Advances in Dielectrics

Series editor

Friedrich Kremer, Leipzig Sachsen, Germany

For further volumes:
http://www.springer.com/series/8283

Aims and Scope

Broadband Dielectric Spectroscopy (BDS) has developed tremendously in the last decade. For dielectric measurements it is now state of the art to cover typically 8–10 decades in frequency and to carry out the experiments in a wide temperature and pressure range. In this way a wealth of fundamental studies in molecular physics became possible, e.g. the scaling of relaxation processes, the interplay between rotational and translational diffusion, charge transport in disordered systems, and molecular dynamics in the geometrical confinement of different dimensionality—to name but a few. BDS has also proven to be an indispensable tool in modern material science; it plays e.g. an essential role in the characterization of Liquid Crystals or Ionic Liquids and the design of low-loss dielectric materials.

It is the aim of "Advances in Dielectrics" to reflect this rapid progress with a series of monographs devoted to specialized topics.

Target Group

Solid state physicists, molecular physicists, material scientists, ferroelectric scientists, soft matter scientists, polymer scientists, electronic and electrical engineers.

Friedrich Kremer
Editor

Dynamics in Geometrical Confinement

Springer

Editor
Friedrich Kremer
Group of Molecular Physics
University of Leipzig
Leipzig
Germany

ISSN 2190-930X ISSN 2190-9318 (electronic)
ISBN 978-3-319-38079-7 ISBN 978-3-319-06100-9 (eBook)
DOI 10.1007/978-3-319-06100-9
Springer Cham Heidelberg New York Dordrecht London

Softcover reprint of the hardcover 1st edition 2014

Printed on acid-free paper

Springer is part of Springer Science+Business Media (www.springer.com)

Preface

Low molecular weight and polymeric molecules can be constrained under the conditions of geometrical confinement having different dimensionalities:

(i) in nanometer thin layers or self-supporting films (one-dimensional),
(ii) in pores or tubes having nanometric diameter (two-dimensional), or
(iii) as micelles embedded in a matrix (three-dimensional), or as nanodroplets.

Within the last two decades, their dynamics under such conditions has been the focus of intense worldwide research activities. Evidence exists that the overall molecular mobility results from a subtle balance between surface and confinement effects. Due to attractive guest/host interactions, the former causes a slowing down of the molecular dynamics, which can be compensated by an appropriate surface coating. The latter is characterized by an increase in the mobility, which becomes more pronounced with decreasing external length scale, e.g., film thickness or pore diameter. In this case, modification of the host/guest interaction by surface treatment has negligible or no effect. Broadband Dielectric Spectroscopy (BDS) plays an essential role in these studies. It is the intention of this second volume in the series "Advances in Dielectrics" to summarize the state of the art in this emerging field, which is also of fundamental importance for nanotechnology.

Friedrich Kremer

Preface

Contents

Dielectric Relaxation of a Polybutadiene Melt at a Crystalline Graphite Surface: Atomistic Molecular Dynamics Simulations

Mathieu Solar, Kurt Binder and Wolfgang Paul

Abstract Dielectric experiments are an indispensable tool to further our understanding of the relaxation behavior of polymers, not only in bulk samples but also in confined situations. A chemically realistic Molecular Dynamics simulation, in which all information about molecular motions is available, can shed light onto the questions of heterogeneity and anisotropy of the underlying molecular relaxation processes which lead to the ensemble averaged experimental dielectric signal. In this contribution, we present a careful analysis of the dielectric response of a weakly polar and confined polymer, 1,4-polybutadiene between graphite walls. The relaxation of the segmental dipole moments was obtained in the time domain and transformed into frequency (Fourier) domain as well as the relaxation time (Debye) domain to highlight the differences between the two types of analysis. A particular bonus of the simulation is that detailed spatially resolved information on structure and dynamics of the confined system is available. We determine the influence of the confinement on the dielectric relaxation and show that for this system the apparent glass transition temperature of a confined film is independent of its thickness even on the scale of a few nanometers.

Keywords Interphase · Molecular Dynamics · Confinement · Coarse-grained model · Atomistic simulations · Dynamic heterogeneity · Polybutadiene · Graphite walls · Dielectric relaxation · Fluctuation-dissipation theorem · Fourier transform · Laplace transform · Dielectric loss spectrum · Relaxation time distribution ·

M. Solar (✉) · W. Paul
Institut für Physik, Martin-Luther-Universitat Halle-Wittenberg, 06099 Halle (Saale), Germany
e-mail: mathieu.solar@physik.uni-halle.de

W. Paul
e-mail: wolfgang.paul@physik.uni-halle.de

K. Binder
Institut für Physik, Johannes-Gutenberg-Universitat, 55099 Mainz, Germany
e-mail: kurt.binder@uni-mainz.de

F. Kremer (ed.), *Dynamics in Geometrical Confinement*, Advances in Dielectrics,
DOI: 10.1007/978-3-319-06100-9_1, © Springer International Publishing Switzerland 2014

Alpha-process · Vogel/Fulcher/Tammann equation · Relaxation map · Glass transition temperature · Torsion stiffness · Layering · Anisotropic relaxation

Abbreviations

DIP	Dipole moment
DSC	Differential Scanning Calorimetry
FFT	Fast Fourier Transform
KWW	Kohlrausch-Williams-Watts
MD	Molecular Dynamics
NMR	Nuclear Magnetic Resonance
PBD	PolyButaDiene
RTD	Relaxation Time Distribution
vdW	van der Waals
VFT	Vogel/Fulcher/Tammann

1 Introduction

Dielectric spectroscopy is an important technique to address the dynamics of macromolecular systems [1]. It measures the polarization current of a (loaded) sample capacitor. Thus, the relaxation of the projection of the electric dipole moment of the sample onto the applied electric field can be measured. For polymeric systems, local dipole moments are carried by some or all of the segments making up the chains, so the total dipole moment is a sum of the segmental ones, $\boldsymbol{\mu}_{\text{tot}} = \sum_i \boldsymbol{\mu}_i$. The macroscopic dipole density in a volume V is then given as $\mathbf{p} = (1/V) \cdot \sum_i \boldsymbol{\mu}_i$. In linear response theory, the dielectric permittivity tensor $\boldsymbol{\varepsilon} = \varepsilon_{\alpha\beta}$ links the polarization to the external electric field $\mathbf{E}$ by the relation $\mathbf{p} = (\boldsymbol{\varepsilon} - \mathbf{1})\varepsilon_0\mathbf{E} = \boldsymbol{\chi}\varepsilon_0\mathbf{E}$, where $\boldsymbol{\chi}$ is the dielectric susceptibility tensor and ε_0 is the dielectric permittivity of vacuum. The dielectric signal in the time domain is then proportional to $\langle[\mathbf{p}(t)\cdot\mathbf{E}(t)][\mathbf{p}(0)\cdot\mathbf{E}(0)]\rangle \propto \langle\mu_{\text{tot,E}}(t)\mu_{\text{tot,E}}(0)\rangle$, where the angular brackets indicate a thermodynamic average and the index, E, denotes the projection onto the applied electric field. In most polymers, the individual segmental dipoles relax essentially independently, so one has $\langle\mu_{\text{tot,E}}(t)\mu_{\text{tot,E}}(0)\rangle \propto \langle\mu_{\text{seg,E}}(t)\mu_{\text{seg,E}}(0)\rangle$, and the experiment probes local segmental relaxation. In so-called type A polymers, the dipole components perpendicular to the chain backbone relax independently, but the parallel components are correlated and add up to a dipole moment parallel to the chain end-to-end vector. In these polymers, dielectric spectroscopy can address local (segmental) as well as global (end-to-end vector) conformational relaxation of polymer chains, and separates into the so-called "segmental" and "normal" modes. Molecular Dynamics (MD) simulations addressing the dielectric response of a polymer melt have been mostly concerned with bulk systems, see [2–4] for example. These

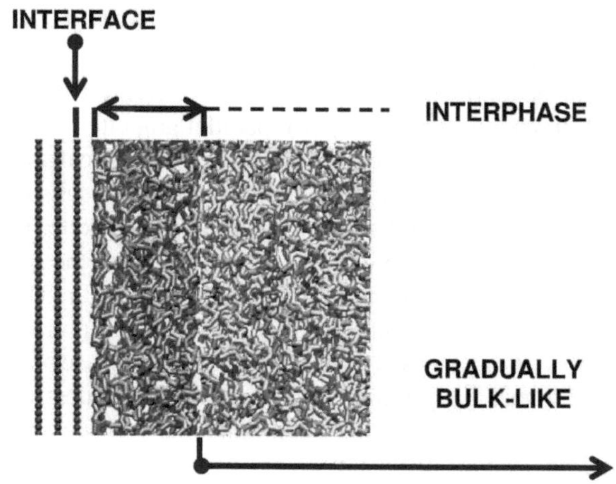

Fig. 1 Snapshot of a polymer melt close to a crystalline surface. The interphase is the region perturbed by the confinement where any physical property gradually evolves to a bulk-like behavior far away from the surface

systems are homogeneous and isotropic and the susceptibility is proportional to the unit tensor, $\boldsymbol{\chi} = \chi \delta_{\alpha\beta}$. Then the dielectric observable is proportional to the autocorrelation function of the dipole vector, $\langle \mu_{\mathrm{tot,E}}(t)\mu_{\mathrm{tot,E}}(0)\rangle \propto \langle \boldsymbol{\mu}_{\mathrm{tot}}(t) \cdot \boldsymbol{\mu}_{\mathrm{tot}}(0)\rangle$. It has been found in experiments [5] as well as in simulations [3] that the assumption of uncorrelated relaxation of different segments discussed above seems to be a good approximation to describe bulk dielectric behavior.

The development of polymer nanocomposite materials required to go beyond the analysis of bulk dielectric behavior and led to the study of confined polymer systems, where the confinement may be one-dimensional (films) or two-dimensional (nanopores). One then has to address the question what changes occur in the polymer dynamics at a solid surface and what may be the characteristic width of the perturbed region in the polymer, the so-called "interphase" (see Fig. 1). Clearly, the relaxation behavior in such a confined system becomes heterogeneous, i.e., it depends on the distance to the surface, as well as anisotropic: an electric field applied perpendicular to the surface probes other molecular motions than a field applied parallel to the surface. For a flat (i.e., atomically corrugated) surface, the susceptibility tensor is written as $\boldsymbol{\chi} = \chi_\alpha \mathbf{e}_\alpha \otimes \mathbf{e}_\alpha$ in a Cartesian basis $\mathbf{e}_\alpha$; $\alpha = x, y, z$. Due to the rotational symmetry, one has $\chi_x = \chi_y \neq \chi_z$ and correspondingly for the dielectric permittivity. The dielectric relaxation of a polymer film was first analyzed carefully for a computer simulation in a study of Peter et al. [6], who considered, however, a coarse-grained bead-spring polymer model supported by a perfectly smooth and flat repulsive interface. This study was motivated by the question, how the glass transition behavior of a polymer melt is changed in a polymer film, a controversial topic in the study of the glass transition phenomenon, which had received much experimental, e.g., [7–9], and simulation attention, e.g., [10–14]. This question is closely related to the problem whether or not the dramatic slowing down of relaxation phenomena in glass forming fluids is associated with the growth of a lengthscale over which motions are strongly correlated [15]. This idea is well established for continuous phase transitions in equilibrium systems ("critical slowing down"), but the extent to which this mechanism can be carried over to glassy freezing is heavily debated [15].

In the following, we discuss the dielectric relaxation of a melt of 1,4-polybutadiene (PBD) chains confined by two parallel graphite surfaces, simulated in a chemically realistic way. We present in detail also the technical aspects of determining the dielectric relaxation behavior for such a chemically realistic simulation. The next section discusses our model and the simulation technique. In Sect. 3, we then present the determination of the dielectric response from the MD simulation trajectory, both in the time domain and in the frequency domain. This section addresses the questions of anisotropy and heterogeneity of the dipolar relaxation. In Sect. 4, we then address the analysis of the temperature dependence of the segmental relaxation and its dependence on film thickness. In addition, the determination of the relaxation time distribution (RTD) from the simulations is discussed in Sect. 5 and finally Sect. 6 presents some conclusions.

2 Atomistic MD Simulations of PBD Between Graphite Walls

We performed MD simulations in the NVT ensemble (i.e., at constant density and temperature) extending for up to approximatively 1 μs in time using the Gromacs package [16]. The simulations were performed for a chemically realistic model of 1,4–polybutadiene (55 % trans, 45 % cis) for which the dielectric relaxation functions in the bulk have been studied before [3]. The polymer melt consisted of N_c chains each of $N_p = 116$ united atom particles representing CH, CH_2 and CH_3 (treated as CH_2) groups, i.e., the chains consisted of 29 repeat units. The force field for the polymer melt may be found in the literature [17]. The graphite model was taken from the literature as well [18] and standard Lorentz-Berthelot combining rules were applied. The glass transition for this polymer[1] was estimated at $T_g = 178$ K from differential scanning calorimetry (DSC) measurements, and at $T_g = 175$ K from rheological measurements [19]. The PBD melt was confined between two graphite walls, which were 10 nm ($T = 353$ K, 323 K, 293 K, 273 K, 253 K and then $N_c = 720$) and 20 nm ($T = 240$ K, 225 K and 213 K and then $N_c = 1440$) apart, respectively. Periodic boundary conditions were applied in the three Cartesian directions to simulate a semi-infinite polymer film confined between two half-spaces of graphite. The temperature of the melt was kept fixed using a Nosé-Hoover thermostat, whereas the carbon atoms were frozen. The Newton equations were integrated using a leap-frog algorithm with a timestep of $\delta t = 1$ fs. Since we are considering a weakly polar polymer at a neutral interface, the influence of the partial atomic charges on the segmental dynamics can be neglected (same as for the bulk [20]), so that the atomic charges

[1] In Ref. [19], a polymer with 7 % of 1, 2, 52 % of 1,4-trans and 41 % of 1,4-cis fractions was considered.

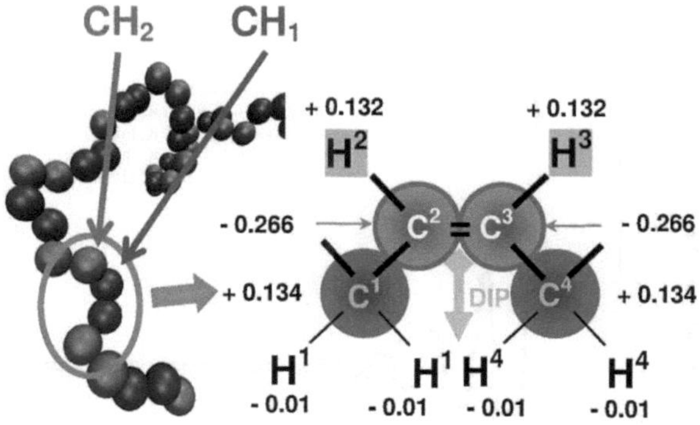

Fig. 2 The *left* picture shows a snapshot of one polymer chain consisting of a repetition of the monomeric unit $CH_2{-}CH = CH{-}CH_2$. A cis conformation is isolated and shown with the real charges on the *right* [3]. The dipole moment (DIP) of the cis segment is computed using the position of the particles. The hydrogen atoms lie on bisections between carbon triplets

were not included in the MD simulations to speed-up the calculations. Thus, the dielectric properties were estimated *a-posteriori* by reinserting partial charges into the stored trajectories as shown in Fig. 2. For the bulk, this procedure gave results which were in quantitatively good agreement with dielectric experiments [3], NMR spin lattice relaxation time measurements [21] and neutron spin echo measurements of the chain dynamics [22]. Each chain in the PBD melt contains 13 cis and 16 trans conformations in a random sequence. A trans conformation has no dipole moment, so that only the segmental relaxation of the cis-conformations is observed in the dielectric spectroscopy experiment.

3 The Dielectric Relaxation of PBD at Graphite

To illustrate the effect that the presence of the graphite wall has on the structure of the melt adjacent to it, we show in Fig. 3 the dependence of the particle density and the chain center of mass density profiles as a function of the distance from the next wall. The crystalline graphite surface attracts the united atoms through a van der Waals attraction. This attraction leads to a strong density layering. Similarly, an adsorbed layer is observable in the center of mass density. With decreasing temperature the layering in the particle density becomes sharper, but the distance over which this layering extends, does not grow significantly in the simulated temperature range. These structural changes influence the relaxation behavior of the polymer segments and lead to a heterogeneous response, depending on the distance to the walls, but only on the scale of a few nanometers, as one would expect from Fig. 3.

Due to the presence of the attraction to the walls, motions parallel (x, y) and perpendicular (z) to the walls are also not equivalent, so one has heterogeneous as

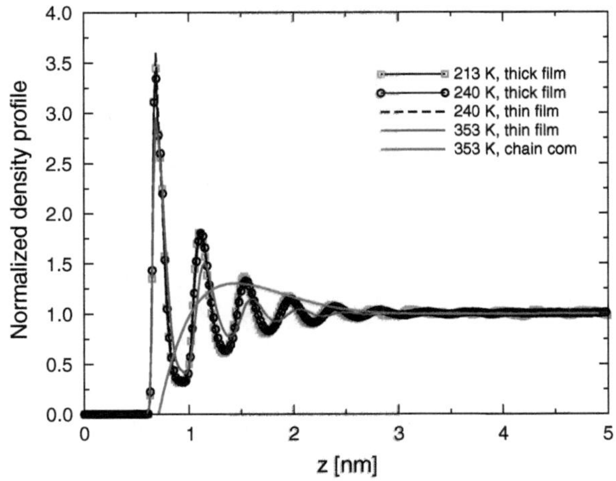

Fig. 3 Particle and chain center of mass density profiles. All densities are normalized to their values in the center of the film, where the particle density is equal to the bulk density at this temperature

well as anisotropic dynamics. In a dielectric experiment on a thin polymer film, the support will always provide one of the electrodes, so that the electric field is oriented perpendicular to the walls. According to the discussion in the introduction, the dielectric experiment is then sensitive to the relaxation of the z-component of the total dipole moment or respectively the segmental dipole. The measured dielectric permittivity is then $\varepsilon = \varepsilon_{zz}$. The most interesting behavior of the segmental relaxation function is observed for a layer of segments directly adjacent to the walls, which is shown in Fig. 4. The dipole relaxation functions in Fig. 4 are not describable as a simple exponential decays $\phi(t) = \exp(-t/\tau_0)$, where t is time and τ_0 the relaxation time constant. After a short time vibrational decay, which is visible in the log-lin plot in the inset and which happens on times below 1 ps, there are two more processes observable.

They can be clearly seen at the higher temperatures but seem to merge at the lower temperatures. The first of these is the structural relaxation (or α-relaxation) of the system, the second one an additional decay process linked to the wall-desorption kinetics [23, 24]. It is known, that the α-process is well described by a stretched exponential decay (Kohlrausch-Williams-Watts, KWW, law) $\phi(t) = \exp[-(t/\tau_{\text{KWW}})^{\beta}]$, where the stretching exponent obeys $0 < \beta < 1$ and τ_{KWW} is the time constant. We empirically fitted the relaxation functions in Fig. 4 by a sum of two stretched exponentials to capture the two relaxation processes present,

$$\phi(t) = a_0\Big[a_1 \exp\left[-(t/\tau_0)^{\beta_0}\right] + (1-a_1)\exp\left[-(t/\tau_1)^{\beta_1}\right]\Big], \qquad (1)$$

where a_0, a_1, τ_0, β_0, τ_1 and β_1 are the fitting parameters. The overall amplitude, a_0, is smaller than one because the short time vibrational decay can not be captured by

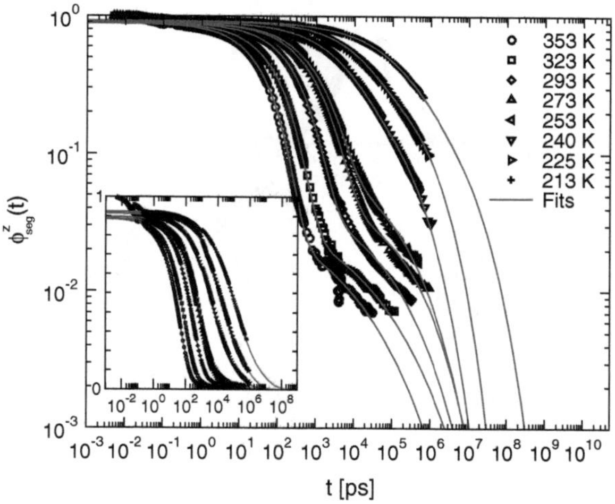

Fig. 4 Main panel: log—log representation of the segmental dipole relaxation functions as a function of temperature. The relaxation functions were computed for a layer of width 1.2 nm next to the walls. The functional form of Eq. (1) was fitted to the data (*grey full lines*). Small panel: log—lin representation of the same data

this fit. For the center region of the film, the final wall-induced relaxation process is absent (data not shown) and a single exponential is enough to fit the autocorrelation functions, in the same way as for the bulk.

3.1 The Dielectric Loss Spectra

The relaxation functions in Fig. 4 span many decades in time and contain processes on very different time scales, from the fast vibrational motions to the long time desorption kinetics. To obtain these functions requires a logarithmic sampling in time, and it is impossible to resolve the whole curve with a resolution adapted to the fastest process. Thus, the raw simulation data were first interpolated and linearly sampled over the full time range of observation, and then a Fast Fourier Transform algorithm was used to obtain the frequency spectrum. In addition, the good fit functions (as shown by the red lines in Fig. 4) at hand were also used to generate the correlation functions on a uniformly spaced time grid of width Δt having 2^{26} time points, which could be used to extend the time range over which the relaxation function was observed compared to the raw data. This was done for all temperatures.

The Shannon-Nyquist sampling theorem then relates the time increment to the largest frequency one can resolve, and the lowest frequency is given by the inverse of the longest time, $(2^{26}\Delta t)^{-1}$, one reaches this way. The time interval Δt depends on temperature because the longest time scale one has to reach increases with decreasing temperature, so the frequency grid points for which one determines the Fourier

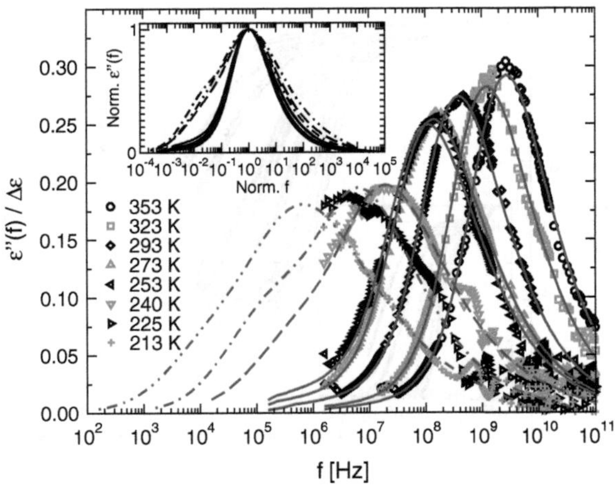

Fig. 5 Main panel: temperature dependence of the dielectric loss for the layer next to the wall. At high temperatures the Fourier transforms of the data and the fit agree, at the lowest temperatures, the Fourier transform of the fit allows for an extrapolation to lower frequencies than obtainable from the data. Inset: scaled representation of the same data showing a broadening of the frequency spectra at low temperatures

transform also depend on temperature. The dielectric spectra shown in Fig. 5 show that the results obtained by transforming the original data and by transforming the fit functions agree with each other for high temperatures. At the lowest temperatures, the relaxation becomes so slow that it can not be completely resolved within the simulation time window. Here the Fourier transforms of the fits (which agree with those of the data at high frequencies) allow for an extrapolation of the spectra to lower frequencies.

4 The Temperature Dependence of the α-Process

The temperature dependence of the position of the maximum in the dielectric loss, which was used in the inset of Fig. 5 to scale the frequencies, gives a measure for the temperature dependence of the α process. When we perform this analysis for different layer widths starting at the graphite surface, we can obtain an idea on the thickness dependence of the glass process in PBD films. This temperature dependence is not described by an Arrhenius law in fragile glass formers to which PBD belongs, rather it is empirically fitted by the Vogel/Fulcher/Tammann (VFT) equation [15]

$$f = f_{\infty} \exp\left[- \frac{E_{\mathrm{a}}}{(T - T_0)} \right], \tag{2}$$

or

Table 1 Results of VFT fits (see Fig. 6) adjusted to the data from the MD simulations

Param.	Bulk	1.2 nm	2.5 nm	3.7 nm	5.0 nm
$\log(f_\infty)$	11.405 ± 0.061	10.758 ± 0.152	10.884 ± 0.128	10.986 ± 0.131	11.027 ± 0.131
E_a	980 ± 53	664 ± 43	691 ± 37	721 ± 39	739 ± 40
T_0	141 ± 5	155 ± 1	153 ± 1	151 ± 1	149 ± 1
T_g	173 ± 7	178 ± 3	176 ± 2	175 ± 3	174 ± 3

The glass transition temperature and its (asymptotic standard) error bar are estimated from the fit parameters using Eq. (3)

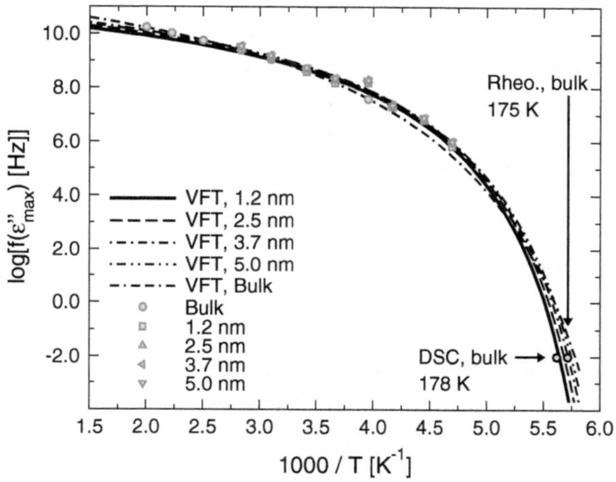

Fig. 6 The relaxation map for various layer thicknesses. The error in the data is smaller than the size of the symbols. The respective thicknesses are given in the legend. The *lines* show empirical VFT fits (see Table 1), extrapolating to the known bulk glass transition temperature within the error bars

$$T = T_0 + \frac{E_a}{[\ln f_\infty - \ln f]}, \tag{3}$$

where f_∞ (frequency as $T \to +\infty$), E_a (activation temperature) and T_0 (so-called ideal glass transition or Vogel temperature) are the fitting parameters. We recall that one model that leads to Eqs. (2) or (3) is based on the idea of regions where relaxations require cooperative rearrangements of the molecules (or monomeric units) that they contain, implying that the size of these regions diverges at T_0. However, the empirical fits of actual data show (see Table. 1) that T_0 always is distinctly smaller than T_g.

The second form allows for the calculation of the glass transition temperature when the VFT parameters are known. The glass transition temperature corresponds to a relaxation rate of $f = 10^{-2}$Hz [25]. Therefore, one may define here the dielectric glass transition by $T_g = T(f = 10^{-2}\text{Hz})$ [1]. When one performs VFT fits to the data in Fig. 6, one obtains the fit parameters shown in Table 1, where also the error bars of the fit parameters are given.

There is an obvious jump in the value of the Vogel temperature T_0 going from the bulk to the film data. The reason for this is the change in the temperature range for which the data are available. The bulk data are taken from Ref. [3] and are given for $253\text{K} \leq T \leq 500\text{K}$, whereas the film data are given for $213\text{K} \leq T \leq 353\text{K}$. This jump reveals the fact that the VFT law actually is not able to cover the complete temperature dependence from the high temperature liquid to the supercooled regime close to T_g. This was worked out in pioneering dielectric experiments by Stickel et al. [26, 27] and was also found for simulations on bulk PBD [20], when the time scale for incoherent neutron scattering was analyzed. Rather than a VFT law, one has to use two Arrhenius laws to capture the crossover from high temperature liquid-like to low temperature solid-like relaxation. However, when we look at the prediction for the glass transition temperature obtainable from these VFT fits, we can conclude that they agree with each other and with the experimentally known value within the error bars inherent in the VFT fitting procedure. There is no indication that the glass transition temperature changes with a reduction in film thickness, similar to what is found in recent dielectric experiments, for example in [28, 29]. This conclusion differs from the one reached in the simulation work by Peter et al. [6]. We explain this difference by the difference between a coarse-grained and a chemically realistic model. In a bead-spring coarse-grained model, the glass transition is mostly determined by density effects alone, so a layering as visible in Fig. 3 leads to strong effects on the glass transition temperature. In a chemically realistic model, the internal rotation barriers of the dihedral angles are much more important and induce a larger dynamic chain stiffness than a bead-spring coarse-grained model possesses, where no torsion stiffness is included. Removing the torsion potential from the simulation shifts the glass transition temperature of PBD (simulated at the realistic densities) from about 175 K to about 60 K [2]. The layering at the walls (or for that purpose, the reduced density at a free surface) therefore has a much smaller effect on T_g in such realistic models and in real confined polymer systems then it has for simulations of a coarse-grained model.

5 The Relaxation Time Distribution

It is astonishing that the two processes clearly visible in the time domain behavior (Fig. 4) do not lead to separate peaks in the frequency spectra (Fig. 5). This smearing out of features in the Fourier analysis can be avoided when one instead determines the relaxation spectrum $\rho(\tau)$ (or distribution of relaxation times) of the time domain signal. Here, the relaxation function is written as an integral over Debye processes, i.e.,

$$\phi(t) = \int_0^{+\infty} \rho(\tau) \exp[-t/\tau] d\tau , \quad (4)$$

where one has to fulfill the normalization condition $\int_0^{+\infty} \rho(\tau)d\tau = 1$. Eq. (4) may be reinterpreted as a Laplace transform if one replaces the real variable t by the complex variable $s = \zeta + i\eta$ (i stands for imaginary unit) and carries out a change of variables $\lambda = 1/\tau$. It leads to Eq. (5), where the function $\overline{\rho}(\lambda) = \rho(1/\lambda)/\lambda^2$ should be reinterpreted as a spectrum of relaxation rate constants, which is simpler to use from a mathematical point of view. From there, the function $\overline{\rho}$ of λ can be calculated as the inverse Laplace transform $\mathscr{L}_{\phi}^{-1}$ of the relaxation function ϕ using the variable s. It follows that

$$\mathscr{L}_{\overline{\rho}}(s) = \int_0^{+\infty} \overline{\rho}(\lambda) \exp[-s\lambda] d\lambda \,, \tag{5}$$

$$\overline{\rho}(\lambda) = \frac{1}{2\pi \mathrm{i}} \int_{\zeta - \mathrm{i}\infty}^{\zeta + \mathrm{i}\infty} \phi(s) \exp[+s\lambda] ds \,. \tag{6}$$

Since the function ϕ, see Eq. (1), as a function of $s = \zeta + i\eta$ belongs to the Hardy space (i.e., ϕ is a holomorphic function for $\Re(s) > 0$ and $\sup_{\zeta > 0} \int_{-\infty}^{+\infty} |\phi(\zeta + i\eta)|^2 d\eta < +\infty$), the Paley-Wiener theorem allows one to set ζ to zero in equation (6), i.e., the inverse Fourier transform of $\phi(i\eta)$ exists, and $\overline{\rho}$ is a square-integrable function (i.e., $\int_{-\infty}^{+\infty} |\overline{\rho}(\lambda)|^2 d\lambda < +\infty$), see [30, 31]. Thus, the inverse Laplace transform may be written as an inverse Fourier transform. The relaxation rate distributions were computed for each model $\phi(t)$ using the FFT algorithm, after the function ϕ was normalized. The computation was done for several integration time windows, so as to cover a large range of relaxation rate constants. We carefully checked that the normalization condition was satisfied, and that the relaxation rate distributions obtained give the model $\phi(t)$ according to Eq. (4). Finally, one can convert the results into distributions of relaxation time constants.

The RTD is able to separate the two processes as shown in Fig. 4 and to reveal their temperature dependence. This dependence is shown in Fig. 7 the layer of width 1.2 nm next to the wall for which the contribution of the desorption process is clearest. For both processes, the typical relaxation times increase as the temperature is lowered, however, the time scale for the α-relaxation increases faster than the one for the desorption process. Thus the two peaks first merge and the relative position of the two contributions even changes. In addition and for a given temperature, in Fig. 7, a big red filled circle corresponds to the frequency of the maximum dielectric loss from Fig. 5. Doing this for the temperature range that was simulated, one obtains a master line, which indicates that the maximum dielectric loss position in frequency is a blend of the two processes. One can also conclude that different physics of the glass transition may be addressed, if one considers the maximum dielectric loss or the maxima in the relaxation rate distributions.

It is interesting to perform a direct comparison of the analysis in terms of a superposition of Debye processes and the analysis in terms of a superposition of harmonic oscillators (Fourier spectrum). We perform this comparison in Fig. 8 for

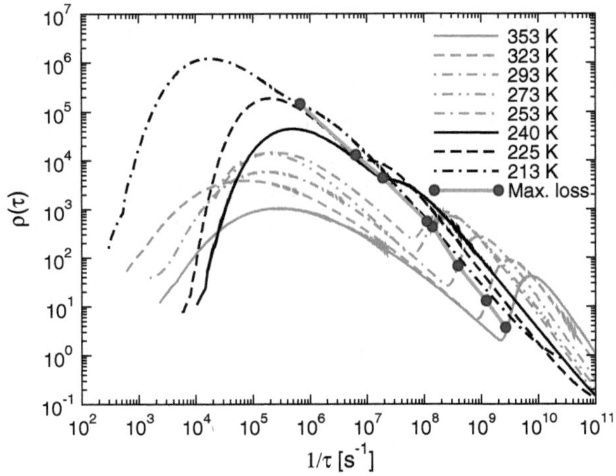

Fig. 7 The temperature dependence of the relaxation time distributions as a function of the relaxation rate constant (equivalent to the inverse of the relaxation time constant) for the layer next to the wall. The big *filled circles* are located at the frequency of maximum dielectric loss from Fig. 5

the relaxation behavior observed in layers of different thicknesses next to the walls and for a temperature of 353 K. Clearly, the Fourier signal shows a single peak which only broadens upon reducing the layer thickness for which the dielectric signal is evaluated (only for the smallest thickness of 1.2 nm there is a small shift of the peak position). Both relaxation processes contribute to this peak and the stronger contribution of the wall process for the smaller layer thicknesses only leads to the observed broadening. In contrast, for the RTD a clear bimodal signal is observed. The peak at lower τ captures the contributions to the α process at this temperature, the peak at larger τ captures the wall-desorption process. The latter one is completely absent when one considers a layer in the center of the film as shown by the dashed line.

6 Conclusion

In this contribution, MD simulations (extending for up to approximatively 1 μs in time) for a chemically realistic model of 1,4–polybutadiene (55 % trans, 45 % cis) confined by graphite walls were performed to address the question of the dielectric relaxation behavior of a polymer in confinement. Since polybutadiene is a weakly polar polymer, the partial charges were included after the simulations of the same but nonpolar polymer. Then, using the fluctuation–dissipation theorem which states that the linear response of a given system to an external perturbation is expressed in terms of fluctuation properties of the system in thermal equilibrium, one is able to estimate the dielectric permittivity from the fluctuations of the dipole moments of the

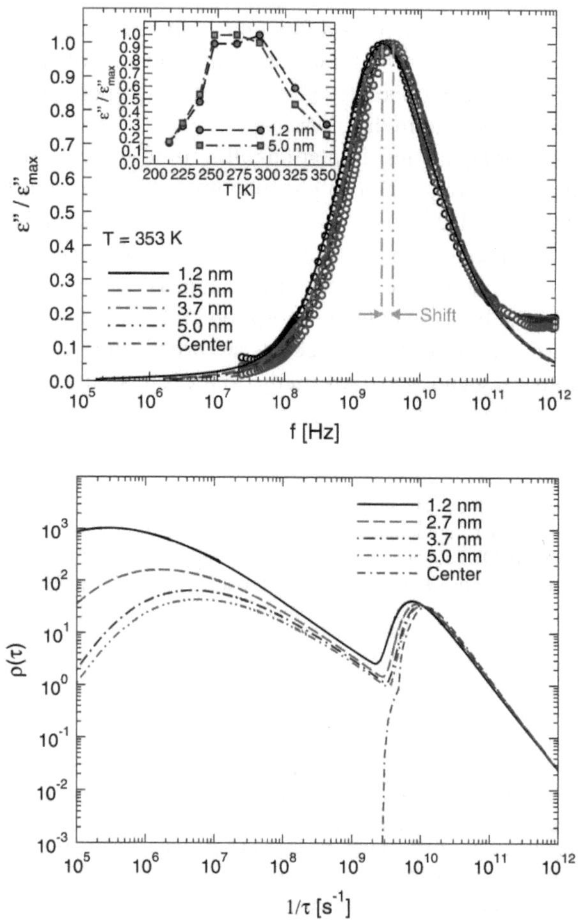

Fig. 8 *Top panel*: frequency dependent dielectric permittivity for $T = 353\,\mathrm{K}$ and several layer thicknesses indicated in the legend. A small shift of the peak position for the thinnest layer of 1.2 nm is indicated. Points are Fourier transforms of the simulation data, lines are transforms of fits, as done in Fig. 4. The inset shows the temperature dependence at two indicated thicknesses at a frequency of $f = 2.25 \cdot 10^8\,\mathrm{Hz}$. *Bottom panel*: the relaxation time distributions as a function of the inverse of the relaxation time constant at $T = 353\,\mathrm{K}$. Data are shown for comparable layer thicknesses as in the *top panel*

cis segments in a polymer molecule. The dielectric relaxation functions in the bulk had been studied before for this polymer. The Fourier spectra and the relaxation time distributions were computed from the raw data and from a model fit capturing the data (and also extrapolating the behavior to long times). This allows one to conclude that the confinement introduces mainly a broadening of the frequency spectrum (Fourier space), which, however, is the result of a sum of two unimodal processes according to the relaxation time distributions. We estimated the glass transition for this confined polymer systems using the activation representation, where the frequency of the maximum dielectric loss is plotted as a function of inverse temperature. When one performs standard VFT fits of the temperature dependence of the maximum loss frequencies, no shift of the extrapolated glass transition temperature with decreasing film thickness is obtained within the error bars.

Since our Molecular Dynamics simulations allow us to study the relaxation behavior spatially well resolved (e.g., focusing on monomers that are no further away from the graphite surface than 1.2 nm), we are able to identify the signals of two distinct slowly relaxing phenomena, both in the time dependence of relaxation functions (see Fig. 4) and in the relaxation time spectrum (see Fig. 8): namely glassy freezing of the same type as in the bulk, and chain desorption kinetics at the graphite wall. The relaxation time distributions of both processes are broad, and due to their somewhat different temperature dependence, they merge in the temperature region of interest. We also show that it is hard to disentangle these processes in the dielectric loss spectra (see Fig. 5).

Acknowledgments The authors are grateful for funding received within the Focused Research Program SPP 1369 of the German Science Foundation. A generous grant of CPU time at the Jülich Supercomputer Center (JUQUEEN and JUROPA) is gratefully acknowledged. We thank F. Kremer and E.U. Mapesa (University of Leipzig, Germany) for stimulating discussions.

References

1. Kremer F, Schönhals A (2003) Broadband dielectric spectroscopy. Springer, Berlin
2. Bedrov D, Smith GD (2005) Molecular dynamics simulation study of the α- and β-relaxation processes in a realistic model polymer. Phys Rev E 71:050801(R)
3. Smith GD, Borodin O, Paul W (2002) A molecular-dynamics simulation study of dielectric relaxation in a 1,4-polybutadiene melt. J Chem Phys 117:10350
4. Eslami H, Muller-Plathe F (2009) Structure and mobility of poly(ethylene terephthalate)- a molecular dynamics simulation study. Macromolecules 42:8241
5. Bahar I, Erman B, Kremer F, Fischer EF (1992) Segmental motions of cis-polyisoprene in the bulk state: interpretation of dielectric relaxation data. Macromolecules 25:816
6. Peter S, Napolitano S, Meyer H, Wübbenhorst M, Baschnagel J (2008) Modeling dielectric relaxation in polymer glass simulations - dynamics in the bulk and in supported polymer films. Macromolecules 41:7729
7. Alcoutlabi M, McKenna GB (2005) Effects of confinement on material behaviour at the nanometre size scale. J Phys Condens Matter 17:R461
8. Ediger MD (2000) Spatially heterogeneous dynamics in supercooled liquids. Ann Rev Phys Chem 51:99
9. Kremer F, Mapesa EU, Tress M (2012) Molecular dynamics of polymers at nanometric length scales: from thin layers to isolated coils. In: Kalmykov YP (ed) Recent advances in broadband dielectric spectroscopy. Springer, Dordrecht
10. Daoulas KC, Harmandaris VA, Mavrantzas VG (2005) Detailed atomistic simulation of a polymer melt / solid interface - structure, density and conformation of a thin polyethylene melt film adsorbed on graphite. Macromolecules 38:5780
11. Harmandaris VA, Daoulas KC, Mavrantzas VG (2005) Detailed atomistic simulation of a polymer melt / solid interface - local dynamics and diffusion of a thin polyethylene melt film adsorbed on graphite. Macromolecules 38:5796

12. Varnik F, Baschnagel J, Binder K (2002) Reduction of the glass transition temperature in polymer films: a molecular-dynamics study. Phys Rev E 65:021507
13. Torres JA, Nealey PF, De Pablo JJ (2000) Molecular simulation of ultrathin polymeric films near the glass transition. Phys Rev Lett 85:3221
14. Hudzinskyy D, Lyulin AV, Baljon ARC, Balabaev NK, Michels MAJ (2011) Effects of strong confinement on the glass-transition temperature in simulated atactic polystyrene films. Macromolecules 44:2299
15. Binder K, Kob W (2011) Glassy materials and disordered solids: an introduction to their statistical mechanics. World Scientific Publishing Company, Singapore
16. Hess B, van der Spoel D, Lindahl E (2008)J Chem Theory Comput 4:435. http://www.gromacs.org
17. Smith GD, Paul W (1998) United atom force field for molecular dynamics simulations of 1,4-polybutadiene based on quantum chemistry calculations on model molecules. J Phys Chem A 102:1200
18. Steele WA (1973) The physical interaction of gases with crystalline solids-i, Gas-solid energies and properties of isolated adsorbed atoms. Surf Sci 36:317
19. Zorn R, Mopsik FI, McKenna GB, Willner L, Richter D (1997) Dynamics of polybutadienes with different microstructures - 2. Dielectric response and comparisons with rheological behavior 107:3645
20. Paul W, Smith GD (2004) Structure and dynamics of amorphous polymers - computer simulations compared to experiment and theory. Rep Prog Phys 67:1117
21. Smith GD, Borodin O, Bedrov D, Paul W, Qiu X, Ediger MD (2001) ^{13} C-NMR spin-lattice relaxation and conformational dynamics in a 1,4-polybutadiene melt. Macromolecules 34:5192
22. Smith GD, Paul W, Monkenbusch M, Richter D (2001) On the non-gaussianity of chain motion in unentangled polymer melts. J Chem Phys 114:4285
23 Yelash L, Virnau P, Binder K, Paul W (2010) Slow process in confined polymer melts: layer exchange dynamics at a polymer solid interface. Phys Rev E 82:050801
24 Yelash L, Virnau P, Binder K, Paul W (2012) Three-step decay of time correlations at polymer-solid interfaces. Europhys Lett 98:28006
25. Binder K, Baschnagel J, Paul W (2003) Glass transition of polymer melts: test of theoretical concepts by computer simulation. Prog Polym Sci 28:115
26. Stickel F, Fischer EW, Richert R (1995) Dynamics of glass-forming liquids: i - temperature derivative analysis of dielectric relaxation data. J Chem Phys 102:6251
27. Stickel F, Fischer EW, Richert R (1996) Dynamics of glass-forming liquids: ii - detailed comparison of dielectric relaxation, dc-conductivity and viscosity data. J Chem Phys 104:2043
28. Mapesa EU, Tress M, Schulz G, Huth H, Schick C, Reichec M, Kremer F (2013) Segmental and chain dynamics in nanometric layers of poly(cis-1,4-isoprene) as studied by broadband dielectric spectroscopy and temperature-modulated calorimetry. Soft Matter. doi:10.1039/C3SM51311D
29. Tress M, Mapesa EU, Kossack W, Kipnusu WK, Reiche M, Kremer F (2013) Glassy dynamics in condensed isolated polymer chains. Science 341:6152
30. Gradshteyn IS, Ryzhik IM (2007) Table of integrals, series, and products. Academic Press, USA
31. Schwartz L (1966) Mathematics for the Physical Sciences. Herman, Paris

Glass Transition of Ultra-Thin Polymer Films: A Combination of Relaxation Spectroscopy with Surface Analytics

Huajie Yin, Sherif Madkour and Andreas Schönhals

Abstract The glass transition behavior of ultra-thin supported polymer films is discussed controversially in the literature for around 20 years. Substantial efforts have been archived to understand it. In this contribution, a combination of methods sensitive to bulk properties of a system, like dielectric or specific heat spectroscopy with surface analytics, for instance, atomic force microscopy (AFM), contact angle measurements, and X-ray photoelectron spectroscopy (XPS) were employed to study the glass transition of ultra-thin supported films. All investigations were carried out on identically prepared and treated samples. Different systems with different complexities going from more or less flexible homopolymers over rigid main chain macromolecules to polymer blends have been studied. For the investigated flexible macromolecules, the dynamic glass transition temperature estimated within the frame of the linear response approach is independent of the film thickness down to several nanometers and identical to the bulk value. For polystyrene it was found the thermal glass transition temperatures can depend on the film thickness. This different behavior is not well understood till now and needs further experimental clarification. For the investigated main chain polymers polycarbonate and polysulfone. Dynamic and thermal glass transition temperature estimated from the dielectric measurements increases with decreasing film thickness. This is discussed in the frame of a strong interaction of the polymer segments with the surface of the substrate. In general for homopolymers, the interaction energy of the polymer segments with the substrate surface cannot be considered as the only parameter, which is responsible for the change in the thermal glass transition with the film thickness. For the investigated miscible blend system of polystyrene/poly(vinyl methyl ether) at a composition of 50/50 wt-% a decrease of the dynamic glass transition temperature with decreasing film thickness is found. This is explained by the formation of a poly(vinyl methyl ether)-rich surface layer with a higher molecular mobility.

H. Yin · S. Madkour · A. Schönhals (✉)
BAM Federal Institute for Materials Research and Testing, Unter den Eichen 87, 12205 Berlin, Germany
e-mail: Andreas.Schoenhals@bam.de

F. Kremer (ed.), *Dynamics in Geometrical Confinement*, Advances in Dielectrics,
DOI: 10.1007/978-3-319-06100-9_2,

Keywords Broadband dielectric spectroscopy · Specific heat spectroscopy · Photoelectron spectroscopy · Contact angle measurements · Polystyrene · Poly (vinyl methyl ether) · Poly(2-vinylpyridine) · Polycarbonate · Polysulfone · Polystyrene/ Poly(vinyl methyl ether) blend

Abbreviations

AFM	Atomic force microscopy
XPS	X-ray photoelectron spectroscopy
PS	Polystyrene
PVME	Poly(vinyl methyl ether)
P2VP	Poly(2-vinylpyridine)
PC	Polycarbonate
PSU	Polysulfone
T_g	Glass transition temperature
T_0	Vogel temperature
M_w	Molecular weight
D	Film thickness
P	Polymer
S	Substrate
γ_{SP}	Polymer/substrate interaction energy
θ	Contact angle
S_iO_2	Silicon dioxide
AlO_x	Aluminum oxide
ω	Radial frequency
f	Technical frequency ($\omega = 2\pi f$)
$J^*(\omega)$	Complex compliance
$G^*(\omega)$	Complex modulus
$\chi^*(\omega)$	Complex dielectric susceptibility
$\varepsilon^*(\omega)$	Complex dielectric function
R	Resistance
HN	Havriliak/Negami
VFT	Vogel/Fulcher/Tammann
f_p	Relaxation rate
$\Delta\varepsilon$	Dielectric strength
AC	Alternating current
min	Minute(s)
K	Kelvin
μW	Microwatt
nm	Nanometer
ΔU	Complex differential voltage
U_R	Real part of complex differential voltage
φ	Phase angle of complex differential voltage
ρ	Density

D	Fragility parameter
ξ	Cooperativity length scale
c_p	Specific heat capacity
δT	Mean temperature fluctuation

1 Introduction

The properties of ultra-thin polymer films are of great interest from both the technological and basic research points of view. The former point is due to the importance of polymeric films in a broad variety of technological fields like coatings, membranes, organic electronic devices, etc, where polymers combine a high tunability, flexibility, and low production costs. For these application fields, several problems like wetting [1], instabilities [2], the viscoelastic properties [3, 4], diffusion processes [5–7], and especially the glass transition behavior (see for instance Refs. [8–45]) have to be considered. This concerns also several chapters of this book [46–49].

The investigation of the glass transition behavior plays a key role, since several material properties change drastically at the glass transition. It has to be noted that the glass transition itself is a topical problem of soft matter physics [50, 51]. From the scientific point of view, ultra-thin polymer films provide an ideal sample geometry for studying the confinement effects on the glass transition of polymers because the confining dimensions (film thickness) can be easily tuned by spin coating [52].

Since the pioneering work of Keddie and Jones [33, 34] the thickness dependence of the glass transition temperature has been controversially discussed in the literature where for the same system (polymer and substrate) divergent results have been published. For polymers supported by a nonattractive substrate (see for instance [34]) a decrease of the thermal glass transition temperature T_g with decreasing film thickness (T_g depression) is often observed. This T_g depression is discussed to originate as a result of a free surface (polymer/air) having a higher molecular mobility than the bulk due to missing polymer–polymer segment interactions [53]. For polymer films supported by substrates having a strong interfacial interaction, T_g may increase with the reduction of film thickness [42, 54], which is explained by the formation of an adsorbed boundary layer in which the polymer segments have a lower molecular mobility; hence, a higher glass transition temperature [55]. There are attempts to correlate depression or increase of the glass transition temperature with the interaction energy between the substrate and the polymer surface γ_{SP} [56]. For values of γ_{SP} smaller than a critical value γ_{c}, the free surface will dominate and a depression of T_g will be observed, while for $\gamma_{\mathrm{SP}} > \gamma_{\mathrm{c}}$ the reduced mobility layer at the substrate dominates and an increase of T_g should be observed. Tsui et al. [26] criticized this concept by concluding that the interaction energy between the polymer segments and the substrate is not the only relevant parameter and the packing of the segments at the interface should be also of importance.

On the one hand, in many investigations on thin films like ellipsometric studies [33], a thermodynamic property (or an associated quantity) is measured during a temperature scanning experiment. A change in the temperature dependence is interpreted as thermal glass transition where a corresponding thermal glass transition temperature T_g can be extracted (dilatometric methods, see for instance [53, 57, 58]). On the other hand, also techniques, which are directly sensing segmental fluctuations like broadband dielectric spectroscopy (see for instance [8]), specific heat spectroscopy (see for instance [59]), or optical photobleaching techniques [60] have been employed to investigate the glass transition of thin polymeric films. These methods are related to the segmental fluctuations and therefore to the dynamic glass transition. For various polymers, e.g., polystyrene [44, 45, 59, 60], poly(methyl methacrylate) [61, 62], or poly(vinyl methyl ether) [63], the dynamic glass transition temperature in the specific frequency range of the employed method was found to be independent of the film thickness down to several nm.

Obviously, the thickness dependence of the glass transition temperature of ultrathin films is influenced by different factors, which may have opposite impacts. Often discussed factors are the interaction of the polymer segments with the substrate, the existence of a free polymer/air interface [64], and the preparation and annealing conditions. In this contribution, recent advances and results for thin films are collected which were obtained by combining methods sensitive to bulk properties like dielectric or specific heat spectroscopy with surface analytics for instance atomic force microscopy (AFM), contact angel measurements, and X-ray photoelectron spectroscopy (XPS). Different systems with different complexities, going from more or less flexible homopolymers over rigid main chain macromolecules to polymer blends, were studied.

2 Methodology

The used methodology is based on the concept that bulk sensitive methods like dielectric relaxation and specific heat spectroscopy are combined with surface analytics. The investigations are carried out on samples, which are prepared and annealed under identical conditions to ensure a comparability of the results of the different methods.

2.1 *Preparation*

As homopolymers polystyrene (molecular weights: $M_w = 50\,\text{kg/mol}$; $M_w = 260$ kg/mol; $M_w = 1408\,\text{kg/mol}$;) poly(vinyl methyl ether) (molecular weights: $M_w = 60\,\text{kg/mol}$), poly(2-vinylpyridine) ($M_w = 1020\,\text{kg/mol}$), polycarbonate ($M_w = 22\,\text{kg/mol}$), and polysulfone ($M_w = 45\,\text{kg/mol}$) were used.

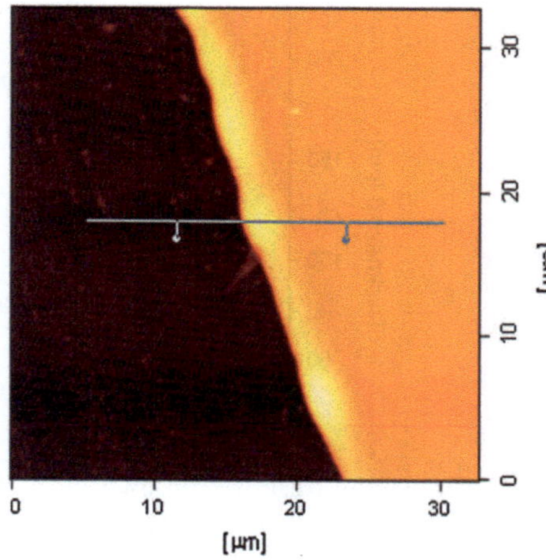

Fig. 1 AFM image of a scratch across a poly(vinyl methyl ether) film (thickness ca. 70 nm) spin coated on a silicon wafer. The figure was adopted from Ref. [63]

All thin films were prepared on a substrate by spin coating from a filtered polymer solution for both the dielectric and the specific heat spectroscopy experiments. For the dielectric measurements, films were prepared between two electrodes (for details see Refs. [42, 45] and inset of Fig. 6). This means for the dielectric experiments the polymer segments are in interaction with an AlO_x surface. In case of the specific heat spectroscopy, films were prepared onto the SiO_2 surface of the calorimeter chip as described below. Before spin coating, the substrates were cleaned by sophisticated procedures as described in Ref. [42], for the dielectric measurements, and in Ref. [43], for the thermal investigations.

After spin coating, films were annealed at temperatures well above the glass transition temperature of the bulk material in an oil-free vacuum for a well-adopted time. The topography of the films was investigated after annealing by AFM. Some examples are shown in Fig. 1. No sign of dewetting was detected even for polymers with glass transition temperatures well below room temperature as in the case of poly(vinyl methyl ether).

The thickness of the prepared films was adjusted by the concentration of the polymer solution. The film thickness was estimated also by AFM by measuring the step height of a scratch, which was made by a sharp needle through the film down to the substrate. As an example, Fig. 2 gives the relation of the thicknesses of polycarbonate films prepared on an AlO_x surface versus the concentration of the solution.

2.2 *Bulk Methods*

2.2.1 Linear Response Approach

In the linear regime, relaxation (or retardation) experiments can be theoretically described by the linear response theory. The general principle is given in Fig. 3.

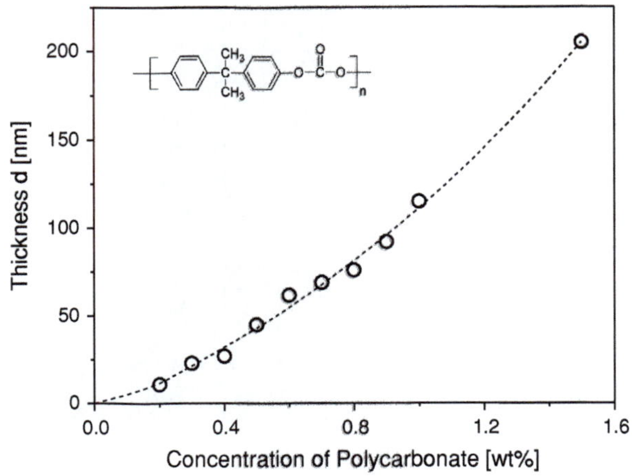

Fig. 2 Film thickness d of a polycarbonate film prepared on an AlO_x surface versus the concentration of polycarbonate dissolved in dichloromethane. The *line* is a guide for the eyes. The *inset* gives the chemical structure of polycarbonate

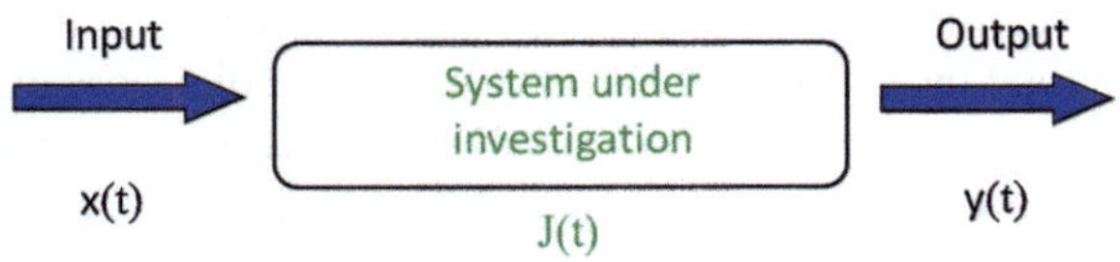

Fig. 3 Scheme of linear response theory

An outer disturbance $x(t)$ is applied to the system (material under investigation), which is considered as a black box with the material function $J(t)$. $x(t)$ provokes a response of the system $y(t)$. Assuming linearity and causality

$$y(t) = y_\infty + \int_{-\infty}^{t} J(t - t') \frac{dx(t')}{dt'} dt' \tag{1}$$

holds. The corresponding material function $J(t)$ (compliance) can be measured as the time dependent response to a step-like disturbance Δx as $J(t) = (y(t) - y_\infty)/\Delta x . y_\infty$ is the response for short times (instantaneous response). The Fourier transformation of Eq. (1) yields

$$y^* (\omega) = J^* (\omega)\, x^* (\omega) \text{ with } J^* (\omega) = J' (\omega) - i J'' (\omega) \tag{2}$$

with ω the radial frequency and $i = \sqrt{-1}$. $J^*(\omega)$ is a generalized compliance with corresponding real $J'(\omega)$ and loss part $J''(\omega)$.

Because of its linearity Eq. (1) can be inverted to

$$x(t) = x_\infty + \int_{-\infty}^{t} G(t - t') \frac{\mathrm{d}y(t')}{\mathrm{d}t'} \mathrm{d}t' \tag{3}$$

where $G(T)$ is a generalized modulus, which reads in the complex form $G^*(\omega) = G'(\omega) + iG''(\omega)$. The relationship between the compliance and the modulus is given by

$$J^*(\omega)G^*(\omega) = 1. \tag{4}$$

Glass transition temperatures deduced in the frame work of the linear response approach are estimated in the thermodynamic equilibrium state (thermodynamic metastable equilibrium, supercooled state) of a system and called dynamic glass transition temperature.

2.2.2 Broadband Dielectric Spectroscopy

In the case of dielectric spectroscopy, the disturbance is the electric field E and the response of the system is the polarization P. The material function is complex dielectric susceptibility $\chi^*(\omega) = \varepsilon^*(\omega) - 1$ where $\varepsilon^*(\omega) = \varepsilon'(\omega) - i\varepsilon''(\omega)$ is the complex dielectric function or the complex permittivity. For a detailed discussion, see Ref. [65]. All presented data were obtained by a high resolution Alpha analyser with an active sample head (Novocontrol®) in the frequency range from 10^{-1} to 10^7 Hz. The temperature was controlled by a Quatro Novocontrol® cryo-system with a temperature stability better than 0.1 K.

Figure 4 shows an example for a dielectric measurement for a polysulfone film with a thickness of 48 nm. A conductivity contribution is observed at low frequencies followed by a peak due to dynamic glass transition (α-relaxation) related to the segmental fluctuations. Both processes shift to higher frequencies with increasing temperature. At frequencies around 10^6 Hz, a further (parasitic) process is evident as a peak. In the case of the thin film capacitors, the resistance R of the thin Al electrodes cannot be neglected. Together with the sample capacity, this resistance leads to an artificial loss peak (electrode peak) on the high frequency side of the spectra with a time constant $\tau_{\mathrm{Res}} = R^* C'$ (C'—sample capacity). Because of the fact that the electrode peak should obey the Debye function, the dielectric data can be analyzed by fitting the model function

$$\varepsilon^*_{\mathrm{Fit}} = \varepsilon^*_{\mathrm{HN}}(\omega) - i\frac{\sigma}{\omega^s \varepsilon_0} + \frac{\Delta\varepsilon_{\mathrm{Res}}}{1 + i\omega\tau_{\mathrm{Res}}} \tag{5a}$$

to the data. The first part is the model function of Havriliak/Negami [66] which reads

$$\varepsilon^*_{\mathrm{HN}}(\omega) - \varepsilon_\infty = \frac{\Delta\varepsilon}{(1 + (i\omega/\omega_0)^\beta)^\gamma} \tag{5b}$$

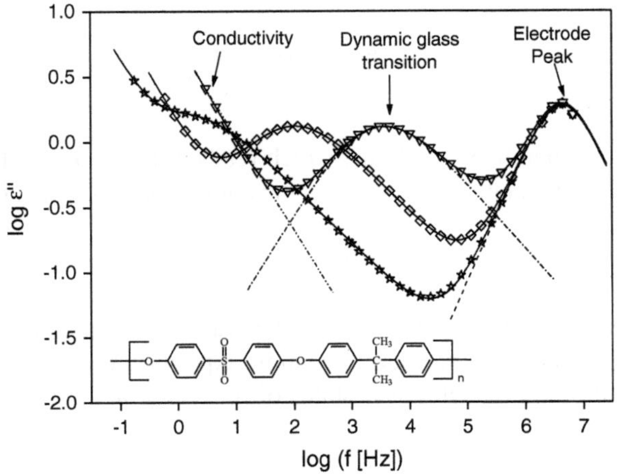

Fig. 4 Dielectric loss versus frequency for a thin polysulfone film with a thickness of 48 nm at different temperatures: *stars* 469 K; *diamonds* 477 K; *triangles* 487 K. The *solid lines* are fits of Eq. 5a to the data. The *dashed lines* are the contributions of the conductivity (*dashed-dotted-dotted*), the α-relaxation (*dashed-dotted*) and the resonance peak (*dashed*) for $T = 477$ K. The figure is redrawn from Ref. [54]. The inset gives the chemical structure for polysulfone

ω_0 is a characteristic frequency related to the frequency of maximal loss f_p (relaxation rate). ε_∞ describes the value of the real part ε' for $f \gg f_0$. β and γ are fractional parameters ($0 < \beta \leq 1$ and $0 < \beta\gamma \leq 1$) characterizing the shape of the relaxation time spectra. $\Delta\varepsilon$ denotes the dielectric strength. The second term in Eq. 5a describes conduction effects where σ is related to the DC conductivity. The parameter s ($0 < \mathrm{s} \leq 1$) describes Ohmic, for $\mathrm{s} = 1$, and non-Ohmic effects, for $\mathrm{s} < 1$, in the conductivity. The third term describes the electrode peak. In the case that the electrode peak is located outside of the experimental frequency window, the Debye function can be expanded, which results in

$$\varepsilon^*_{\mathrm{Fit}} = \varepsilon^*_{\mathrm{HN}}(\omega) - i\frac{\sigma}{\omega^s \varepsilon_0} - i\,\mathrm{const}\,\omega \tag{5c}$$

where const is a fitting parameter mainly due to τ_{Res}. Figure 4 demonstrates that the data measured for ultra-thin polysulfone films [54] can be well described by this approach.

2.2.3 Specific Heat Spectroscopy

Broadband dielectric and specific heat spectroscopy provide different windows to look at the glass transition phenomena. The formulation of the specific heat spectroscopy in the frame work of linear response theory was done by Schawe [67].

A detailed discussion can be found in Chap. "Dynamic calorimetric glass transition in thin polymer films" of this book [48]. In this case, the disturbance is an oscillation of the temperature and the reaction of the system is entropy (or enthalpy). The corresponding generalized compliance is the complex specific heat capacity , which can be directly compared with the complex dielectric function.

Specific heat spectroscopy is carried out by employing AC chip calorimetry. On the AC calorimeter chip, heaters as well as thermopiles are arranged [68]. The heat capacity of the system is measured by the temperature change sensed by the thermopiles in a lock-in approach (complex voltage). The differential setup developed by Schick et al. [59] is employed where the calorimeter chip XEN 39390 (Xensor Integration, Nl) is used as the measuring cell. The differential approach will minimize the contribution of the heat capacity of the empty sensor to the measured data. In the approximation of thin films (submicron), the heat capacity of the sample C_S is then given by

$$C_S = i\omega\overline{C}(\Delta U - \Delta U_0)/P_0 S \tag{6}$$

where $\overline{C} \equiv C_0 + G/i\omega$ describes the effective heat capacity of the empty sensor where $G/i\omega$ is the heat loss through the surrounding atmosphere, S is the sensitivity of the thermopile, P_0 is the applied heating power, ΔU is the complex differential thermopile signal for an empty and a sensor with a sample, and ΔU_0 is the complex differential voltage measured for two empty sensors. For identical sensors, $\Delta U_0 = 0$ holds.

The polymer film was directly prepared onto the SiO_2 surface of the sensor. The thickness of the thin films cannot be directly measured on the sensor. Therefore, a second set of films was prepared under identical conditions on a silicon wafer with a native SiO_2 layer to estimate the film thickness. Since the silicon wafer has similar surface properties to the sensor, it is assumed that under identical spin coating conditions the film on the silicon wafer has the same thickness as that on the sensor.

In the thin film approximation, absolute values of the complex specific heat capacity [69] can be deduced from the measured complex differential voltage using calibration techniques. Here the real part of the complex differential voltage and the phase angle are considered. Moreover, the temperature-scan-mode was employed, which means that the frequency was kept constant while the temperature was scanned with a heating/cooling rate of 1–2.0 K/min depending on the used frequency. During the measurements, the heating power for the modulation was kept constant at about 25 μW, which ensures that the amplitude of the temperature modulation is less than 0.5 K. Figure 5 shows an example of an AC chip calorimetry measurement on a polycarbonate film with a thickness of 38 nm [43].

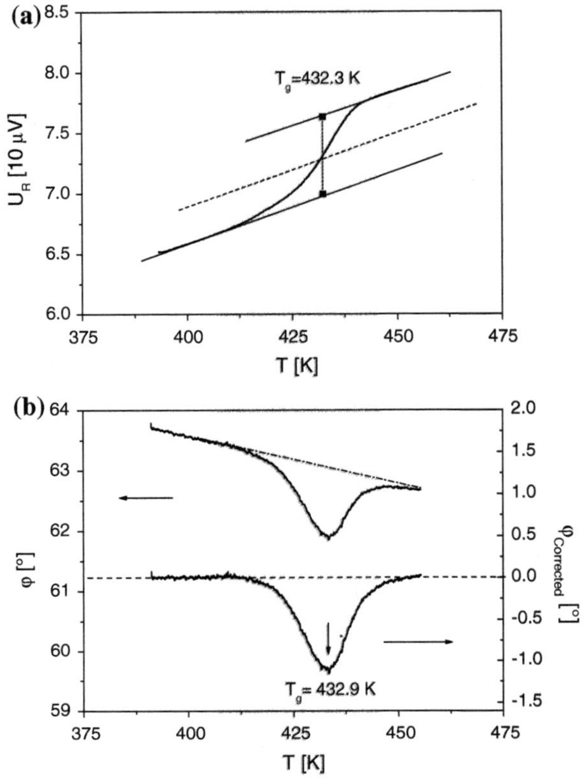

Fig. 5 Real part (**a**) and phase angle (**b**) of the complex differential voltage as measure of the specific heat capacity for a polycarbonate film (38 nm) at a frequency of 160 Hz. The contribution of the underlying step in heat capacity in the raw data of the phase angle (*upper panel* of Fig. 5b) was subtracted from the overall *curve* (*lower panel* of Fig. 5b). The heating rate was 1 K/min. Data were taken from Ref. [43]

2.2.4 Dielectric Expansion Dilatometry

Dielectric expansion dilatometry relies on the analysis of the temperature dependence of the dielectric permittivity $\varepsilon'(T)$ (or real part of the capacity $C'(T)$) under the assumption that no dielectric active processes take place in the selected frequency and temperature range. A detailed discussion of its theoretical basis is provided in Ref. [8]. Generally, the real part of the permittivity can be expressed by $\varepsilon'(f_e, T) = \varepsilon_\infty(T) + \Delta\varepsilon(f_e, T)$; where $\Delta\varepsilon$ is related to a dielectric dispersion, ε_∞ is the real part of the permittivity in the high frequency limit, and f_e is the detection frequency. For a thin polymer film capped between two thin aluminum electrodes (see inset Fig. 6), the geometrical capacitance $C_0(T)$ of the sample capacitor depends on temperature due to the thermal expansion of the film in a temperature scan experiment. In a linear approximation,

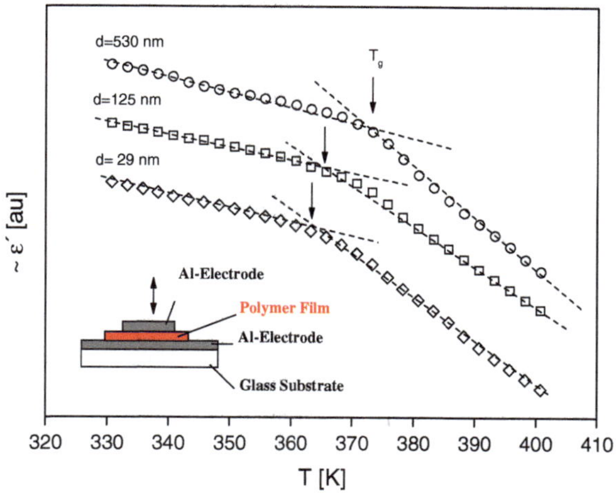

Fig. 6 Temperature dependence of the real part ε' of the complex dielectric function at a frequency of 1.7*10^5 Hz for polystyrene (M_W = 1408 kg/mol) at the indicated thicknesses. The *curves* are shifted on the y-scale for sake of clearness. *Lines* are linear fits to the data in the different temperature regions. Data were taken from Ref. [70]. The *scheme* in the *inset* shows the principle of dielectric expansion dilatometry

$$C_0(T) = \varepsilon_0 \frac{Ar}{d(T)} \sim \varepsilon_0 \frac{Ar}{d_R(1 + \alpha(T)\Delta T)} \tag{7}$$

where d is the thickness of the film, Ar is the electrode area, ΔT is the temperature change with regard to a reference temperature $T_R(T = T_R + \Delta T)$. $\alpha(T)$ is the thermal expansion coefficient normal to the film surface. For frequencies f_e, where no relaxation process is present in the experimental window and/or weakly polar polymers ($\Delta\varepsilon \approx 0$), $\varepsilon'(T) = \varepsilon_\infty(T) \sim \varepsilon_\infty(T_R)(1 - \alpha(T)\Delta T)$ holds. A thermal glass transition temperature can be estimated for nonpolar and/or weakly polar polymers from the change in the temperature dependence of ε' at sufficiently high frequencies. Figure 6 gives an example of dielectric expansion dilatometry for thin polystyrene films [70].

In difference to the dynamic glass transition temperature, which is estimated under equilibrium conditions a thermal glass transition temperature measured for instance by dielectric expansion dilatometry is accompanied by equilibrium/out-of-equilibrium transition of the system.

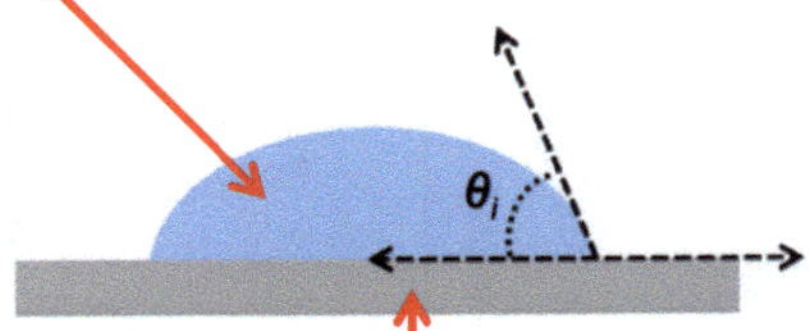

Fig. 7 Principle of contact angle measurements

2.3 *Surface Analytics*

2.3.1 Contact Angle Measurements

In Ref. [56], a correlation between the change of the thermal glass transition temperature with the film thickness and the interaction energy between the substrate and the polymer surface γ_{SP} is suggested. Contact angle measurements can be employed to estimate interaction energy between the substrate and the surface. In the Fowkes-van Oss-Chaudry-Good (FOCG) model [71], the surface tension γ^{Total} is given as the sum of the dispersive γ^{LW} and the polar component γ^{P} by

$$\gamma^{Total} = \gamma^{LW} + \gamma^{P} = \gamma^{LW} + 2\sqrt{\gamma^{+}\gamma^{-}} \tag{8}$$

where γ^{+} is the electron-acceptor and γ^{-} is the electron-donor component [71, 72]. The equation of Young and Dupré [71] can be applied to estimate γ^{LW}, γ^{+} and γ^{-} from the measured contact angels θ_i

$$(1 + \cos\theta_i)\gamma_1 = 2\lfloor(\gamma_X^{LW}\gamma_1^{LW})^{1/2} + (\gamma_X^{+}\gamma_1^{-})^{1/2} + (\gamma_X^{-}\gamma_1^{+})^{1/2}\rfloor \tag{9}$$

where the index 1 indicates a test liquid i and X symbolizes the polymer P or the substrate S. For the principle see Fig. 7. The Good-Girifalco-Fowkes [73] combining rule

$$\gamma_{SP} = \left(\sqrt{\gamma_S^{LW}} - \sqrt{\gamma_P^{LW}}\right)^2 + 2\left[(\gamma_S^{+}\gamma_S^{-})^{1/2}(\gamma_P^{+}\gamma_P^{-})^{1/2} - (\gamma_S^{+}\gamma_P^{-})^{1/2} - (\gamma_S^{-}\gamma_P^{+})^{1/2}\right] \tag{10}$$

allows to estimate γ_{SP} from the experimental data.

2.3.2 X-Ray Photoelectron Spectroscopy

With the method of X-ray photoelectron spectroscopy (XPS), the chemical composition of a surface can be analyzed with a penetration depth of several nm (3–7 nm) [74]. In these investigations, a polymeric surface is irradiated with X-rays. The X-rays generate photoelectrons, which can be analyzed with regard to their frequency of occurrence at certain energy. The value of the energy will provide information about the chemical bonding of an atom within a molecule and therefore about the composition.

3 Results

3.1 Flexible Polymers

3.1.1 Polystyrene

Broadband dielectric spectroscopy, dielectric expansion dilatometry and specific heat spectroscopy were combined to compare the thickness dependence of the dynamic glass transition temperature (broadband dielectric and specific heat spectroscopy) with that of the thermal glass transition temperature sensed by dielectric expansion dilatometry for polystyrene samples. The different molecular weights have been selected to discuss the influence of the molecular weight on the thermal glass transition of thin supported films. It should be explicitly noted that the data obtained by broadband dielectric spectroscopy and dielectric expansion dilatometry were measured simultaneously on the same sample.

Dynamic Glass Transition Temperature

Figure 8a gives the normalized dielectric loss at a fixed frequency ($f = 6$ kHz) versus temperature for polystyrene with a molecular weight of 1408 kg/mol (PS1408) for several film thicknesses. The data treatment was discussed in detail in Ref. [70]. Figure 8a shows that that the loss peaks for all film thicknesses collapse into a single curve. This means that the segmental dynamics measured by broadband dielectric spectroscopy is independent of the film thickness down to the lowest film thickness. This is first in agreement with literature results [20, 44] and secondly for all molecular weights studied [70].

The relaxation map can be deduced (see Fig. 9a) from the maximum position of the dielectric loss. The data are curved when plotted versus $1/T$ as expected for glassy dynamics. Such a dependence can be described by the Vogel/Fulcher/Tammann-(VFT-) equation [75–77]

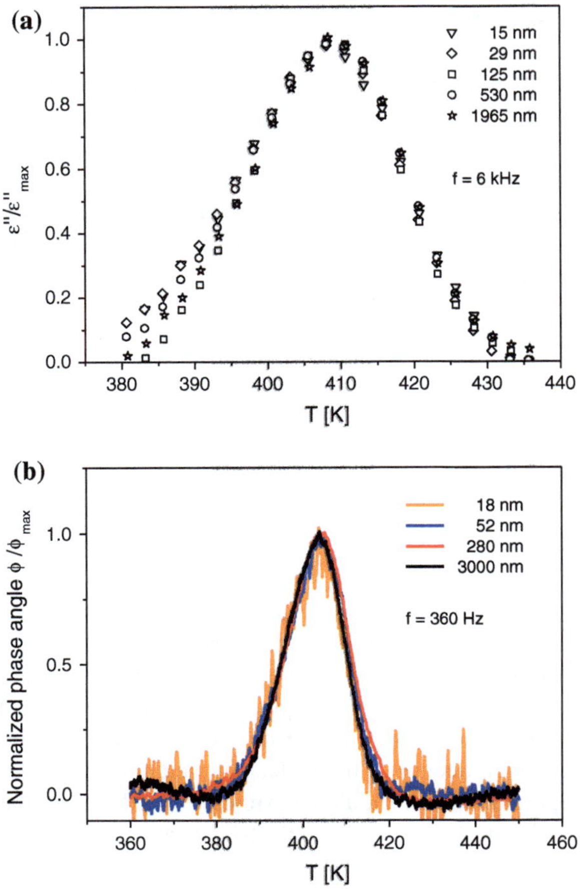

Fig. 8 **a** Normalized dielectric loss versus temperature at a frequency of 6 kHz for thin polystyrene films ($M_W = 1408$ kg/mol) at the indicated film thicknesses. All data are taken from Ref. [70]. **b** Normalized phase angle measured by specific heat spectroscopy for thin polystyrene films ($M_W = 1408$ kg/mol) with a frequency of 360 Hz at the indicated film thicknesses. All data are taken from Ref. [70]

$$\log f_p = \log f_\infty - \frac{A}{T - T_0} = \log f_\infty - \frac{\ln(10) D\, T_0}{T - T_0} \tag{11}$$

where f_∞ and A are parameters. T_0 is called Vogel or ideal glass transition temperature, which is found to be 30–70 K below the thermal glass transition temperature. D is the so-called fragility parameter. Among others [78, 79] it can be used as a quantitative measure of "fragility", which provides a useful scheme to classify glass-forming materials. Glass formers are called "fragile" if their $f_p(T)$ dependence deviates strongly from an Arrhenius-type behavior (high values of D) and "strong" if $f_p(T)$ is close to the latter (low values of D). As it is already obvious from the raw

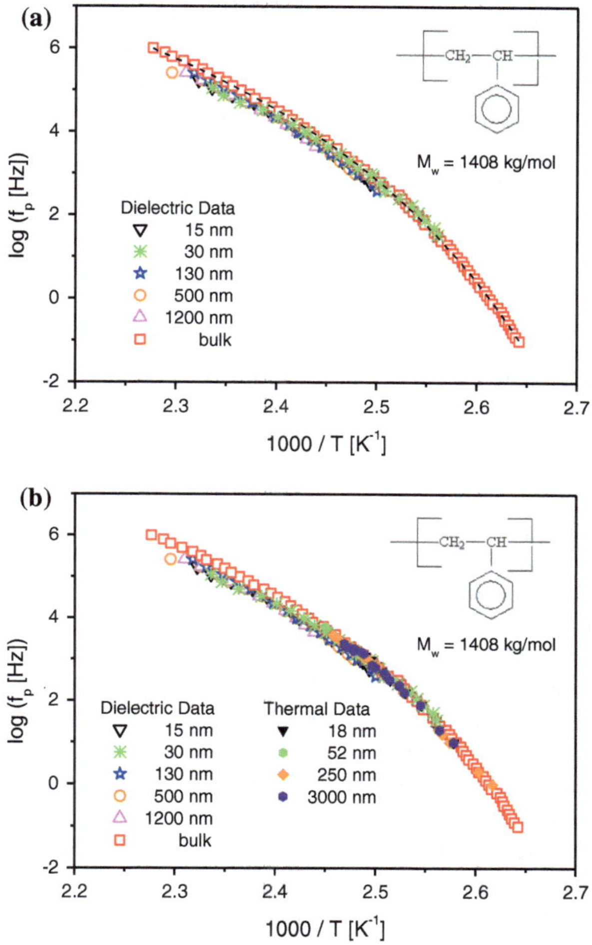

Fig. 9 **a** Dielectric relaxation rate f_p versus $1/T$ for thin polystyrene films ($M_W = 1408$ kg/mol) at the indicated film thicknesses. All data are taken from Ref. [70]. The *line* is a fit of the VFT-equation to the bulk data. **b** Dielectric and thermal relaxation rates f_p versus $1/T$ for thin polystyrene films ($M_W = 1408$ kg/mol) at the indicated thicknesses

data presented in Fig. 8a the temperature dependence of the relaxation rate measured for all film thicknesses is identical. This means that the dynamic glass transition temperature measured by dielectric spectroscopy in equilibrium does not depend on the film thickness. Similar results were obtained for thin films prepared from polystyrene samples with molecular weights of 260 and 50 kg/mol.

The segmental dynamics of thin polystyrene films were further investigated by means of specific heat spectroscopy. The real part U_R and the phase angle of the complex differential voltage were obtained as a measure of the complex heat capacity. Figure 8b gives the temperature dependence of the normalized phase angle measured at a frequency of 360 Hz at the given film thicknesses. Similarly to the dielectric data, all curves collapse into one chart indicating that the segmental dynamics, and therefore also the dynamic glass transition, studied by specific heat spectroscopy, is independent of the film thickness, analogously to the findings obtained by broadband dielectric spectroscopy.

The specific heat spectroscopy data can be further analyzed by constructing the relaxation map (Fig. 9b). First, Fig. 9b shows that the relaxation rates measured by specific heat spectroscopy have the same temperature dependence for all considered film thicknesses. Second, the temperature of the relaxation rates measured by specific heat and dielectric spectroscopy are identical.
In conclusion, for polystyrene, the dynamic glass transition temperature measured in equilibrium in the framework of linear response approach is independent of the film thickness down to values of d of several nanometers. This agrees with literature results [44, 48].

Decoupling of Dynamic and Thermal Glass Transition Temperature

Before staring the discussion on thin films, one has to state that most researchers will agree that the dynamic and thermal glass transition should have a similar (or more strongly an identical) dependence on the structure of the sample. The difference between the thermal and the dynamic glass transition is that the latter is estimated under equilibrium (or metastable equilibrium) conditions where the former one is related to an out-of-equilibrium transition.

Figure 6 depicts results of dielectric expansion dilatometry for polystyrene with a molecular weight of 1408 kg/mol for different thicknesses. All samples are well annealed in an oil-free vacuum at $T = 433$ K for 4 days. From the kink-like change of the temperature dependence of the real part of the complex permittivity, a thermal glass transition temperature can be retrieved. With decreasing film thickness, the thermal glass transition temperature decreases.

This result, measured by dielectric expansion dilatometry for the thermal glass transition, is contrast to the conclusion drawn from dielectric relaxation and specific heat spectroscopy, where a thickness independent dynamic glass transition temperature was found. This means that for thin films the dynamic and thermal glass transition temperature can have a different dependence on thickness. In a first conclusion, this implies that there is a decoupling of the thickness dependence of the dynamic and thermal glass transition temperature [45, 70] for thin films. (Please also not that in the dielectric case dielectric expansion dilatometry and dielectric relaxation spectroscopy were measured simultaneously on the same sample.) The same effect was observed for other molecular weights [70] and also polystyrene nanospheres [80].

One attempt to discuss these controversial results is based on the fact that the dynamic T_g (segmental dynamics) is measured under equilibrium conditions, whereas the measurement of the thermal T_g involves a transition from an equilibrium to a nonequilibrium state. Recently, a free volume hole diffusion model (see for instance [81–84]) based on ideas of Simha et al. [85] was used to rationalize this experimental finding [45]. It can describe simultaneously a T_g depression and an invariant segmental dynamics. In the frame of this model, a polymer can maintain equilibrium when cooling down by diffusing free volume holes out of the available interfaces of the system. Thus, the efficiency of maintaining equilibrium is determined by the velocity of the diffusion of the free volume holes, which is related to both the molecular mobility of the polymer and the amount of interface. The latter is supposed to scale with the inverse of the film thickness. A much deeper discussion can be found in Chap. "Equilibrium and Out-of-Equilibrium Dynamics in Confined Polymers and Other Glass Forming Systems by Dielectric Spectroscopy and Calorimetric Techniques" of this book [49].

Molecular Weight Dependence of the Thermal Glass Transition Temperature

The thickness dependence of the thermal glass transition temperature for thin polystyrene films with a molecular weight of 1408 kg/mol (PS1408) is given in Fig. 10. As evident from the data given in Fig. 6, the thermal glass transition temperature T_g decreases with decreasing film thickness. These data are compared with results of freestanding polystyrene films of the same molecular weight [45] measured also by dielectric expansion dilatometry. The freestanding films show a much stronger depression of T_g with decreasing film thickness than those capped between Al-electrodes. These data demonstrate the influence of the deposited electrodes, which will lead to the formation of a boundary layer with a reduced mobility. The formation of such a boundary layer by adsorption processes was in detail discussed for polystyrene on AlO_x in Ref. [86]. This boundary layer with a reduced mobility would lead to weaker thickness dependence for a film capped between the AlO_x electrodes than for freestanding films.

In addition to experiments carried out on thin polystyrene films with a molecular weight of 1408 kg/mol, measurements were done on samples with molecular weights of 260 kg/mol (PS260) and 50 kg/mol (PS50). All samples were well annealed for 4 days in vacuum above the glass transition temperature of the bulk (for details see [70]). Similar to the results presented in Fig. 6, for the films of the molecular weight of 1408 kg/mol, Fig. 11 depicts data for polystyrene films of a molecular weight of 50 kg/mol. In contrast to PS1408, for polystyrene films with a molecular weight of 50 kg/mol, T_g increases with decreasing film thickness (see also Fig. 10). In addition, the data for the sample PS260 are in-between those of PS1408 and PS50. This means for PS260, T_g decreases with decreasing film thickness d yet still much weaker than that of PS1408 (see Fig. 10). (It should be noted that a similar dependence is also observed for polystyrene films by varying the cooling rate [87].) Therefore, it can be concluded that, under the same annealing conditions, the thickness dependence of the thermal glass transition temperature of thin polystyrene films depends on the molecular weight. The results presented here agree with literature data [88–90], although it is also reported that the glass transition temperature of ultra-thin polystyrene films is independent of the molecular weight [9, 10, 91].

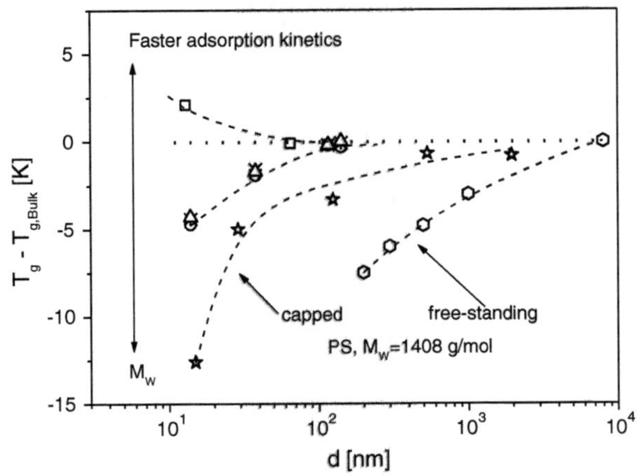

Fig. 10 Thermal glass transition temperature for thin polystyrene films versus film thickness d estimated by dielectric expansion dilatometry: *stars* $M_w = 1408$ kg/mol, *triangles*, *asterisks*, *circles* $M_w = 260$ kg/mol (different annealing times), *squares* $M_w = 50$ kg/mol. All samples were annealed at 433 K for 4 days. All data are taken from Ref. [70]. *Diamonds* represents data for free-standing polystyrene films with a molecular weight of $M_w = 1408$ kg/mol taken from Ref. [45]. *Lines* are guides for the eyes

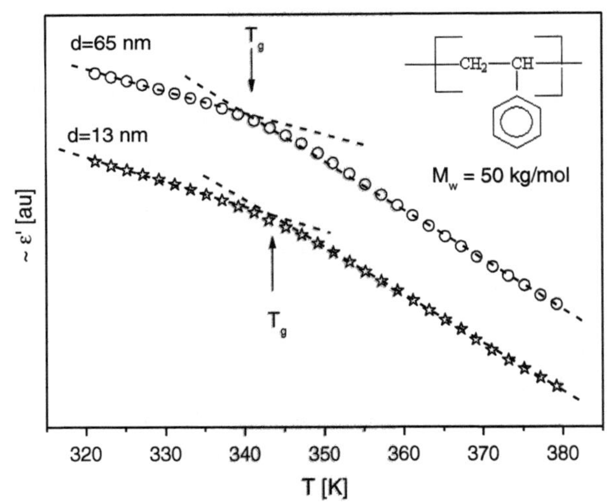

Fig. 11 Temperature dependence of the real part ε' of the complex dielectric function versus temperature at frequency of $1.7 * 10^5$ Hz for polystyrene ($M_W = 50$ kg/mol) at the indicated thicknesses. The *curves* are shifted on the *y*-scale for sake of clearness. *Lines* are linear fits to the data in the different temperature regions. Data were taken from Ref. [70]

The observed dependence of the thermal glass transition temperature on the molecular weight is also in contradiction to the assumption that the interaction energy between the polymer segments and surface of the substrate is the only parameter, which determines the value of the glass transition temperature of ultra-thin polystyrene films. This will be discussed in more detail below.

The molecular weight dependence of the thermal glass transition can be explained in terms of the effect of an irreversibly adsorbed layer onto the Al substrate of PS segments melt formed by annealing the film at a temperature higher than the glass transition temperature of the bulk [21, 25, 86]. Chain arrangements at the interface between the polymer and the surface of the substrate result in a local densification of chain segments, which will reduce the free surface effect. These chain arrangements will require a given chain mobility, which depends of course on the molecular weight of the polymer. At the same temperature, the molecular mobility is higher for lower molecular weight samples than that of polymers with higher molecular weight. This implies that the formation and densification of the irreversibly adsorbed surface layer is more advanced for a lower molecular weight polymer than that of a higher molecular weight one. Because of the fact that the interfacial layer with a reduced mobility will compensate the free surface effect on the depression of the thermal glass transition temperature of thin films (evidenced for the free standing films), the T_g depression of thin films should be smaller for a lower molecular weight polymer than for a macromolecule with a higher molecular weight. This is observed experimentally (see Fig. 10). For the lowest molecular weight of 50 kg/mol the higher densification of the surface layer with a reduced mobility will even lead to an increase of the thermal glass transition temperature (see Fig. 10) indicating that the effect of the free polymer/air interface is overcompensated by the reduced mobility layer. This line of argumentation agrees with the consideration given in Ref. [86], where an increase of T_g with increasing annealing time above T_g was reported.

To analyze the results obtained here in details, a model first proposed by Tsui et al. was modified and used [26]. The deviation of the glass transition temperature of a film with the thickness d compared to the bulk value can be described by

$$\Delta T_g = T_g - T_{g,\,\mathrm{Bulk}} = \frac{\delta T_g}{\delta \rho}\frac{2\xi}{d}(\rho_i - \rho). \tag{12}$$

where ρ_i is the density of the polymer over a distance ξ from the AlO_x substrate surface and the remaining film with a thickness of $d - 2\xi$ having the bulk density ρ. $\delta T_g/\delta_\rho$ is the variation of T_g due to a change in the mass density ρ. For polystyrene a value of $\delta T_g/\delta_\rho$ of $1.35 \times 10^3 \mathrm{cm}^3$ K/g was taken from Ref. [26]. For film of PS1408 with the thickness of 15 nm, a 13 K reduction in T_g is observed for $2\xi(\rho_i - \rho)$ with a value of 1.67×10^{-8} g/cm^2. If $\xi = 3$ nm (several Kuhn lengths) is assumed for the calculation, $\rho_i - \rho = -0.028\,\mathrm{g/cm}^3$ is obtained. This means that only a ca. 3 % decrease of the density at both Al interfaces is required for a 13 K reduction in the glass transition temperature for a film thickness of 15 nm. For a 14-nm-thin film of PS260 a 5 K decrease in T_g requires a ca. 1 % reduction of the density. For the case $\rho_i - \rho = 0$ the glass transition temperature of a film will be independent of the film thickness. When $\rho_i - \rho$ is positive, it results in an increase in T_g. This model describes the combined effects on the T_g deviation of thin polymer films. Moreover, it provides a quantitative measure for the perturbations on the overall segment density.

For the sake of completion, the model of the diffusion of free volumes holes proposed by Boucher et al. [81–84] can be employed to describe the molecular weight dependence of $T_g(d)$ for ultra-thin polymer films.

3.1.2 Poly(vinyl methyl ether)

The thickness dependence of the dynamic glass transition temperature for thin films of poly(vinyl methyl ether) were investigated by specific heat spectroscopy [63]. Figure 12a shows the temperature dependence of the normalized phase angle for PVME measured at a frequency of 640 Hz. Down to the lowest film thickness of 12 nm, all datasets collapse into one chart. This indicates that the dynamic glass transition temperature for poly(vinyl methyl ether) films is independent of the film thickness [63], similarly to polystyrene (see Fig. 8b). Similar information is obtained from the real part of the complex differential voltage, which is normalized to $U_R/(U_{R,T=210k} - U_{R,T=330\,k}) - U^N_{R,Tg}$ (see Fig. 12b). For all film thicknesses, the rescaled data collapse into one chart. This indicates that the temperature corresponding to half of the amplitude step, as well as that of maxima of the phase angle of the complex differential voltage, is independent of the film thickness down to 12 nm.

The dynamic glass transition temperatures obtained for different frequencies are used to construct the relaxation map with the film thickness as a parameter (see Fig. 13). The relaxation rates for poly(vinyl methyl ether) measured by broadband dielectric spectroscopy are also included. The calorimetric data are close to the dielectric ones. At a given frequency, the data measured for different film thicknesses are located in a quite narrow temperature range of ± 2 K, which is within the uncertainty of the AC calorimeter measurement. The VFT-equation is fitted to all thermal data. This fit describes the calorimetric data quite well and the deviation of the experimental result from the fit is within the experimental error limit (± 2 K) even for the thinnest film thicknesses and the lowest frequencies.

In conclusion, the dynamic glass transition temperature for thin poly(vinyl methyl ether) films is independent of the film thickness.

A systematic shift between the data estimated from specific heat and dielectric spectroscopy is observed. The trace of the calorimetric points is shifted by about one order of magnitude to lower frequencies compared to the dielectric data. This shift is much larger than the experimental errors and consistent with the behavior reported in Ref. [92]. Jacobsen et al. [93] provide similar results for two simple glass-formers although the differences of the different datasets are small. On the other hand, it was also reported for some other polymers that the dielectric and calorimetric data collapse into one chart (see for instance Fig. 9b and Ref. [44, 94, 95]). From this literature survey, one can concluded that specific heat and dielectric spectroscopy detect the dynamic glass transition process, while each method provides a different window to look at the underlying phenomena. Dielectric spectroscopy is sensitive to fluctuations of dipoles, while specific heat spectroscopy detects entropy (or enthalpy) fluctuations. The reason for this shift is not yet understood till now.

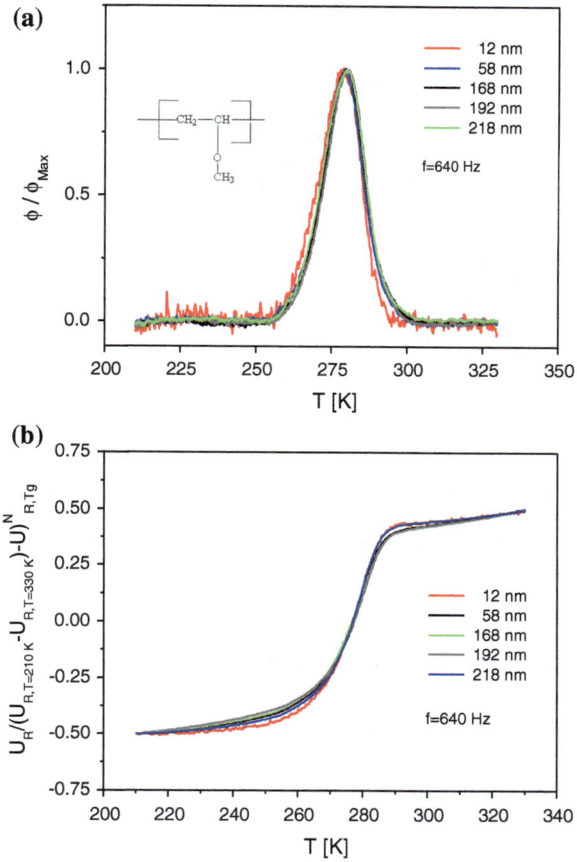

Fig. 12 Normalized phase angle (**a**) and rescaled real part (**b**) of the complex differential voltage measured by specific heat spectroscopy for thin poly(vinyl methyl ether) films at a frequency of 640 Hz at the indicated film thicknesses. The heating rate was 1 K/min. All data are taken from Ref. [63]

3.1.3 Poly(2-vinylpyridine)

Besides polystyrene and poly(vinyl methyl ether) thin films of poly(2-vinylpyridine) (P2VP) were investigated as a further example for flexible macromolecules by specific heat spectroscopy [96]. Recently, a dielectric study about the dynamics of condensed (semi-) isolated poly(2-vinylpyridine) chains have been published [46, 97]. A thermal expansion study of thin P2VP films [25] on cleaned SiO_2 have shown strong increase of the thermal glass transition temperature with decreasing film thickness.

The inset of Fig. 14 gives the normalized phase angle of the complex differential voltage versus temperature measured at a frequency of 160 Hz by specific heat spectroscopy. Within the experimental error, the data measured for the different film

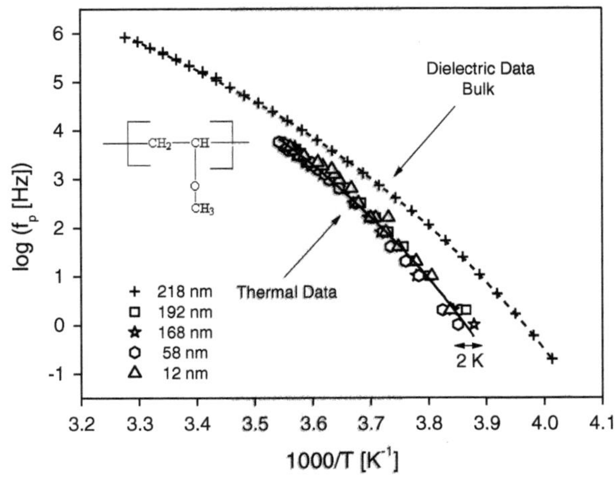

Fig. 13 Relaxation rate f_p versus $1/T$ for thin poly(vinyl methyl ether) films at the indicated film thicknesses. The *solid line* is a fit of the VFT-equation to all thermal data. *Crosses* represent the dielectric data measured for a bulk sample. The *dashed line* is a fit of the VFT-equation to the dielectric data. All data are taken from Ref. [63]

thicknesses overlay. Therefore, the dynamic glass transition is independent of the film thickness down to 10 nm.

This is further confirmed by the relaxation map (Fig. 14). All data points estimated for the different frequencies and film thicknesses collapse more or less into one chart. Within the experimental error, the data for the different film thicknesses can be described by a single VFT fit.

3.1.4 Conclusions: Flexible Polymers

In conclusion, for the investigated flexible macromolecules, the dynamic glass transition temperature estimated by broadband dielectric spectroscopy and/or specific heat spectroscopy is independent of the film thickness. In other words, within the experimental error of the methods, the dynamic glass transition temperature is equal to the bulk value down to film thicknesses of ca. 10 nm.

For polystyrene it was found that the thermal glass transition temperature could depend on the film thickness. The different thickness dependence of the dynamic and thermal glass transition temperature is not expected from the bulk behavior and is not well understood till now. It might be related to the fact that the thermal glass transition is related to an out-of-equilibrium transition where the dynamic glass transition temperature is measured under equilibrium conditions.

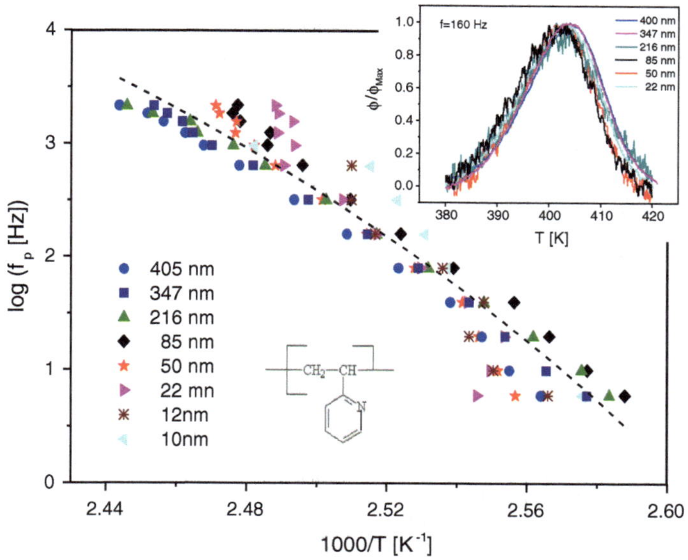

Fig. 14 Relaxation rate f_p versus $1/T$ for thin poly(2-vinylprydine) films at the indicated film thicknesses. The *dashed line* is a fit of the VFT-equation to all data. The *inset* gives the normalized phase angle of the complex differential voltage measured by specific heat spectroscopy for thin poly(2-vinylpyridine) films at a frequency of 160 Hz at the indicated film thicknesses. The heating rate was 2 K/min. All data were taken from Ref. [96]

3.2 Main Chain Polymers

The inset of Fig. 15 shows results of the dielectric expansion dilatometry for the main chain polymer poly(bisphenol A carbonate) (polycarbonate, PC) [42]. The normalized real part of the complex dielectric function decreases with increasing temperature. A thermal glass transition temperature can be estimated at a given frequency, whereas the temperature dependence of ε' changes. In Fig. 15, the difference in the glass transition temperature for thin films and the bulk is plotted versus the film thickness d. With decreasing film thickness, T_g increases strongly for film thicknesses smaller than 20 nm. This behavior implies an interaction of polycarbonate with the aluminum substrates (electrodes), which leads to the formation of an adsorbed boundary layer with reduced mobility. This will be discussed in detail later.

Besides dielectric expansion dilatometry, dielectric relaxation spectroscopy was carried out simultaneously on the identical polycarbonate samples. The estimated dielectric relaxation rates are plotted for the different thicknesses versus inverse temperature (relaxation map) in Fig. 16.

All data sets are curved in such a representation as found typical for glassy dynamics and can be described by the VFT-equation. To analyze the temperature dependence of the relaxation rate in more detail, a derivative method can be employed [98]. Using Eq. (11), it holds

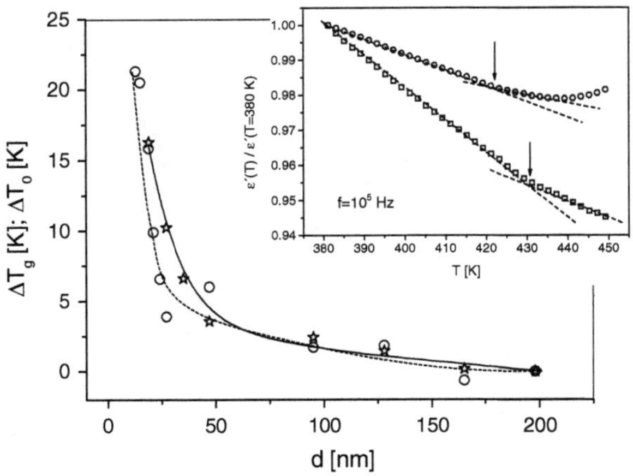

Fig. 15 Dependence of the glass transition temperature (*circles*) and the Vogel temperature (*stars*) on the film thickness for polycarbonate. The *dashed* and the *solid lines* are guides for the eyes respectively. The *inset* gives an example for dielectric expansion dilatometry for thin polycarbonate films at thicknesses of 13 nm (*squares*) and 47 nm (*circles*). *Lines* are linear regressions to the data in the corresponding temperature regions. Data were taken from Ref. [42]

$$\left[\frac{\mathrm{d}\log f_p}{\mathrm{d}T}\right]^{-1/2} = A^{-1/2}(T - T_0). \tag{13}$$

Therefore, a dependence in accordance to the VFT equation shows up as a straight line in the derivative plot. This technique was applied to the data of thin PC films (see inset of Fig. 16). For all thicknesses, the data can be well described by straight lines. From the intersection with $y = 0$, the Vogel temperature T_0 can be extracted. Figure 15 compares the thickness dependence of T_0 with that of T_g estimated from the dielectric expansion dilatometry. Both temperatures estimated independently from each other have a similar dependence on the film thickness indicating that the same process is sensed.

In addition to the dielectric data, Fig. 16 depicts also the relaxation rates as a function of inverse temperature obtained for thin polycarbonate films by specific heat spectroscopy [43]. Compared to the dielectric data, the results obtained by thermal spectroscopy are shifted slightly to lower temperatures.

Similar to the dielectric data, the temperature dependence of the relaxation rates estimated from specific heat capacity is curved when plotted versus $1/T$. All data can be described by a common set of VFT parameters. This means down to the lowest film thickness no dependence of the glass transition temperature on the film thickness is observed within the experimental uncertainty of ca. 3 K.

The different thickness dependence of the polycarbonate films measured by dielectric and specific heat spectroscopy could be explained by the different interaction of the polymer segments with the substrate. Figure 17 gives the dependence of the glass

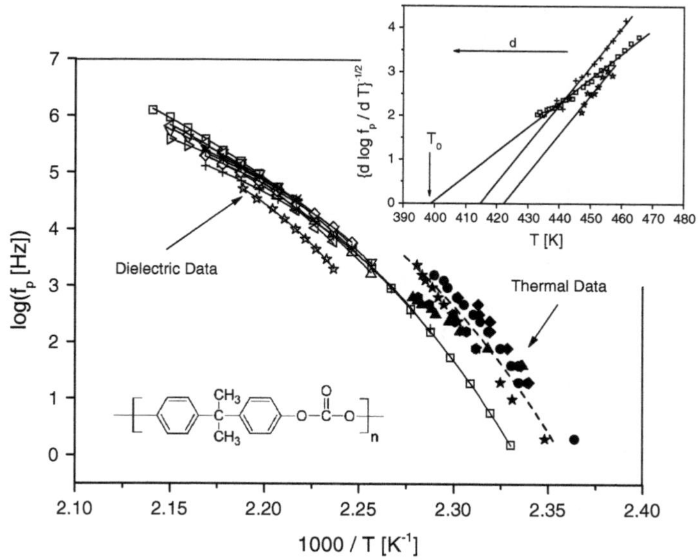

Fig. 16 Relaxation rate f_p versus $1/T$ for thin polycarbonate films. Open data points are dielectric data for the film thicknesses: *squares* bulk, *triangle down sided* 198 nm, *triangle left sided* 165 nm, *triangle right sided* 128 nm, *triangle up sided* 95 nm, *diamonds* 47 nm, *crosses* 35 nm, *stars* 27 nm, *open stars* 19 nm. Data are taken from Ref. [42]. *Solid* symbols are thermal data for the film thicknesses: *hexagons* 10 nm, *diamonds* 13 nm, *circles* 38 nm, *stars* 41 nm, *triangles* 55 nm. Data are taken from Ref. [43]. *Lines* are fits of the VFT-equation to the different datasets. The *inset* gives $\{d \log f_p/dT\}^{-1/2}$ versus T for selected thicknesses: *squares* bulk, *crosses* 35 nm, *stars* 19 nm. *Lines* are linear fits to the different datasets

transition temperature versus the interaction energy between the substrate and the polymer segments γ_{SP} for ca. 20 nm thin polystyrene and polycarbonate films according to Ref. [56], where the number of polar groups of an octodecyltrichlorsilane/SiO_2 surface was varied by X-ray radiation. A correlation between the change of the T_g of the film and the interaction energy was observed. For polycarbonate on AlO_x γ_{SP} is 2.51 mJ/m^2 [42], while for PC on SiO_2 $\gamma_{SP} = 2.15$ mJ/m^2 is estimated [43]. These two values agree with the correlation line found in Ref. [56], which might support the assumption that the change in the glass transition temperature of thin films is due to the interaction of the polymer segments with the substrate.

The presence of a reduced mobility layer with a profile of mobility affecting the segmental dynamics of the film [99, 100] should be also reflected in the relaxation time distribution of the segmental dynamics, this means in the width of the α-peak. Figure 18 shows that the loss peak related to segmental dynamics broadens with decreasing film thickness. This broadening is due to heterogeneity in the molecular dynamics, related to the introduction of slower modes in the distribution of relaxation times. At the polymer/metal interface, the mobility of chains is hindered down to the segmental motion, due to less available space [101] and the favorable interactions with the AlO_x surface. The perturbations into the chain conformations responsible

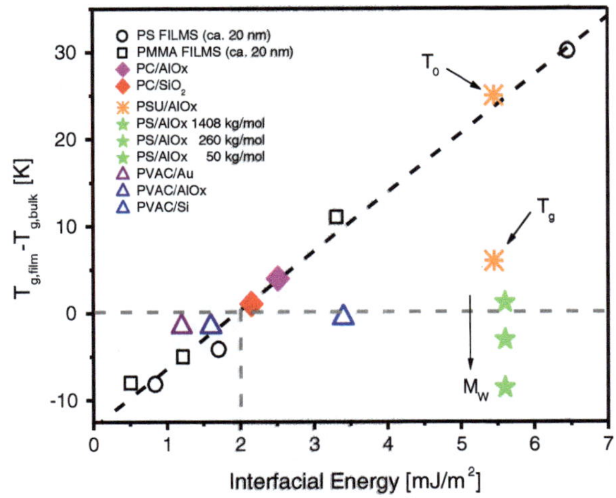

Fig. 17 Difference of the glass transition temperature for 20-nm-thick films and the bulk value versus the interfacial energy. The *open symbols* for poly(methyl methacrylate) and polystyrene are taken together with the *correlation line* from Ref. [56]. The data (*diamonds*) for polycarbonate on AlO_x are from [42] and those on the SiO_2 from Ref. [43]. The *asterisks* represent data for polysulfone taken from Ref. [54]. The glass transition temperatures (*stars*) are taken from literature source [70] where the corresponding interaction energy between polystyrene and AlO_x is taken from Ref. [86]. The data for poly(vinyl acetate) prepared on the indicated surfaces are taken from Ref. [104]

for such deviation from bulk dynamics vanish after exceeding a dynamic length scale λ [42].

Figure 19 gives the relaxation map for polysulfone for the dynamic glass transition. The derivative method described above is employed to estimate the parameters of the VFT equation (Eq. (11); for details see Ref. [54]). From the VFT parameters as a measure for T_g a dielectric glass transition temperature, T_g^{Diel} can be calculated and plotted versus the film thickness, in the inset of Fig. 20. Besides the absolute values, the qualitative behavior is more or less the same as that for polycarbonate. With decreasing film thickness, T_g increases strongly for film thicknesses smaller than 20 nm.

Like for polycarbonate the interaction energy between the segments and the surface was estimated and a value of $\gamma_{SP} = 5.45$ mJ/m^2 [54] was obtained. This value is much larger than the value of 2.51 mJ/m^2 [42] measured for polycarbonate where a similar increase of the glass transition temperature of about 5 K (for a ca. 20-nm-thickfilm) was found. This means by using the glass transition temperature the corresponding point for polysulfone is located far away from the correlation line between the polymer/substrate interaction energy and the change in the glass transition temperature (see Fig. 17). This result might provide some evidence that the interaction energy between the polymer segments and the surface of the substrate is not the only relevant parameter for the glass transition behavior of thin films.

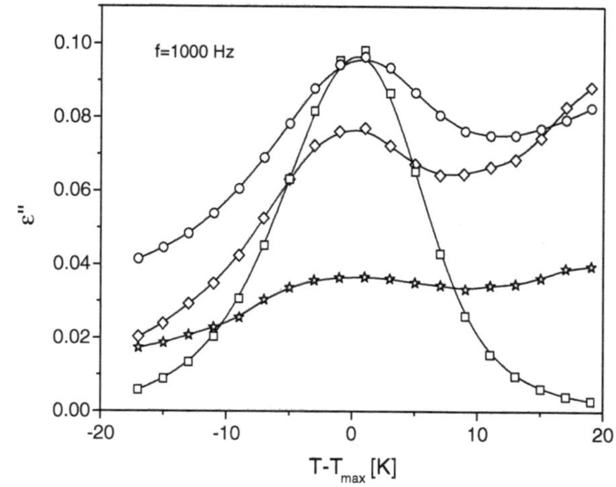

Fig. 18 Dielectric loss versus normalized temperature at 1 kHz for polycarbonate at different film thicknesses: *squares* bulk, *circles* 95 nm, *diamonds* 47 nm, *full stars* 19 nm. The *lines* are guides for the eyes. The data were taken from Ref. [42]

Figure 20 compares the thickness dependence of the Vogel temperature T_0 for polycarbonate and polysulfone. As a result, for polysulfone $T_0(d)$ is much stronger than that of polycarbonate. This is probably directly related to the higher interaction energy between the polysulfone segments and the AlO_x substrate. Moreover, by taking T_0 as measure for the thermal glass transition temperature, the correlations between the change in the glass transition temperature and the polymer substrate / interaction energy [56] is fulfilled (see Fig. 17).

The Debye theory of dielectric relaxation generalized by Kirkwood and Fröhlich [102] predicts for the temperature dependence of the dielectric relaxation strength

$$\Delta\varepsilon = \frac{1}{3\varepsilon_0} g \frac{\mu^2}{k_B T} \frac{N}{V} \tag{14}$$

where μ is the mean dipole moment of the process under consideration and N/V is the number density of dipoles involved. g is the so-called Kirkwood/Fröhlich correlation factor, which describes static correlation between the dipoles. The Onsager factor is omitted for sake of simplicity. The inset of Fig. 21 gives the temperature dependence of the dielectric relaxation strength estimated from the fit of the HN-function to the polysulfone data for different film thicknesses. With decreasing thickness $\Delta\varepsilon$ decreases strongly. $\Delta\varepsilon$ is taken at a temperature of 485 K and plotted versus inverse layer thickness in Fig. 21. As it can be already seen from the raw data, $\Delta\varepsilon$ decreases strongly but follows a linear dependence when plotted versus inverse layer thickness. Because the dipole moment is a molecular property it should not depend on the film thickness. Further, assuming that the correlation factor g weakly depends on the film thickness, it is concluded that the strong decrease of $\Delta\varepsilon$ is due to a strong reduction

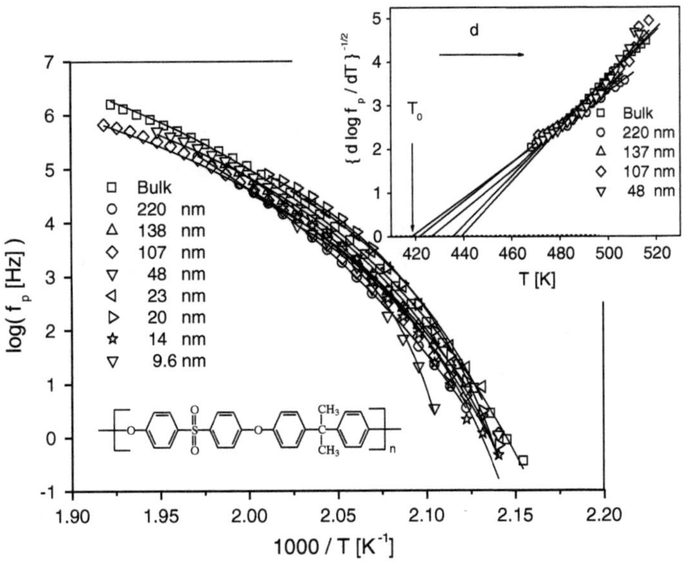

Fig. 19 Relaxation rate f_p versus $1/T$ for thin polysulfone films at the indicated thicknesses measured by dielectric spectroscopy. The *lines* are fits of the VFT-equation as described in the text. The *inset* gives $\{d\log f_p/\mathrm{d}T\}^{-1/2}$ versus T for selected labeled thicknesses. All data were taken from Ref. [54]

of the number density of fluctuating dipoles with decreasing film thickness. This can be understood by employing the reduced mobility layer model discussed above. Due to the high interaction energy between the polysulfone segments and AlO_x surface, the segments close to the electrodes get adsorbed at the substrate. Because of this interaction, these segments have a strongly reduced mobility, in comparison to the time scale of the dynamic glass transition of bulk polysulfone. For polycarbonate [42] as well as for other polymers [11, 16, 22], a corresponding behavior is observed. In general, a gradient of molecular mobility starting from the substrate into the middle of the film as suggested in Ref. [55] has to be discussed.

Equation (11) defines the fragility parameter D, which is plotted for polycarbonate and polysulfone versus the reciprocal film thickness in Fig. 22. For both polymers, a similar dependence of D on the film thickness is observed. First, it decreases with decreasing film thickness and reaches a more or less constant value for the thinnest films. This value seems to be similar for both main chain polymers. This decrease of the fragility parameter with decreasing film thickness corresponds to change of the system from a fragile to a stronger one. In Ref. [103], it was discussed that the fragility of the bulk system is one key parameter for the thickness dependence of the glass transition temperature for thin polymer films. Although the fragility was not estimated in dependence of the film thickness, this hypothesis, however, seems to disagree with the experimental results presented here.

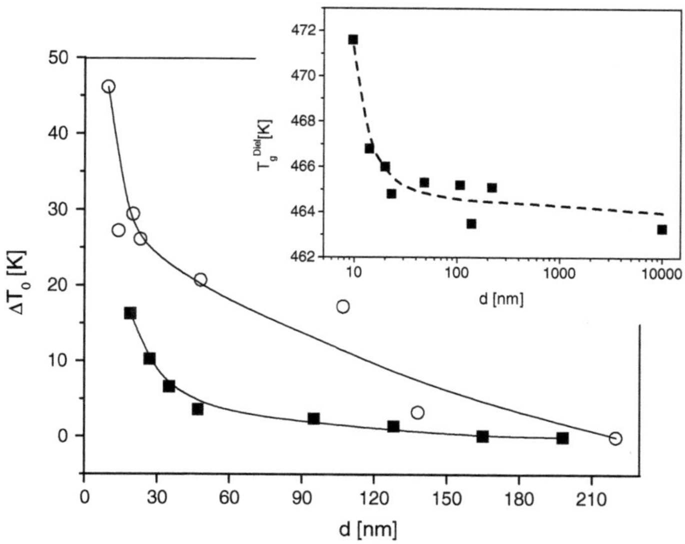

Fig. 20 Thickness dependence of the Vogel temperature with respect to the bulk value for polycarbonate (*solid squares*) [42] and polysulfone (*open circles*) [53]. The *lines* are guides for the eyes. The *inset* gives the thickness dependence of the dielectric glass transition temperature for polysulfone estimated for $f = 0.1$ Hz. The *line* is a guide for the eyes. Data were taken from Ref. [54]

3.3 Proof of the Concept of the Polymer/Surface Interaction Energy

In Ref. [56], attempts to correlate the change in the glass transition temperature with the interaction energy between the substrate and the polymer surface γ_{SP} are made. Figure 17 the data for all the polymers discussed here are summarized. For the both main polycarbonate and polysulfone, it is already discussed above that this correlation is partly fulfilled. In Fig. 17, the data for the different polystyrene samples are added, whereas the interaction energy between the AlO_x surface of the substrate and the polystyrene segments is taken from Ref. [86]. The corresponding points for the different polystyrene samples are located far away from the correlation line between the polymer/substrate interaction energy and the change in the glass transition temperature. The observed dependence of the thermal glass transition temperature on the molecular weight is also in contradiction to the assumption that the interaction energy between the polymer segments and surface of the substrate is the only parameter, which determines the value of the glass transition temperature of ultra-thin polystyrene films. Moreover, data for poly(vinyl acetate) taken from Ref. [104] prepared on different surfaces were added to Fig. 17. Also these data do not support the correlation between the polymer/substrate interaction energy and the change in the glass transition temperature. In conclusion, the polymer/substrate interaction

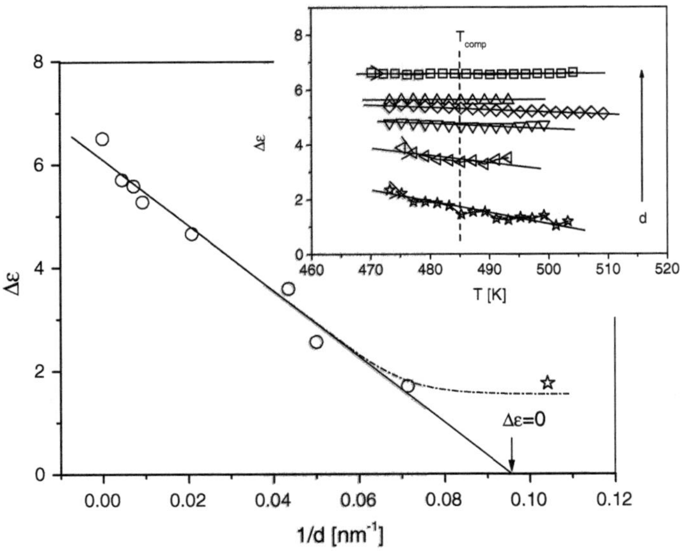

Fig. 21 Dielectric relaxation strength $\Delta\varepsilon$ for polysulfone versus inverse film thickness at $T = 485\,\mathrm{K}$. The *dashed dotted line* is a guide for the eyes. The *solid line* is a linear regression to the data except the point for the lowest layer thickness. The *inset* gives $\Delta\varepsilon$ for polysulfone versus temperature for different film thicknesses: *squares* bulk, *up pointed triangles* 138 nm, *diamonds* 107 nm, *down pointed triangles* 48 nm, *left pointed triangles* 23 nm, *stars* 14 nm. *Lines* are linear regressions to the corresponding data. All data were taken from Ref. [54]

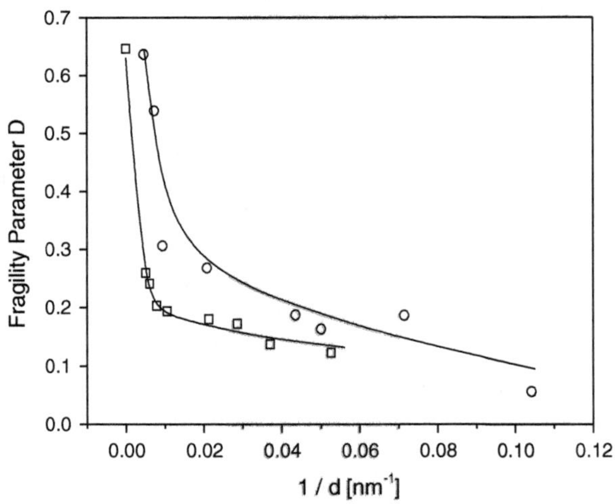

Fig. 22 Fragility parameter D versus inverse film thickness d: *circles* polysulfone, *squares* polycarbonate. *Lines* are guides for the eyes

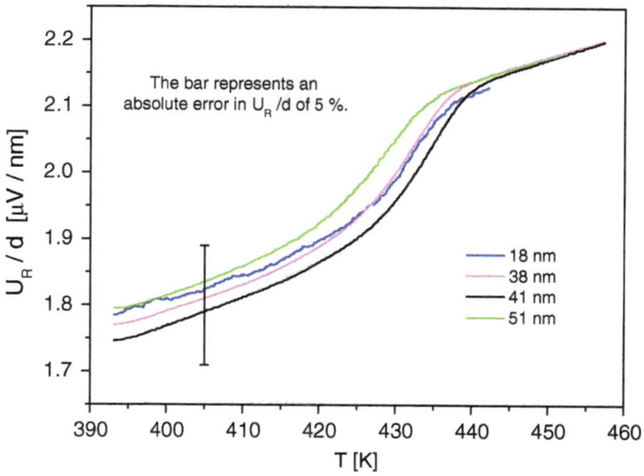

Fig. 23 Real part of the complex differential voltage divided by the thickness of the film for polycarbonate for the given different values of d. The *bar* gives an absolute error of 5 % at 405 K. The measuring frequency is 160 Hz

energy is not the only parameter, which is responsible for the change in the thermal glass transition temperature with the film thickness. Packing effects and/or densifications of the reduced mobility layer at the surface as discussed above and also in Refs. [54, 86] plays a role as well.

3.4 Length Scale of Cooperativity at the Glass transition

Ultra-thin films provide an ideal sample geometry to analyze the confinement effect on the cooperativity length scale ξ or the corresponding volume V_{CRR} at the dynamic glass transition by varying the film thickness. The cooperativity approach to the glass transition by Adam and Gibbs [105] was extended by the fluctuation theory of Donth [106]. Within this approach, a correlation length ξ (or volume V_{CRR}) of dynamically correlated segments at the glass transition can be estimated to

$$\xi^3 \sim V_{\mathrm{CRR}} = \frac{k_B T_g^2 \Delta(1/c_p)}{\rho \delta T^2} \tag{15}$$

T_g is the dynamic glass transition temperature, ρ is the density at T_g, and $\Delta(1/c_p)$ is the step of the reciprocal specific heat capacity at the glass transition, where $c_V \approx c_p$ was assumed. δT is the width of the glass transition and can be extracted experimentally from the temperature dependence of the specific heat capacity [107, 108]. Moreover, δT can be estimated from the phase angle of specific heat spectroscopy measurements [109–112]. Two quantities, the specific heat capacity c_p and the mean temperature fluctuation δT, will determine the extent of cooperativity. According to

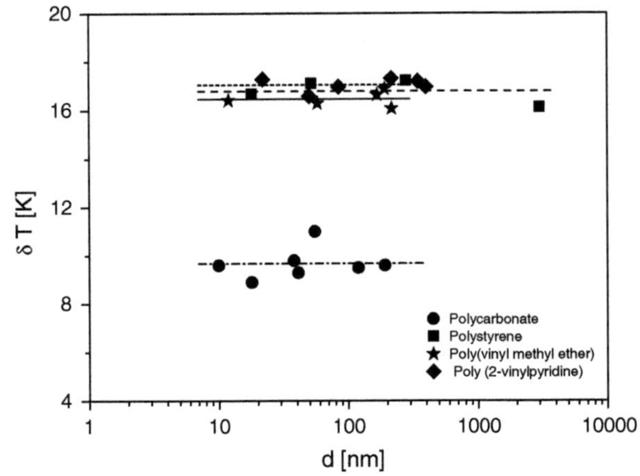

Fig. 24 Thickness dependence of the mean temperature fluctuation δT for the indicated polymers. The data for polycarbonate and poly(2-vinylpyridine) were measured with a frequency of 160 Hz, those of polystyrene with 320 Hz and that of poly(vinyl methyl ether) at 640 Hz. The different *lines* indicate the mean values of δT for the different polymers

Eq. (6), for identical chips ($\Delta U_0 = 0$), the specific heat capacity of the material can be related to the measured heat capacity of the system C_S by

$$c_p = \frac{C_S}{m} = i\omega\frac{\overline{C}U_R}{m\,S\,P_0} = i\omega\frac{\overline{C}U_R}{\rho d\,Ar\,S\,P_0} \sim \frac{U_R}{d}. \tag{16}$$

where m is the sample mass, ρ the density, Ar is the heated area of the sensor and d is the film thickness. $\overline{C}$ and S can be estimated from the frequency dependence of the thermopile voltage of a single sensor [69]. Figure 23 gives the real part of the complex differential voltage divided by the film thickness d for polycarbonate for different values of d. Within the experimental error, all data collapse onto a single chart. Similar results were obtained for polystyrene and poly(vinyl methyl ether). Furthermore in the literature, identical results have been found [69]. This means c_p in both the glassy and the liquid state is independent of the film thickness within the experimental error limit.

The mean temperature fluctuation δT is the remaining quantity to estimate the extent of the cooperativity according to Eq. (15). It can be estimated by fitting Gaussians to the data of the phase angle (see for instance [43, 63]). Figure 24 depicts the thickness dependence of δT for polystyrene, polycarbonate, poly(vinyl methyl ether), and poly(2-vinylpyridine). For all four polymers with a quite different structure, δT is independent of the film thickness down to the lowest values of d. In conclusion, this means that firstly, the cooperativity length scale at the glass transition should be smaller than ca. 10 nm and secondly, there is no confinement effect on the cooperativity length scale.

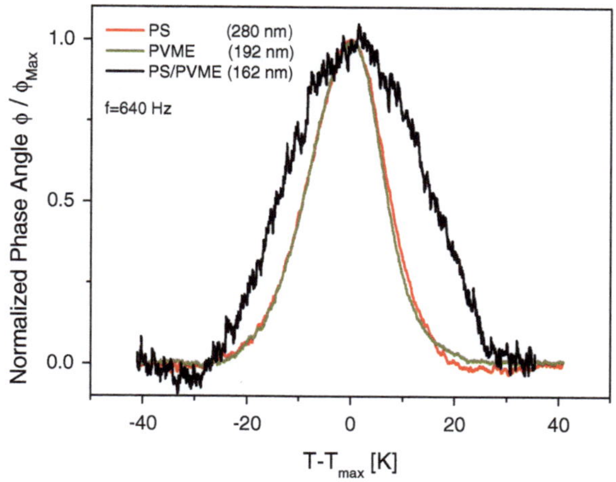

Fig. 25 Normalized phase angle of the complex differential voltage measured by specific heat spectroscopy for thin films of the homopolymers (PS and PVME) as well as for a blend at 50/50 wt-% composition for the given thicknesses. Data were taken from Ref. [119]

3.5 Thin Films of Miscible Polymer Blends

By investigating thin films of blends, where the both components are miscible in the bulk state, the interaction of the polymer segments with the surface of the substrate and/or the formation of a surface layer with higher molecular mobility can be studied. Corresponding experiments were carried out by specific heat spectroscopy for the system polystyrene (PS)/poly(vinyl methyl ether) (PVME) at a concentration of 50/50 wt-%. Recently, some results for this system have been published [113–118].

PS/PVME blend of 50/50 weight ratio was dissolved in toluene to form a solution with different overall polymer concentrations. After spin coating, all samples were dried in an oil-free vacuum for 24 h at room temperature and further annealed at 50 °C($T = Tg$, Bulk + 45 K) for at least 72 h in order to remove the residual solvent and the stress induced by the spin coating procedure. AFM topography images before and after the annealing procedure further show homogeneous films with a low roughness even for the thinnest films. No sign of dewetting was observed.

Figure 12a shows the temperature dependence of the normalized phase angle for $PVME_{Bulk}$ measured at a frequency of 640 Hz and Fig. 13 presents the corresponding relaxation map. Down to the lowest film thickness of 12 nm all datasets collapse into one chart indicating that the dynamic glass transition temperature for poly(vinyl methyl ether) films is independent of the film thickness [63], like for polystyrene (see Fig. 8b).

Figure 25 compares the temperature dependence of the normalized phase angle of a film of a PVME/PS blend with that of the corresponding homopolymers [119]. Similarly to the corresponding bulk system [120], the characteristic broadening of the relaxation function for the blends is also observed for the cases of thin films. Figure 26 gives the normalized phase angle for the blend PVME/PS at a concentration of 50/50 wt-% for different film thicknesses. In contrast to the behavior of

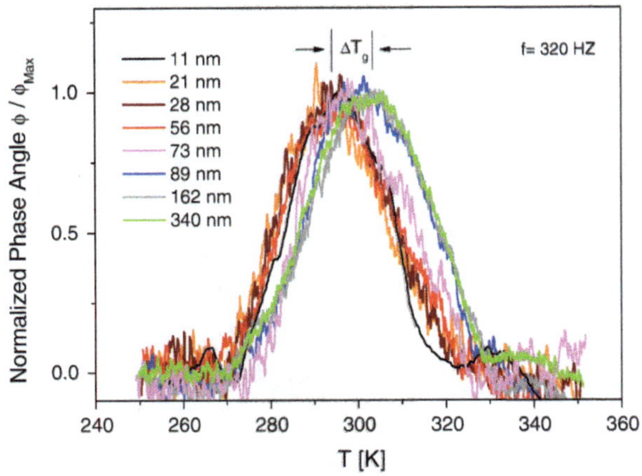

Fig. 26 Normalized phase angle of the complex differential voltage measured by specific heat spectroscopy for thin films of the polymer blend PS/PVME (50/50 wt-%) at a frequency of 320 Hz for the indicated film thicknesses. The data were taken from Ref. [119]

both homopolymers, polystyrene (see Fig. 8b) and poly(vinyl methyl ether) (see Fig. 12a), where the dynamic glass transition temperature does not depend on the film thickness, the loss peak for the blends shifts to lower temperatures with decreasing film thickness.

From the peak position of the phase angle, a corresponding dynamic glass transition temperature can be estimated and together with measuring frequency the relaxation map can be constructed (Fig. 27). As already evident from the raw data (Fig. 12a) for pure PVME within the experimental uncertainty (approximately 2 K) the data for all film thicknesses collapse into one chart (see also Ref. [63]). The same is true for the specific heat spectroscopy data of pure polystyrene [45]. This is different for the blend system. With decreasing film thickness the data shifts to lower temperatures.

To discuss this in more detail, Fig. 28 gives the dynamic glass transition temperature measured by specific heat spectroscopy at a frequency of 320 Hz. For both homopolymers, the dynamic glass transition temperature is independent of the film thickness (see inset of Fig. 28). For larger thicknesses of the blend film (d > 100 nm), the glass transition temperature is also independent of d similarly to the corresponding homopolymers. At a thickness of around 80 nm, T_g drops down with decreasing film thickness by around 10 K followed by a further decrease of the dynamic glass transition temperature with decreasing film thickness.

To discuss these results, a structural model for thin films of polymer blends had to be developed. PVME has a lower surface tension, with respect to air, than polystyrene. Therefore, it is possible that poly(vinyl methyl ether) segments segregates at the air/polymer surface of the film [121] forming a PVME-rich surface layer. Therefore, the following model can be assumed to discuss the thickness dependence of the

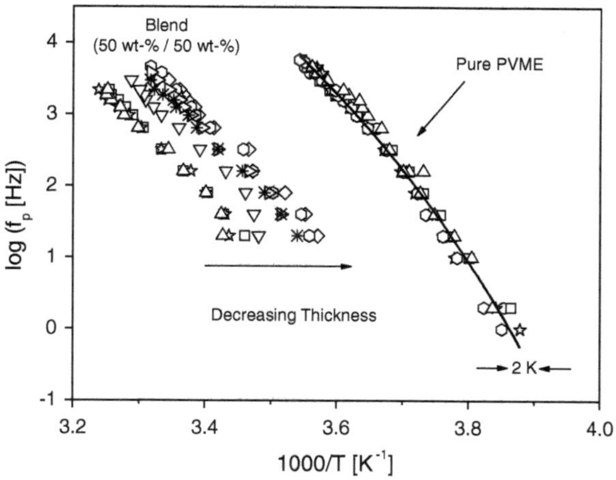

Fig. 27 Relaxation map for pure poly(vinyl methyl ether) and for the PVME/PS blend at 50/50 wt-% overall composition for different film thicknesses. PVME: *triangles* 12 nm, *hexagons* 58 nm, *stars* 168 nm, *squares* 192 nm, *circles* 218 nm. The *solid line* is a fit of the VFT-equation to all data of pure PVME. PVME/PS blend: *diamonds* 11 nm, *hexagon* 21 nm, *right-sited triangles* 28 nm, *asterisks* 56 nm, *down-sited triangles* 73 nm, *squares* 89 nm, *stars* 162 nm, *up-sited triangles* 340 nm. The data are taken from Ref. [119]

dynamic glass transition temperature of thin polymer blend films. At the SiO_2 surface of the sensor, a polystyrene-rich layer with a reduced mobility is formed. In the middle of the film, a more or less bulk-like behavior is present. Whereas at top of the film, the polymer/air surface, a poly(vinyl methyl ether)-rich layer is developed, due to the mentioned segregation process. In contrast to polystyrene, PVME contains oxygen. Therefore, the concentration of poly(vinyl methyl ether) in a surface layer (ca. 3–7 nm) can be estimated by XPS to evidence the existence of PVME-rich surface layer. Figure 29 depicts the XPS-spectrum of a PS/PVME blend (50/50 wt%) film, with a thickness of 200 nm. The two observed peaks correspond to the C–H as well as C–O bonds as indicated [74]. The areas of both peaks ($I_C - O$ and $I_C - H$) can be estimated by fitting Gaussians to the data. For the weight fraction w of PVME in a surface layer of 3–7 nm thickness one obtains

$$\frac{I_{C-O}}{I_{C-H}} = \frac{2w/M_{\mathrm{VME}}}{8(1-w)/M_S + 3w/M_{\mathrm{VME}}} \tag{17}$$

where M_{VME} and M_S are the molecular weight of the repeating units of PVME and PS, respectively. This calculation leads to the results, that the concentration of PVME in the surface layer is 84 wt-%, in contrast to the formulation which is 50/50 wt-%. For thick films of the polymer blend, the bulk-like middle layer of the film dominates the glass transition behavior. Therefore, the dynamic glass transition temperature is independent of the film thickness for $d \approx> 100$ nm. For decreasing film thicknesses, the PVME-rich surface layer with a higher molecular mobility

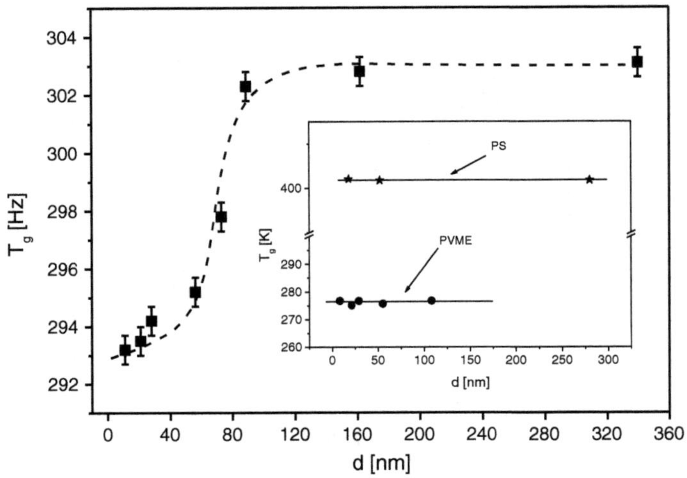

Fig. 28 Dynamic glass transition temperature measured at 320 Hz by specific heat spectroscopy versus film thickness *d* for PVME/PS blend films at a nominal composition of 50/50 wt-%. The *line* is guide for the eyes. The *inset* gives the same data for the homopolymers polystyrene and poly(vinyl methyl ether). The *lines* represent the average values. The data are taken from Ref. [119]

becomes more and more important for the overall glass transition behavior of the film. Therefore, the observed decrease of the dynamic glass transition temperature with decreasing film thickness for the blend (see Fig. 28) systems can be assigned to the increasing influence of the PVME-rich surface layer on the glass transition of thin polymer blend films of polystyrene and poly(vinyl methyl ether). In general, these results provide evidence that a mobile surface layer is important for the glass transition phenomena of thin polymer films [64]. At the lowest film thicknesses, the thickness dependence of the dynamic glass transition is due to a counter play of the adsorbed layer at the surface of the substrate with a lower molecular mobility and the surface layer at the polymer/air interface with a high molecular mobility.

4 Summary, Conclusions, Outlook

Although the glass transition behavior of ultra-thin supported polymer films is discussed controversially in the literature for around 20 years, substantial efforts have been archived to understand its molecular dynamics and especially the glass transition phenomena.

On one side, there are many experimental investigations that the polymer segments undergo an attractive interaction with the surface of the substrate. These interactions will give rise to the formation of a layer with a reduced molecular mobility, which will lead, in general, to an increase in the thermal glass transition temperature.

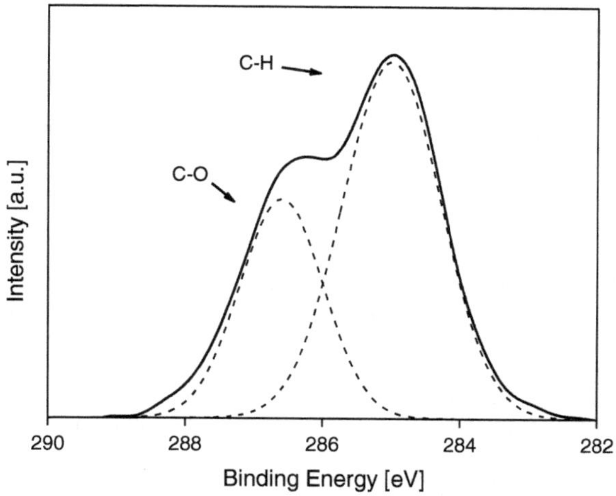

Fig. 29 Core-level XPS spectrum of a PS/PVME (50/50 wt-%) thin film with a thickness of 200 nm. The *dashed lines* represent the fitted Gaussians for the C–O and the C–H peak. Data were taken from [119]

The strength of these interactions will depend on both the chemical nature of the surface, including its roughness, or the chemical reactivity and the structure of the macromolecule. It should be noted that a simple modeling of the substrate polymer interaction by the molecular interaction energy between the polymer segments and the surface of the substrate would not be valid in the general case. Other effects like the packing density of the polymer segments close to the surface of the substrate, including its change with annealing above the glass transition temperature or even the formation of real chemical bonds, will have an impact onto the glass transition behavior.

On the other side, in most cases, thin polymer films will have a free polymer/air interface. Due to the lack of molecular interactions, which are present in the bulk state, the free polymer/air interface will lead to a formation of a surface layer with a higher molecular mobility than the bulk state. In principle, this highly mobile surface layer will lead to a decrease of the thermal glass transition temperature compared to the bulk value.

The actual value of the thermal glass transition temperature of a polymer film with a given thickness will be due to a delicate balance of the reduced mobility layer at the substrate surface, the bulk-like middle layer and the highly mobile surface layer at the polymer/air interface.

The recent results of the decoupling of the thickness dependence of the thermal and dynamic glass transition temperature need further experimental and theoretical investigations.

Acknowledgments The authors gratefully acknowledge the assistance of Mr. D. Neubert, Mrs. G. Hidde, and Dr. R.-D. Schulze for experimental help. Moreover we thank for fruitful discussions with Prof. Dr. Ch. Schick and Dr. H. Huth (University Rostock). The financial support from the German Science Foundation (Deutsche Forschungsgemeinschaft, SCHO-470/20-1) is highly acknowledged.

References

1. de Gennes PG (1985) Wetting: statics and dynamics. Rev Modern Phys 57:827–863
2. Besancon BM, Green PF (2004) Moving fronts in entangled polymeric films. Phys Rev E 70:051808
3. Alsten JV, Granick S (1990) Shear rheology in a confined geometry: polysiloxane melts. Macromolecules 23:4856–4862
4. Hu H-W, Granick S (1992) Viscoelastic dynamics of confined polymer melts. Science 258:1339–1342
5. Frank B, Gast AP, Russell TP, Brown HR, Hawker C (1996) Polymer mobility in thin films. Macromolecules 29:6531–6534
6. Zheng X, Sauer BB, Van Alsten JG, Schwarz SA, Rafailovich MH, Sokolov J et al (1995) Reptation dynamics of a polymer melt near an attractive solid interface. Phys Rev Lett 74: 407–410
7. Zheng X, Rafailovich MH, Sokolov J, Strzhemechny Y, Schwarz SA, Sauer BB et al (1997) Long-range effects on polymer diffusion induced by a bounding interface. Phys Rev Lett 79:241–244
8. Hartmann L, Fukao K, Kremer F (2002) Molecular dynamics in thin polymer films. In: Kremer F, Schönhals A (eds) Broadband dielectric spectroscopy. Springer, Berlin
9. Fukao K, Miyamoto Y (1999) Glass transition temperature and dynamics of α-process in thin polymer films. Euro Phys Lett 46:649
10. Fukao K, Miyamoto Y (2000) Glass transitions and dynamics in thin polymer films: dielectric relaxation of thin films of polystyrene. Phys Rev E 61:1743–1754
11. Priestley RD, Broadbelt LJ, Torkelson JM, Fukao K (2007) Glass transition and α-relaxation dynamics of thin films of labeled polystyrene. Phys Rev E 75:061806
12. Serghei A, Kremer F (2003) Confinement-induced relaxation process in thin films of cis-polyisoprene. Phys Rev Lett 91:165702
13. Serghei A, Mikhailova Y, Huth H, Schick C, Eichhorn KJ, Voit B et al (2005) Molecular dynamics of hyperbranched polyesters in the confinement of thin films. Eur Phys J E 17: 199–202
14. Serghei A, Mikhailova Y, Eichhorn KJ, Voit B, Kremer F (2006) Discrepancies in the characterization of the glass transition in thin films of hyperbranched polyesters. J Polym Sci Part B: Polym Phys 44:3006–3010
15. Serghei A, Kremer F (2006) Unexpected preparative effects on the properties of thin polymer films. In: Grundke K, Stamm M, Adler H-J (eds) Characterization of polymer surfaces and thin films, vol 132. Springer, Berlin Heidelberg, pp 33–40
16. Serghei A, Tress M, Kremer F (2006) Confinement effects on the relaxation time distribution of the dynamic glass transition in ultrathin polymer films. Macromolecules 39:9385–9387
17. Serghei A, Kremer F (2006) Broadband dielectric spectroscopy on ultrathin organic layers having one free (upper) interface. Rev Sci Instrum 77:116108
18. Serghei A, Kremer F (2008) Broadband dielectric studies on the interfacial dynamics enabled by use of nanostructured electrodes. Rev Sci Instrum 79:026101
19. Lupaşcu V, Picken SJ, Wübbenhorst M (2006) Cooperative and non-cooperative dynamics in ultra-thin films of polystyrene studied by dielectric spectroscopy and capacitive dilatometry. J Non-Cryst Solids 352:5594–5600
20. Lupaşcu V, Huth H, Schick C, Wübbenhorst M (2005) Specific heat and dielectric relaxations in ultra-thin polystyrene layers. Thermochimica Acta 432:222–228

21. Napolitano S, Wübbenhorst M (2007) Dielectric signature of a dead layer in ultrathin films of a nonpolar polymer. J Phys Chem B 111:9197–9199
22. Napolitano S, Prevosto D, Lucchesi M, Pingue P, D'Acunto M, Rolla P (2007) Influence of a reduced mobility layer on the structural relaxation dynamics of aluminum capped ultrathin films of poly(ethylene terephthalate). Langmuir 23:2103–2109
23. Orts WJ, van Zanten JH, Satija SK (1993) Observation of temperature dependent thicknesses in ultrathin polystyrene films on silicon. Phys Rev Lett 71:867–870
24. Wallace WE, van Zanten JH, Wu WL (1995) Influence of an impenetrable interface on a polymer glass-transition temperature. Phys Rev E 52:R3329–R3332
25. van Zanten JH, Wallace WE (1996) Effect of strongly favorable substrate interactions on the thermal properties of ultrathin polymer films. Phys Rev E 53:R2053–R2056
26. Tsui OKC, Russell TP, Hawker CJ (2001) Effect of interfacial interactions on the glass transition of polymer thin films. Macromolecules 34:5535–5539
27. Kanaya T, Miyazaki T, Watanabe H, Nishida K, Yamano H, Tasaki S et al (2003) Annealing effects on thickness of polystyrene thin films as studied by neutron reflectivity. Polymer 44:3769–3773
28. Roth CB, McNerny KL, Jager WF, Torkelson JM (2007) Eliminating the enhanced mobility at the free surface of polystyrene:? Fluorescence studies of the glass transition temperature in thin bilayer films of immiscible polymers. Macromolecules 40:2568–2574
29. Sills S, Overney RM, Chau W, Lee VY, Miller RD, Frommer J (2004) Interfacial glass transition profiles in ultrathin, spin cast polymer films. J Chem Phys 120:5334–5338
30. Forrest JA, Dalnoki-Veress K, Stevens JR, Dutcher JR (1996) Effect of free surfaces on the glass transition temperature of thin polymer films. Phys Rev Lett 77:2002–2005
31. Forrest JA, Dalnoki-Veress K, Dutcher JR (1997) Interface and chain confinement effects on the glass transition temperature of thin polymer films. Phys Rev E 56:5705–5716
32. Dalnoki-Veress K, Forrest JA, Murray C, Gigault C, Dutcher JR (2001) Molecular weight dependence of reductions in the glass transition temperature of thin, freely standing polymer films. Phys Rev E 63:031801
33. Keddie JL, Jones RAL, Cory RA (1994) Interface and surface effects on the glass-transition temperature in thin polymer films. Faraday Discuss 98:219–230
34. Keddie JL, Jones RAL, Cory RA (1994) Size-dependent depression of the glass transition temperature in polymer films. Euro Phys Lett 27:59
35. Grohens Y, Brogly M, Labbe C, David M-O, Schultz J (1998) Glass transition of stereoregular poly(methyl methacrylate) at interfaces. Langmuir 14:2929–2932
36. Kawana S, Jones RAL (2001) Character of the glass transition in thin supported polymer films. Phys Rev E 63:021501
37. Pham JQ, Green PF (2002) The glass transition of thin film polymer/polymer blends: interfacial interactions and confinement. J Chem Phys 116:5801–5806
38. Pham JQ, Green PF (2003) Effective Tg of confined polymer-polymer mixtures. Influence of molecular size. Macromolecules 36:1665–1669
39. Pham JQ, Mitchell CA, Bahr JL, Tour JM, Krishanamoorti R, Green PF (2003) Glass transition of polymer/single-walled carbon nanotube composite films. J Polym Sci Part B: Polym Phys 41:3339–3345
40. Kim JH, Jang J, Zin W-C (2001) Thickness dependence of the glass transition temperature in thin polymer films. Langmuir 17:2703–2710
41. Torres JA, Nealey PF, de Pablo JJ (2000) Molecular simulation of ultrathin polymeric films near the glass transition. Phys Rev Lett 85:3221–3224
42. Yin H, Napolitano S, Schönhals A (2012) Molecular mobility and glass transition of thin films of poly (bisphenol a carbonate). Macromolecules 45:1652–1662
43. Yin H, Schonhals A (2012) Calorimetric glass transition of ultrathin poly (bisphenol a carbonate) films. Soft Matter 8:9132–9139
44. Tress M, Erber M, Mapesa EU, Huth H, Müller J, Serghei A et al (2010) Glassy dynamics and glass transition in nanometric thin layers of polystyrene. Macromolecules 43:9937–9944

45. Boucher VM, Cangialosi D, Yin H, Schönhals A, Alegria A, Colmenero J (2012) T_g depression and invariant segmental dynamics in polystyrene thin films. Soft Matter 8:5119–5122
46. Tress M, Kremer F (2014) Molecular dynamics of condensed (semi)-isolated polymer chains. In: Kremer F (ed) Dynamics in confinement. Springer, Berlin
47. Mapesa E, Kremer F (2014) Molecular dynamics of poly (cis-1,4-isoprene in) in 1- and 2-dimensional confinement. In: Kremer F (ed) Dynamics in confinement. Springer, Berlin
48. Schick C (2014) Dynamic calorimetric glass transition in thin polymer films. In: Kremer F (ed) Dynamics in confinement. Springer, Berlin
49. Cangialosi D (2014) Equilibrium and out-of-equilibrium dynamics in confined polymers by dielectric spectroscopy and calorimetric techniques. In: Kremer F (ed) Dynamics in confinement. Springer, Berlin
50. Angell CA (1995) Formation of glasses from liquids and biopolymers. Science 267: 1924–1935
51. Sastry S, Debenedetti PG, Stillinger FH (1998) Signatures of distinct dynamical regimes in the energy landscape of a glass-forming liquid. Nature 393:554–557
52. Hall DB, Underhill P, Torkelson JM (1998) Spin coating of thin and ultrathin polymer films. Polym Eng Sci 38:2039–2045
53. Ellison CJ, Torkelson JM (2003) The distribution of glass-transition temperatures in nanoscopically confined glass formers. Nat Mater 2:695–700
54. Labahn D, Mix R, Schönhals A (2009) Dielectric relaxation of ultrathin films of supported polysulfone. Phys Rev E 79:011801
55. Rotella C, Wübbenhorst M, Napolitano S (2011) Probing interfacial mobility profiles via the impact of nanoscopic confinement on the strength of the dynamic glass transition. Soft Matter 7:5260–5266
56. Fryer DS, Peters RD, Kim EJ, Tomaszewski JE, de Pablo JJ, Nealey PF et al (2001) Dependence of the glass transition temperature of polymer films on interfacial energy and thickness. Macromolecules 34:5627–5634
57. Efremov MY, Kiyanova AV, Last J, Soofi SS, Thode C, Nealey PF (2012) Glass transition in thin supported polystyrene films probed by temperature-modulated ellipsometry in vacuum. Phys Rev E 86:021501
58. Wang J, McKenna GB (2013) A novel temperature-step method to determine the glass transition temperature of ultrathin polymer films by liquid dewetting. J Polym Sci Part B: Polym Phys 51:1343–1349
59. Huth H, Minakov AA, Schick C (2006) Differential AC-chip calorimeter for glass transition measurements in ultrathin films. J Polym Sci Part B: Polym Phys 44:2996–3005
60. Paeng K, Swallen SF, Ediger MD (2011) Direct measurement of molecular motion in freestanding polystyrene thin films. J Am Chem Soc 133:8444–8447
61. Serghei A, Huth H, Schick C, Kremer F (2008) Glassy dynamics in thin polymer layers having a free upper interface. Macromolecules 41:3636–3639
62. Huth H, Minakov AA, Serghei A, Kremer F, Schick C (2007) Differential AC-chip calorimeter for glass transition measurements in ultra thin polymeric films. Eur Phys J Spec Top 141: 153–160
63. Yin H, Schönhals A (2013) Calorimetric glass transition of ultrathin poly (vinyl methyl ether) films. Polymer 54:2067–2070
64. Ediger MD, Forrest JA (2013) Dynamics near free surfaces and the glass transition in thin polymer films: a view to the future. Macromolecules 47:471–478
65. Schönhals A, Kremer F (2002) Theory of dielectric relaxation. In: Kremer F, Schönhals A (eds) Broadband dielectric spectroscopy. Springer, Berlin
66. Havriliak S, Negami S (1967) A complex plane representation of dielectric and mechanical relaxation processes in some polymers. Polymer 8:161–210
67. Schawe JEK (1995) A comparison of different evaluation methods in modulated temperature DSC. Thermochimica Acta 260:1–16
68. http://www.xensor.nl/pdffiles/sheets/nanogas3939.pdf

69. Zhou D, Huth H, Gao Y, Xue G, Schick C (2008) Calorimetric glass transition of poly (2,6-dimethyl-1,5-phenylene oxide) thin films. Macromolecules 41:7662–7666
70. Yin H, Cangialosi D, Schönhals A (2013) Glass transition and segmental dynamics in thin supported polystyrene films: the role of molecular weight and annealing. Thermochimica Acta 566:186–192
71. Van Oss CJ, Chaudhury MK, Good RJ (1988) Interfacial Lifshitz-van der Waals and polar interactions in macroscopic systems. Chem Rev 88:927–941
72. Lee L-H (1996) Correlation between Lewis acid-base surface interaction components and linear solvation energy relationship solvatochromic α and β parameters. Langmuir 12: 1681–1687
73. Good RJ, Girifalco LA (1960) A theory for estimation of surface and interfacial energies. III. Estimation of surface snergies of solids from contact angle aata. J Phys Chem 64:561–565
74. Beamson G, Briggs H (1992) High resolution XPS of organic polymers: the scienta ESCA 300 database. Wiley, Chichester
75. Vogel H (1921) The temperature dependence law of the viscosity of fluids. Phys Z 22:645–646
76. Fulcher GS (1925) Analysis of recent measurements of the viscosity of glasses. J Am Ceram Soc 8:339–355
77. Tammann G, Hesse W (1926) The dependancy of viscosity on temperature in hypothermic liquids. Z Anorg Allg Chem 156:245–257
78. Angell CA (1991) Relaxation in liquids, polymers and plastic crystals– strong/fragile patterns and problems. J Non-Cryst Solids Part 1 131-133:13–31
79. Angell CA (1997) Entropy and fragility in supercooling liquids. J Res Natl Inst Stan 102: 171–185
80. Zhang C, Boucher VM, Cangialosi D, Priestley RD (2013) Mobility and glass transition temperature of polymer nanospheres. Polymer 54:230–235
81. Boucher VM, Cangialosi D, Alegría A, Colmenero J (2014) Accounting for the thickness dependence of the Tg in supported PS films via the volume holes diffusion model. Thermochimica Acta 575:233–237
82. Boucher VM, Cangialosi D, Alegría A, Colmenero J (2012) Enthalpy recovery in nanometer to micrometer thick polystyrene films. Macromolecules 45:5296–5306
83. Napolitano S, Cangialosi D (2013) Interfacial free volume and vitrification: reduction in Tg in proximity of an adsorbing interface explained by the free volume holes diffusion model. Macromolecules 46:8051–8053
84. Cangialosi D, Boucher VM, Alegria A, Colmenero J (2013) Physical aging in polymers and polymer nanocomposites: recent results and open questions. Soft Matter 9:8619–8630
85. Curro JG, Lagasse RR, Simha R (1982) Diffusion model for volume recovery in glasses. Macromolecules 15:1621–1626
86. Napolitano S, Wübbenhorst M (2011) The lifetime of the deviations from bulk behaviour in polymers confined at the nanoscale. Nat Commun 2:260
87. Fakhraai Z, Forrest JA (2005) Probing slow dynamics in dupported thin polymer films. Phys Rev Lett 95:025701
88. Sharp JS, Forrest JA (2003) Free surfaces cause reductions in the glass transition temperature of thin polystyrene films. Phys Rev Lett 91:235701
89. Forrest JA, Dalnoki-Veress K (2001) The glass transition in thin polymer films. Adv Colloid Interface Sci 94:167–195
90. Tsui OKC, Zhang HF (2001) Effects of chain ends and chain entanglement on the glass transition temperature of polymer thin Films. Macromolecules 34:9139–9142
91. Ellison CJ, Mundra MK, Torkelson JM (2005) Impacts of polystyrene molecular weight and modification to the repeat unit structure on the glass transition-nanoconfinement effect and the cooperativity length scale. Macromolecules 38:1767–1778
92. Weyer S, Huth H, Schick C (2005) Application of an extended Tool-Narayanaswamy-Moynihan model. Part 2. Frequency and cooling rate dependence of glass transition from temperature modulated DSC. Polymer 46:12240–12246

93. Jakobsen B, Hecksher T, Christensen T, Olsen NB, Dyre JC, Niss K (2012) Communication: identical temperature dependence of the time scales of several linear-response functions of two glass-forming liquids. J Chem Phys 136:081102
94. Weyer S, Hensel A, Korus J, Donth E, Schick C (1997) Broad band heat capacity spectroscopy in the glass-transition region of polystyrene. Thermochimica Acta 304-305:251–255
95. Schönhals A, Schick C, Huth H, Frick B, Mayorova M, Zorn R (2007) Molecular dynamics in glass-forming poly (phenyl methyl siloxane) as investigated by broadband thermal, dielectric and neutron spectroscopy. J Non-Cryst Solids 353:3853–3861
96. Madkour S, Yin H, Schönhals A (2014) Calorimetric glass transition of ultrathin poly(2-vinylpyridine) films. To be published.
97. Tress M, Mapesa EU, Kossack W, Kipnusu WK, Reiche M, Kremer F (2013) Glassy dynamics in condensed isolated polymer chains. Science 341:1371–1374
98. Kremer F, Schönhals A (2002) The scaling of the dynamics of glasses and supercooled liquids. In: Kremer F, Schönhals A (eds) Broadband dielectric spectroscopy. Springer, Berlin
99. Lipson JEG, Milner ST (2009) Percolation model of interfacial effects in polymeric glasses. Eur Phys J B 72:133–137
100. Milner ST, Lipson JEG (2010) Delayed glassification model for free-surface suppression of Tg in polymer glasses. Macromolecules 43:9865–9873
101. Napolitano S, Rotella C, Wübbenhorst M (2011) Is the reduction in tracer diffusivity under nanoscopic confinement related to a frustrated segmental mobility? Macromol Rapid Commun 32:844–848
102. Schönhals A, Kremer F (2002) Theory of dielectric relaxation. In: Kremer F, Schönhals A (eds) Broadband dielectric spectroscopy. Springer, Berlin
103. Evans CM, Deng H, Jager WF, Torkelson JM (2013) Fragility is a key parameter in determining the magnitude of Tg-confinement effects in polymer films. Macromolecules 46:6091–6103
104. Nguyen HK, Labardi M, Capaccioli S, Lucchesi M, Rolla P, Prevosto D (2012) Interfacial and annealing effects on primary α-relaxation of ultrathin polymer films investigated at nanoscale. Macromolecules 45:2138–2144
105. Adam G, Gibbs JH (1965) On the temperature dependence of cooperative relaxation properties in glass-forming liquids. J Chem Phys 43:139–146
106. Donth E (2001) The glass transition: relaxation dynamics in liquids and disordered materials. Springer, Berlin
107. Schneider K, Schönhals A, Donth E (1981) Über die Größe der kooperativen Umlagerungsbereiche am thermischen Glasübergang amorpher Polymere. Acta Polymerica 32:471–475
108. Donth E (1982) The size of cooperatively rearranging regions at the glass transition. J Non-Cryst Solids 53:325–330
109. Donth E, Hempel E, Schick C (2000) Does temperature fluctuate? Indirect proof by dynamic glass transition in confined geometries. J Phys: Condens Matter 12:L281
110. Donth E, Huth H, Beiner M (2001) Characteristic length of the glass transition. J Phys-Condens Matter 13:L451–L462
111. Saiter A, Delbreilh L, Couderc H, Arabeche K, Schönhals A, Saiter JM (2010) Temperature dependence of the characteristic length scale for glassy dynamics: combination of dielectric and specific heat spectroscopy. Phys Rev E 81:041805
112. Hempel E, Hempel G, Hensel A, Schick C, Donth E (2000) Characteristic length of dynamic glass transition near Tg for a wide assortment of glass-forming substances. J Phys Chem B 104:2460–2466
113. Gotzen N-A, Van Assche G, Van Mele B (2011) Construction of the state diagram of polymer blend thin films using differential AC chip calorimetry. Polymer 52:4277–4283
114. Gotzen N-A, Huth H, Schick C, Van Assche G, Neus C, Van Mele B (2010) Phase separation in polymer blend thin films studied by differential AC chip calorimetry. Polymer 51:647–654
115. Yang H, Green PF (2013) Role of spatial compositional heterogeneity on component dynamics in miscible bulk and thin film polymer/polymer blends. Macromolecules 46:9390–9395
116. Valiquette D, Pellerin C (2011) Miscible and core-sheath PS/PVME fibers by electrospinning. Macromolecules 44:2838–2843

117. Lin Y, Tan Y, Qiu B, Cheng J, Wang W, Shangguan Y et al (2013) Casting solvent effects on molecular dynamics of weak dynamic asymmetry polymer blend films via broadband dielectric spectroscopy. J Membr Sci 439:20–27
118. Frieberg B, Kim J, Narayanan S, Green PF (2013) Surface layer dynamics in miscible polymer blends. ACS Macro Lett 2:388–392
119. Yin H, Schönhals A (2014) To be published.
120. Colmenero J, Arbe A (2007) Segmental dynamics in miscible polymer blends: recent results and open questions. Soft Matter 3:1474–1485
121. Tanaka K, Yoon J-S, Takahara A, Kajiyama T (1995) Ultrathinning-induced surface phase separation of polystyrene/poly(vinyl methyl ether) blend film. Macromolecules 28:934–938

Molecular Dynamics of Condensed (Semi-) Isolated Polymer Chains

Martin Tress, Emmanuel Urandu Mapesa, Wilhelm Kossack, Wycliffe Kiprop Kipnusu, Manfred Reiche and Friedrich Kremer

Abstract While structure and conformation of condensed (semi-) isolated low molecular weight and polymeric molecules is well explored, nothing is known about their molecular dynamics as measured in a broad spectral range (1 Hz–1 MHz) and at widely varying temperature (200–400 K). This is achieved in this study by employing broadband dielectric spectroscopy (BDS) with nanostructured electrode arrangements, enabling electrode separations down to 35 nm. The approach offers, additionally, to determine in parallel the structure and conformation of the identical sample by scanning force microscopy (AFM). For the case of high molecular weight poly (2-vinylpyridine), it is demonstrated that even condensed (semi-) isolated polymer chains perform glassy dynamics, well comparable to the bulk state with the characteristic Vogel-Fulcher-Tammann temperature dependence; only $\sim$10$-$15 % of the mobile segments are weakly slowed down. The results are in full accord with the understanding that the length scale of the dynamic glass transition is $\sim$2 $-$ 3 polymer segments. Complementarily, AFM and infrared (IR) spectroscopy are employed to examine the chain conformation and to analyze the interfacial interactions, respectively. The former yields a condensed coil conformation, the average volume of which agrees within limits of $\pm$10 % with that calculated for a single chain assuming bulk density. A fraction of $\sim$30 % of the segments is found to be in a 0.4-nm-layer being directly in contact with the solid substrate. From IR investigations it is deduced that only half of them are influenced by the substrate presumably due to Coulombic interactions.

M. Tress (✉) · E. U. Mapesa · W. Kossack · W. K. Kipnusu · F. Kremer
Universität Leipzig, Leipzig, Germany
e-mail: martintress@gmx.de

M. Reiche
Max-Planck-Institut für Mikrostrukturphysik Halle, Halle, Germany

W. K. Kipnusu
e-mail: kipnusu@physik.uni-leipzig.de

F. Kremer
e-mail: kremer@physik.uni-leipzig.de

F. Kremer (ed.), *Dynamics in Geometrical Confinement*, Advances in Dielectrics,
DOI: 10.1007/978-3-319-06100-9_3, © Springer International Publishing Switzerland 2014

Keywords Glassy dynamics · Isolated polymers · Condensed chain · Collapsed coil · Broadband dielectric spectroscopy · Nanostructured electrodes · Interfacial interaction · Poly(2-vinylpyridine)

Abbreviations

AFM	Atomic force microscope
BDS	Broadband dielectric spectroscopy
HF	Hydrofluoric acid
IR	Infrared
P2VP	Poly(2-vinylpyridine)
PDI	Poly dispersity index
PS	Polystyrene
VFT	Vogel-Fulcher-Tammann

1 Introduction

Polymers are one of the major basic materials due to their low cost of production, facile processability and wide range of tunable properties enabling the fabrication of tailor-made materials for applications as diverse as lightweight construction elements, functional surface coatings, and organic semi-conductors. Being present in the vast majority of modern everyday-life appliances, polymers are vital in the ongoing demand for miniaturization in recent technology. The perspectives of this development are as appealing as flexible displays realized by ultra-thin functional layers and sensors or completely new devices as promised by molecular motors. Regardless of the particular application, manufacturing at the nanometric length scale down to single molecules requires a detailed understanding of the potential changes of the materials properties due to finite size effects.

The impact of such effects on one key quantity of polymeric materials, the glass transition temperature (T_g), is in the focus of intense research for about two decades now [1–30]. Publication of partly contradicting results has stirred up a controversial discussion; recent ideas aiming for a comprehensive explanation suggest that techniques probing the nonequilibrium state deliver results which are different from those obtained using equilibrium methods [30–37] (see also Chap. "Equilibrium and Out-of-Equilibrium Dynamics in Confined Polymers and Other Glass Forming Systems by Dielectric Spectroscopy and Calorimetric Techniques").

In this chapter, we go beyond the investigation of thin layers and examine glassy dynamics of isolated condensed polymer chains [38]. It will be shown that even the ultimate case of the single building unit of a polymer—the isolated macromolecule—conducts a dynamic glass transition. The challenge of an exiguous amount of sample

material, which would inhibit a successful measurement by any other experimental method because of insufficient signal intensity, is overcome by employing broadband dielectric spectroscopy (BDS) in combination with a refined nanostructured electrode arrangement. This is because a reduction of the thickness of the probe capacitor, and accordingly the probe volume, results in an increase of signal intensity. Furthermore, the (macroscopic) response of BDS has a clear-cut microscopic assignment and its broadband character enables examination of molecular motion in a wide spectral and temperature range [39].
Structural information of these samples on the nanoscale is gained by atomic force microscopy (AFM), which has already demonstrated its abilities to unravel the conformation of isolated polymer chains [40–47]. Additionally, it turned out enlightening to explore the type and especially the extent of polymer-surface interactions by means of infrared (IR) spectroscopy due to its chemical specificity.

2 Sample Preparation

The investigation of isolated polymer chains by means of BDS requires a unique probe setup. This is because the sample material does not cover the electrode completely. Therefore, evaporation of electrodes, as conventionally used to study thin polymeric layers [14, 18, 48–50], is not applicable. Instead, a recently developed technique employing highly conductive ultra-flat silicon electrodes with strongly insulating silica nanostructures as spacers [51] has been refined as described in the following to enable such measurements. This work is complemented by results obtained using IR spectroscopy on the polymer in nanopores, the fabrication of which is briefly described afterward.

2.1 Capacitors with Nanostructured Electrodes

As electrodes, highly doped silicon wafer dice with a specific resistance of <10 mΩ cm are employed. The bottom electrode, which also serves as support for the sample, has a root mean square roughness of <0.3 nm.[1] After a cleaning procedure which comprises rinsing with acetone, a plasma treatment and finally a surface purification with a snow jet gun,[2] isolated polymer nanodroplets are deposited by spin-coating from a very dilute solution ($c \sim 0.001$ wt%). Due to its strong segmental

[1] The root mean square roughness (R_q) refers to an area of $1 \times 1\,\mu\text{m}^2$ scanned with a digital resolution of 512×512 pixels which corresponds to an absolute resolution of $2 \times 2\,\text{nm}^2$. It is calculated from the mean height $\bar{z} = 1/N\ \Sigma_j^N z_j$ of the respective image as $R_q = \sqrt{1/N\ \Sigma_j^N \left(z_j - \bar{z}\right)}$ where N is the number of pixels.

[2] A snow jet gun expands super-critical CO_2 through a nozzle which generates a fast jet of CO_2 gas carrying CO_2 ice crystals. The latter polish the surface and by that remove contaminating particles.

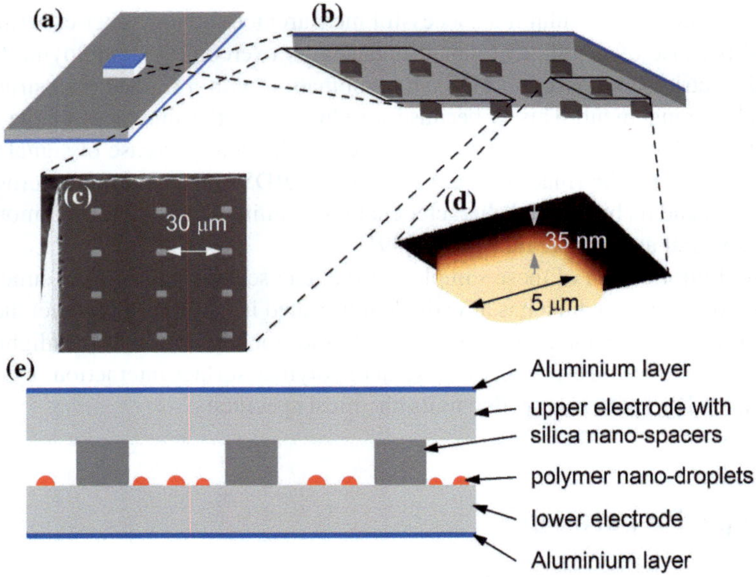

Fig. 1 **a** Sketch of the $4 \times 10\,\text{mm}^2$ large supporting electrode covered by the $1 \times 1\,\text{mm}^2$ sized counter electrode. **b** Schematic enlargement of the counter electrode tilted to show the spacer array which faces the sample on the bottom electrode. **c** Scanning electron microscopy image of the regular array of nanostructures and **d** 3D animation of an AFM height image of a single nanostructure. **e** Scheme of the sample cross section (dimensions not to scale)

dipole moment, poly(2-vinylpyridine) (P2VP) is a suitable candidate for dielectric measurements of extremely small amounts of sample material. As solvent, chloroform is adapted due to its particular interaction with P2VP, which favors the formation of collapsed coils[3] (for details see Sect. 3.1.1).

The final assembly of the capacitor requires some sort of spacer to avoid an electrical connection between the electrodes since the isolated nanodroplets exhibit no continuous surface coverage. For that, the counter electrode bears a regular matrix of strongly insulating nanostructures made of SiO_x, which act as spacers. With a quadratic base area of $5\,\mu\text{m}$ edge length and an inter-spacer separation of $45\,\mu\text{m}$, these nanostructures occupy an area of ~1 % in the assembled capacitor (Fig. 1). Even though the height of the spacers, which has been decreased particularly for these experiments to only 35 nm, is considerably smaller than in previous studies [22–24, 51] the polymer nanodroplets account for only about 1 % of the probe volume. Clearly, such a sensitive preparation requires high standards regarding purity and freedom from dust, for which reason it must be performed in a flow box or clean room.

[3] Notwithstanding the nomenclature used in the field of dissolved proteins for states of conformation, e.g., extended coils and collapsed globules, we will denote the conformation of polymer chains deposited in droplets as *condensed coils*.

Before the actual measurement, a severe annealing at a temperature of ~50 K + T_g (~150 K) for at least 24 h in a high, oil-free vacuum (10^{-6} mbar) is carried out. This procedure has been developed to combine best possible removal of solvent [52, 53], prevention of chemical degradation [54], and equilibration of the polymer to avoid metastable states [55]. Further, AFM checks of the surface topography are essential to verify the conformation and status of agglomeration of the chains. After the sample has been measured by means of BDS, the counter electrode is removed in order to directly investigate the surface of the actual probe area. In contrast, before the BDS measurement, such checks are restricted to spots nearby the top electrode since its removal would allow contaminating particles to vitiate the polymer surface.

2.2 Nanoporous Membranes

Isolated polymer chains forming nanodroplets exhibit a huge surface to volume ratio, therefore it is expected that interactions at the boundary play an important role in determining their structure and dynamics. Since the droplets are supported on a substrate, a considerable fraction of the boundary is composed of the P2VP-silica interface, which is reported to exhibit attractive interactions [56, 57]. However, details about this interaction like length scale or contributing entities are not easily found. For the latter, one may infer from coordination chemistry that the pyridine ring of the P2VP segments establishes some kind of bond with the terminal OH group of the native silica layer of the silicon substrate, but their range or quantity is unclear.[4]

To analyze this in detail, IR spectroscopy has been applied (Sect. 3). Since attempts of such an analysis on the flat silicon substrates employed for the BDS investigations have not been successful, systems with a much larger surface to volume ratio were desired. For that, nanoporous silica membranes turned out to be very useful (Chap. "Rotational Diffusion of Guest Molecules Confined in Uni-directional Nanopores").

These membranes are synthesized from conductive silicon wafers by electrochemical etching [60]. In principle, the wafer is immersed into a hydrofluoric acid (HF) solution and an applied current drives the formation of unidirectional pores of a few nanometers in diameter. The particular pore-diameter and porosity are controlled by the material and process parameters like wafer doping, HF concentration, current density, and anodization time [61]. Afterward, the porous silicon matrix is passivated by thermal oxidation. Details on the preparation can be found elsewhere [60, 62–64]. Immediately before filling the polymer into the pores, the membrane is annealed at 573 K in high vacuum (10^{-6} mbar, oil-free) for 24 h to remove any

---

[4] Although some studies claim to provide insight into the interaction mechanism based on the measurement of the glass transition temperature by means of X-ray reflectometry and fluorescence intensity detection [4, 9, 58, 59], these are very indirect observations, which try to deduce a molecular mechanism from a macroscopic quantity whose molecular origin still remains unclear [31–33].

water contained in this porous material. Subsequently and under argon atmosphere, the P2VP powder is placed on top of the membrane and heated to 443 K under high vacuum (10^{-6} mbar, oil-free) and kept there for 87 h to allow for the infiltration of the pores.

3 Experimental Results

The present investigation of isolated polymer chains contains three complementary experimental methods. These are (i) a comprehensive analysis of the surface by AFM, (ii) measurements of the segmental dynamics by means of BDS, and (iii) a detailed examination of the polymer-substrate interactions at the interface. Subsequently, a molecular picture containing all these results is given.

3.1 Surface Characterization

The characterization of the surface topography is inevitable to verify the existence of isolated chains. Therefore, AFM is a suitable technique which has been applied in several studies to identify the conformation of chains [40–47], but can also be employed to probe many other aspects. In the present investigation, we use AFM to determine (i) the (type of) conformation of the chains, (ii) the volume of the nanodroplets from which the number of enclosed chains is calculated, (iii) the spatial distribution of segments with respect to the substrate, and (iv) the radius of curvature in order to consider the internal pressure in such droplets.

3.1.1 Chain Conformation

Most studies which investigate the conformation of single polymer molecules on solid substrates focus on expanded chains, which are strongly adsorbed at a surface. Such samples would not be suitable to study dynamics since a strong immobilization of the adsorbed segments is expected. In contrast, condensed droplets contain only a certain fraction of segments potentially adsorbed at the substrate while the remaining proportion can be anticipated to exhibit a noticeable mobility.

It is well known that the dimension of polymer chains in solution strongly depends on the solvent, i.e., the interaction between solvent molecules and the segments [65]. This predominantly affects the topology of polymer chains deposited from solution onto solid substrates as evident, e.g., from block copolymers prepared from selective solvents (Fig. 2c) [40, 43]. In the case of P2VP, a sharp transition of the conformation between extended coils and collapsed globules is induced, e.g., by increasing the pH or the ion concentration in aqueous solution (Fig. 2b) [41, 47, 66, 67]. Similar effects have been studied in mixed polymer brushes of P2VP and polystyrene (PS) where

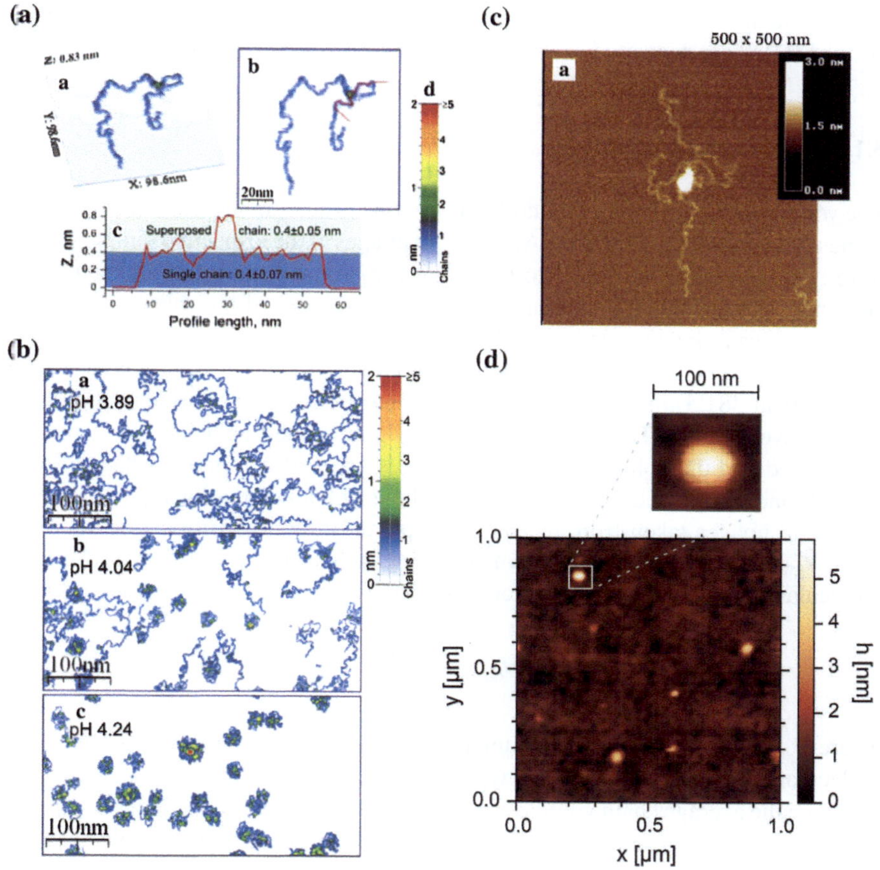

Fig. 2 **a** AFM images **a** & **b** with *height* scale bar **d** and profile **c** of a P2VP chain adsorbed on mica as extended coil, and **b** of P2VP chains on mica in different pH dependent conformations showing the transition between extended coils and collapsed globules (both reprinted with permission from [47]. Copyright 2005 American Chemical Society). **c** AFM image of an agglomerate of three diblock copolymer PS-b-PMMA chains deposited from aqueous solution; the selective solvent enables the PMMA blocks adsorbing on the mica support in an extended conformation while the PS blocks are collapsed (adapted with permission from [43]. Copyright 2003 American Chemical Society). **d** AFM image of nanodroplets of the collapsed P2VP coils investigated in this study (the *inset* shows an enlargement of one of the coils)

the chains were selectively stretched by swelling with different organic solvents [68]. These experiments reveal that in the presence of ethanol, a good solvent for P2VP, the chains undergo a large extension, while swelling with chloroform (which is also a solvent for P2VP), leads to collapsed chains.

Consequently, P2VP is dissolved in the moderate solvent chloroform favoring the formation of collapsed coils already in the solution, from which they are cast onto the substrate. In a sufficiently dilute solution, this also prevents the formation of entanglements with neighboring chains. As a result, isolated noninteracting collapsed coils of P2VP are deposited as visualized in Fig. 2d.

3.1.2 Volume of Isolated Coils

The volume of the isolated P2VP coils enables one to determine the number of chains in these nanodroplets. Therefore, AFM height images of the respective sample surface have been examined according to the following protocol. First, all grains above a certain height threshold, which must cover the roughness of the substrate or rather its peak-to-peak height, are masked. In the present case, the threshold value has been chosen to be 1.3 nm. To exclude potential spikes, any of these masked areas smaller than 381.5 nm^2 (corresponding to a diameter of 22 nm) are excluded from further investigation. Then, for the remaining grains, the Laplacian background is determined and the volume encased by this background, the grain surface and the mask boundary is calculated.[5] The grain volumes obtained by this type of analysis of several images taken from the same sample at different positions yield a mean value $\hat{V}_{\text{grain}}$. As reference, the volume of a single chain V_{chain} is estimated from the molecular weight and the assumption of bulk mass density $\rho = 1.14\,\text{kg/m}^3$ [69]

$$V_{\text{chain}} = \frac{M_n}{\rho N_A} \tag{1}$$

where N_A represents Avogadro's number. For this calculation, the number averaged molecular weight M_n is used since volume scales with the number of segments and therefore the average chain volume is proportional to M_n. On a first glance, assuming bulk density may appear questionable on the level of individual macromolecules, but nevertheless this supposition has been substantiated in previous studies [40, 43, 45].

Table 1 summarizes the average droplet volumes and calculated chain volumes for several polymer portions resulting in distinct distribution regimes of different molecular weights. Two of the regimes show droplet volumes which are about a factor of five larger than the estimated single chain volume. This indicates agglomerates of about five chains in one nanodroplet, which we refer to as semi-isolated chains. In samples cast from even more dilute solutions, the droplets have an average volume comparable to the calculated volume of a single chain, revealing individual well-separated polymer chains. The spread of the detected volumes is assigned to the relatively large molecular weight distribution.

[5] For the described routine the free software Gwyddion, version 2.25 (GNU General Public License, downloaded from http://www.gwyddion.net, 2012) has been applied.

Table 1 Characteristic properties of the used polymers and volumetric analysis of the respective isolated nanodroplets: Weight averaged and number averaged molecular weight M_W and M_n, respectively, polydispersity index *PDI*, calorimetric glass transition temperature of bulk T_g, estimated volume of a single chain V_{chain} and the average volume of the grains/ nanodroplets $\hat{V}_{\text{grain}}$

Property	P2VP 1	P2VP 2	
M_W (kg/mol)	1,020	2,250	
M_n (kg/mol)	766	1,510	
PDI	1.33	1.49	
T_g (K) by DSC	371 (±1)	375 (±1)	
V_{chain} (nm^3) calculated	1,142	2,251	
Chain distribution	Semi-isolated	Semi-isolated	Isolated
$\hat{V}_{\text{grain}}$ (nm^3)$\pm SE$	8,864 (±2,500)	11,564 (±3,500)	2,033 (±500)
$V_{\text{chain}}/\hat{V}_{\text{grain}} \pm SE$	7.8 (±2.2)	5.1 (±1.6)	0.9 (±0.2)
Curvature r (nm) $\pm SE$	325 (±54)	753 (±67)	92 (±12)
Droplet density (μm^{-2})	37.5	10.0	4.0
Area analyzed (μm^2)	8	8	13.75
# droplets analyzed	300	80	55

3.1.3 Distribution of Segments

Especially in view of interfacial interactions which are examined in Sect. 3.3, the distribution of segments with respect to their distance to the supporting substrate is of importance. Interfacial interactions are, depending on their range, likely to alter dynamics of the segments within that specific distance. Combined with information from the dynamics, knowledge of the spatial distribution thus enables an estimation of the range of specific interfacial interactions.

Therefore, height profiles of the nanodroplets are taken from the AFM images. For each droplet, four profiles along different axes (horizontal, vertical and both diagonals) coincide to a large extent which reveals a high degree of radial symmetry. This feature is explained in detail in Sect. 3.1.4. The width of a P2VP chain segment which has been determined to be 0.4 nm [47] (Fig. 2a) is then used to divide these profiles into slices of chain segments. Considering the radial symmetry yields a volume fraction of $\sim 15-20\,\%$ and $\sim 25-30\,\%$ for the very first segment layer at the surface in a droplet of semi-isolated and isolated chains, respectively (Fig. 3).

3.1.4 Internal Pressure

The previous section demonstrates a remarkable symmetry among the isolated nanodroplets, which supports the assumption of a driving mechanism generating this particular shape. In fact, surface tension of liquids controls the shape of small droplets. For spherical droplets, the relation between surface tension γ and radius of curvature r is given by the Young-Laplace equation for the internal pressure p of the droplet

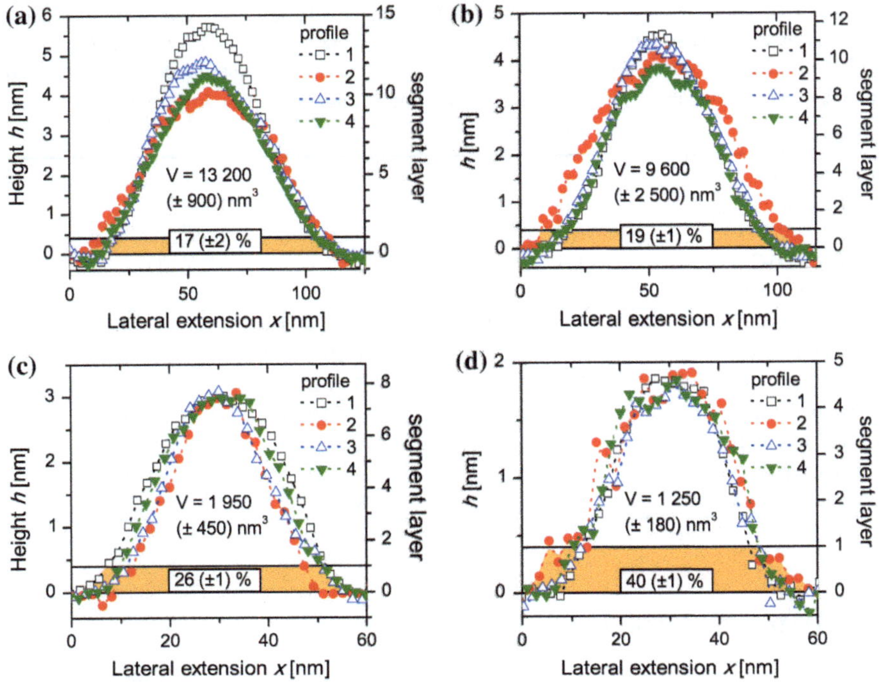

Fig. 3 Profiles of exemplary P2VP nanodroplets extracted from AFM scans for **a** and **b** semi-isolated ($M_W = 1{,}020$ kg/mol) and **c** and **d** isolated chains ($M_W = 2{,}250$ kg/mol). Each graph depicts four profiles taken along four different directions (1-horizontal, i.e. in AFM scan line direction, 2-vertical, i.e. perpendicular to scan line direction, 3-diagonal from top left to bottom right, 4-diagonal from top right to bottom left) of the same droplet. The dashed lines are interpolations of the data and act as a guide to the eye while the colored areas underneath the curves indicate the fraction of the volume within a 0.4-nm-thick layer at the bottom, which represents the first layer of segments at the substrate

$$p = \frac{2\gamma}{r} \tag{2}$$

Insertion of the radii of curvature (Table 1), which range from almost 1 μm to ~100 nm for semi-isolated and isolated chains, respectively (Fig. 4), and an estimated[6] surface tension of ~40 mN/m yields an internal pressure between ~10^5 Pas and ~10^6 Pas. These values correspond to about 1–10 times the ambient pressure which is reasonable in terms of a counterbalance of internal and external pressure and furthermore explains the high degree of radial symmetry of the droplets. However, as evident from Fig. 4 the larger droplets (semi-isolated chains) are not adequately

[6] Sauer and Dee [70] give values for the surface tension of P2VP only in the temperature range of 450–500 K (35–32 mN/m). Extrapolating these values to temperatures of 300 (room temperature as employed during the AFM investigation) and 430 K (representing the annealing temperature) yields approximate surface tensions of 44 and 37 mN/m, respectively.

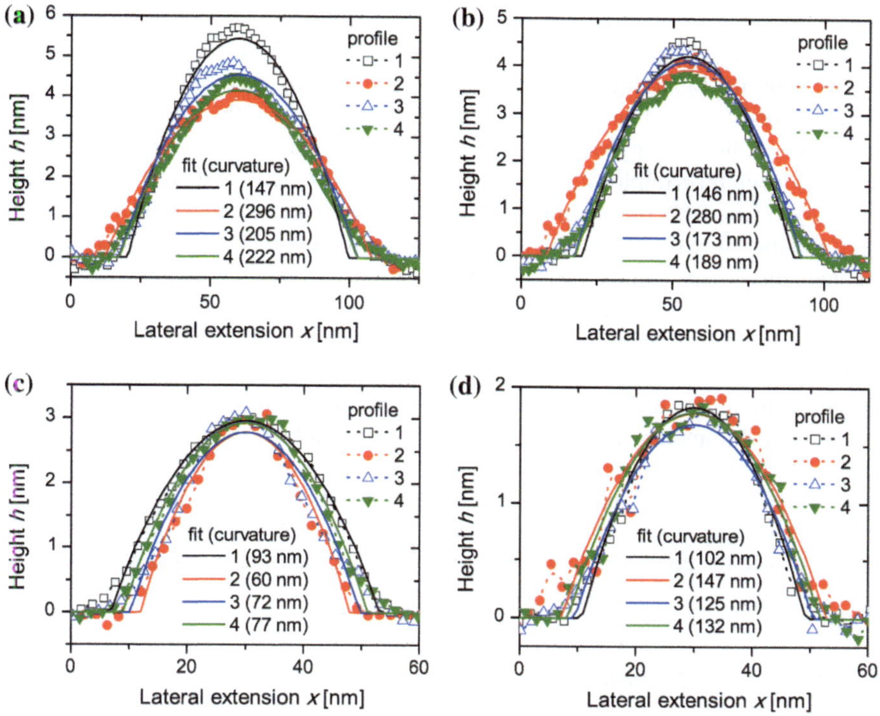

Fig. 4 Same profiles as shown in Fig. 3 of exemplary P2VP nanodroplets for **a** and **b** semi-isolated ($M_W = 1{,}020$ kg/mol) and **c** and **d** isolated chains ($M_W = 2{,}250$ kg/mol). The *dashed lines* are interpolations of the data and act as a guide to the eye while the *solid lines* represent fits describing the profiles as segments of a circle with a curvature as indicated

described by a single radius of curvature. Hence, their curvature is rather close to ~100 nm as well which, in consequence, corroborates an estimated internal pressure of about 10 times the ambient pressure. This does not match the counterbalance reasoning, but a specific interpretation of this value requires a detailed consideration of the thermodynamic history of the samples, which is difficult to comprehend as discussed on a qualitative basis in the following.

In solution the extension of the chains is driven by the interaction between solvent molecules and chain segments and results in a particular expansion which can be probed, e.g., in terms of the hydrodynamic radius r_h. The spin-coating process is expected to introduce a flattened and laterally extended conformation in which the chains are deposited onto the substrate while the solvent volatilizes. This removal however is certainly incomplete and therefore the chains will remain in a swollen state as nanodroplets which consequently are likely to have a reduced surface tension. At this stage, the counterbalance of external pressure and surface tension will cause the formation of droplets with a particular radius of curvature. Since subsequently the solvent continuously evaporates, the surface tension increases resulting in a larger radius of curvature. During the following annealing procedure, the solvent is removed even more rigorously whereas in contrast the increased temperature would decrease the surface tension. Although it is not clear which of these two effects predominates, it is evident that the reduction of the external pressure by up to nine decades (from ambient pressure of ~1 bar to $\sim 10^{-6}\ mbar$) exceeds both of them by far and would cause extraordinarily large radii of curvature. After the annealing, the effect may be reversed when the sample is transferred to room temperature and ambient pressure. In the following BDS measurement, elevated temperatures decrease the surface tension, and therefore the curvature, while the external pressure corresponds (nearly) to ambient conditions. During the AFM investigations, the chain is exposed to ambient pressure and room temperature.

Overall, external pressure, temperature and, related but not exclusive to the latter, surface tension change substantially during the sample preparation and it is by no means clear, up to which stage the radius of curvature of the coils is driven by thermodynamic parameters and when it remains stable and resists due to a sufficiently high surface tension. It should be noted that the values for the latter are extrapolated from the melt at high temperatures, and consequently may be remarkably underestimated, especially for room temperature. Any further interpretation is therefore difficult since the chains undergo several changes in their thermodynamic history.

3.2 Molecular Dynamics

After extensive characterization of the surface topography which proves the existence of semi-isolated and isolated chains in a collapsed coil conformation, the molecular dynamics in such mono-molecular nanodroplets is to be focused. P2VP enables, due to sits strong segmental dipole moment, the investigation of its segmental or glassy dynamics by means of BDS. Figure 5 depicts spectra of the dielectric loss ε'' (the imaginary part of the complex dielectric function $\varepsilon^* = \varepsilon' - i\,\varepsilon''$ with the imaginary unit $i = (-1)^{1/2}$) of P2VP which clearly show the α-relaxation in bulk but also in semi-isolated as well as isolated chains. Consequently, even isolated macromolecules in a condensed coil conformation contain mobile segments.

To analyze the relaxation processes and extract their mean relaxation time τ_c (the inverse of which is the mean relaxation rate and indicates the maximum position of the relaxation peak), the empirical Havriliak-Negami function [71] has been employed

$$\varepsilon^*_{HN}(\omega) = \varepsilon_\infty + \frac{\Delta\varepsilon}{(1 - (i\omega\tau_{HN})^\alpha)^\beta} \tag{3}$$

where ε_∞ and $\Delta\varepsilon$ denote the permittivity of the unrelaxed medium and the dielectric strength, respectively. The latter has not been investigated since, in the used sample arrangement, the reduced filling fraction causes a systematic diminution. Even considering the sample geometry is insufficient to reliably extract $\Delta\varepsilon$ while the other parameters are not affected at all as examined in the appendix. α and β are parameters

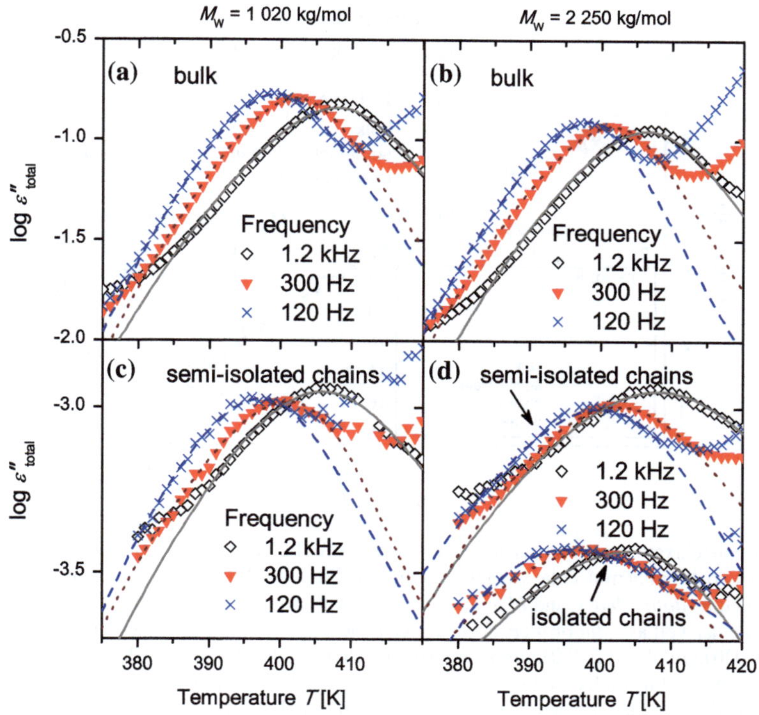

Fig. 5 Dielectric loss versus temperature for P2VP in **a** & **b** bulk and **c** & **d** in semi-isolated as well as **d** isolated chains for two different molecular weights and several frequencies as indicated. The *dashed*, *dotted* and *solid lines* are Havriliak-Negami functions (extended by a Vogel-Fulcher-Tammann temperature dependence of the relaxation time) which have been fitted to the data. At high temperatures (T > 410 K) some curves exhibit a conductivity contribution. Adapted from [38]. Reprinted with permission from AAAS

accounting for symmetric and asymmetric broadening, and τ_{HN} is a relaxation time which can be converted into τ_c by the relation

$$\tau_c = \tau_{HN} \sin\left(\frac{\alpha\beta\pi}{2+2\beta}\right)^{1/\alpha} \sin\left(\frac{\alpha\pi}{2+2\beta}\right)^{-1/\alpha} \tag{4}$$

Analyzing the temperature dependence of the mean relaxation time $\tau_c(T)$ in the semi-isolated and isolated P2VP chains (Fig. 6) reveals a trend described by the Vogel-Fulcher-Tammann (VFT) equation [72–74]

$$\tau\,(T) = \tau_0 \exp\left(-\frac{DT_V}{T-T_V}\right) \tag{5}$$

with the constants τ_0, D and the Vogel temperature T_V. This relation is typically applied to describe the temperature dependence of the mean relaxation rate as shown

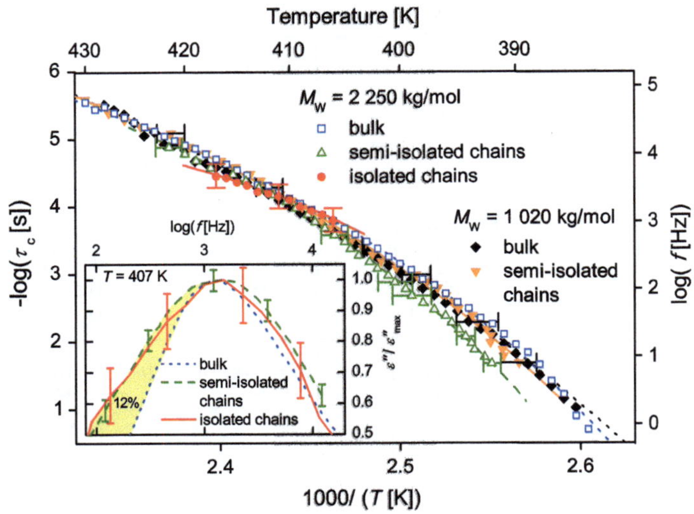

Fig. 6 Mean relaxation time versus inverse temperature for P2VP in bulk, semi-isolated and isolated chains for two different molecular weights as indicated. Fits to the Vogel-Fulcher-Tammann equation are included as *dotted black* and *blue lines* for bulk with M_W of 1,020 kg/mol and 2,250 kg/mol, respectively as well as *solid orange*, *dashed green* and *solid red line* for semi-isolated chains with M_W of 1,020 kg/mol and 2,250 kg/mol and isolated chains with M_W of 2,250 kg/mol, respectively. (*Inset*) Corresponding dielectric loss spectra normalized with respect to the peak maximum of bulk, semi-isolated and isolated chains for a M_W of 2,250 kg/mol at a temperature of 407 K. Adapted from [38]. Reprinted with permission from AAAS

in Fig. 6 and represents a major characteristic of glassy systems. Furthermore, it can be used to introduce temperature as free variable in the Havriliak-Negami function (as used in Fig. 5) by inserting Eq. 5 in Eq. 3, namely by the substitution $\tau_{HN} = \tau(T)$. Beyond the fact that the mean relaxation rate of semi-isolated and isolated chains exhibits a VFT temperature dependence, it even coincides with the curve for the bulk sample. This means that a majority of the mobile segments exhibits bulk-like glassy dynamics.

A detailed inspection of the loss spectra reveals a broadening of the α-relaxation in the case of semi-isolated and isolated chains (Fig. 6 inset). In a representation normalized with respect to the peak maximum, the excessive area in the spectra of the nanodroplets, which occurs on the low frequency side, can be extracted. It amounts to $\sim$12 % of the whole peak area. Accordingly, $\sim$12 % of the mobile segments in the (semi-) isolated chains relax slightly slower than the remaining bulk-like fraction.

This finding apparently does not match the result of Sect. 3.1.3. There, for isolated coils, a fraction of $\sim$30 % of the segments is determined to be in direct contact with the surface. This mismatch cannot be explained so far but is presumably interrelated with the specific polymer-surface interactions for which reason they are addressed in the following section.

3.3 Interfacial Interactions

To explore the interfacial interactions of P2VP with the silica surface, IR spectroscopy is a suitable technique due to its chemical specificity (a detailed description of IR spectroscopy can be found elsewhere [75]). As only material in a very small range near the interface is expected to show alterations, the sample should contain a large fraction of that particular region to provide enough signal strength. Appropriate systems for such investigations are membranes with nanometric unidirectional pores. In silica nanopores of 10 nm diameter, the IR spectrum of P2VP[7] shows a general shift to higher wave numbers by 1.5–2 cm^{-1}. Such effects are well known from other materials in pressure dependent measurements and can be assigned to density effects due to packing defects [76]. Furthermore, an absorption band at 1,590 cm^{-1} specific for the stretching vibration of the pyridine ring [77–79] shows a shoulder which is not present in bulk (Fig. 7). Subtraction of the bulk spectrum, after correction for the density effect, reveals a remaining absorption band which is assigned to a modified stretching vibration of the bound pyridine ring. However, an analysis of the peak area is not reasonable in this case since the transition moment of the bound pyridine ring possibly differs from the nonbound and hence the areas are not comparable any more. In contrast, the silica surface as counterpart of the bond possesses an attribute which enables a quantitative analysis. Namely, the terminal OH group at the silica surface shows an absorption band at ~3,750 cm^{-1} assigned to its stretching vibration in the nonbound case [60, 79, 80]. If the membrane is filled with P2VP, it is observed that the intensity of this band is reduced and its area decreased by 50 % (Fig. 8). This means that about half of the terminal OH groups bind to the pyridine ring.

A comparison of the mean distance (of ~0.45 nm) between the terminal OH groups [60] and the size (~0.4 nm) of a P2VP chain segment [47] reveals a similar surface coverage for both. Considering that only 50 % of the OH groups establish bonds, leads to the conclusion that about half of the P2VP segments in direct contact with the surface are bound. The reason for that remains unclear but is possibly related to tacticity in general and steric afflictions of neighboring segments with respect to the orientation of the pyridine ring in particular.

3.4 Combining the Results: A Molecular Picture

The previous sections present results obtained by three very different methods, that is, AFM, BDS, and FTIR, and reveal a wealth of information which now merges to a consistent and detailed picture (Fig. 9). Single P2VP chains in a collapsed coil conformation put onto a silica surface take the shape of a flat spherical calotte. This is driven by the surface tension and the internal pressure which roughly balances with the ambient pressure. In that shape, ~30 % of the segments directly contact the

---

[7] Note that, to ease the infiltration, the P2VP filled into the pores has a M_W of 22 kg/mol which is much lower than that of the P2VP deposited as isolated chains.

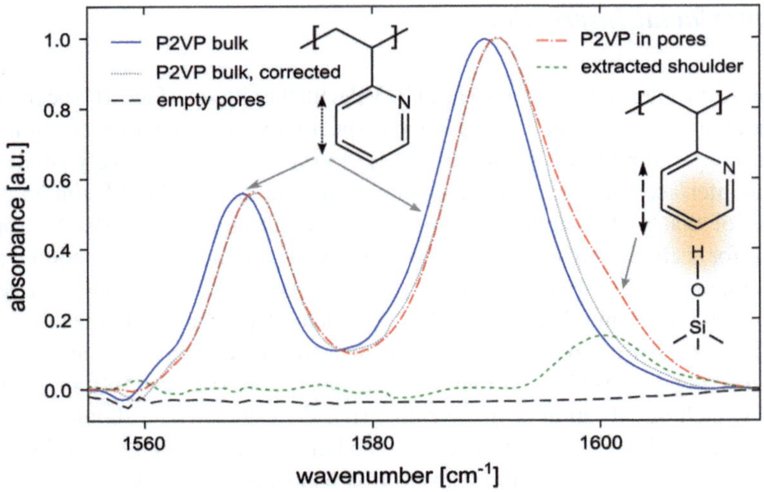

Fig. 7 IR absorption bands of the pyridine ring stretching vibration (*scheme on the left*) of bulk P2VP (*solid line*) and P2VP in silica nanopores of 10 nm diameter (*dash-dotted line*). After the P2VP bulk spectrum has been shifted (*dotted line*) to take into account a density effect (see text for details) it has been subtracted from the spectrum of P2VP in pores. A remaining peak (*short-dashed line*) is assigned to the stretching vibration of pyridine rings bound to terminal OH groups of the silica surface (*scheme on the right*). For comparison, the spectrum of an empty silica membrane of the same type is shown (*long-dashed line*). Adapted from [38], supplementary materials. Reprinted with permission from AAAS

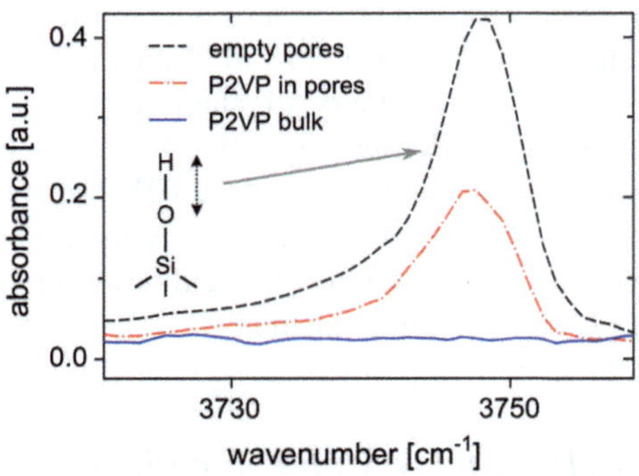

Fig. 8 IR absorption band of the stretching vibration of free (nonbound) OH groups of silica (*scheme on the left*) measured in an empty nanoporous silica membrane (*dashed line*) and of a P2VP filled nanoporous membrane (*dash-dotted line*). The IR signal of bulk P2VP (*solid line*) in the same spectral region is shown for comparison. Adapted from [38], supplementary materials. Reprinted with permission from AAAS

substrate and about half of these are bound to terminal OH groups. About $\sim 12\,\%$ of the mobile segments show a slowed dynamics which is assigned to interactions with the interface. To explain the fact that only about half of the segments in direct

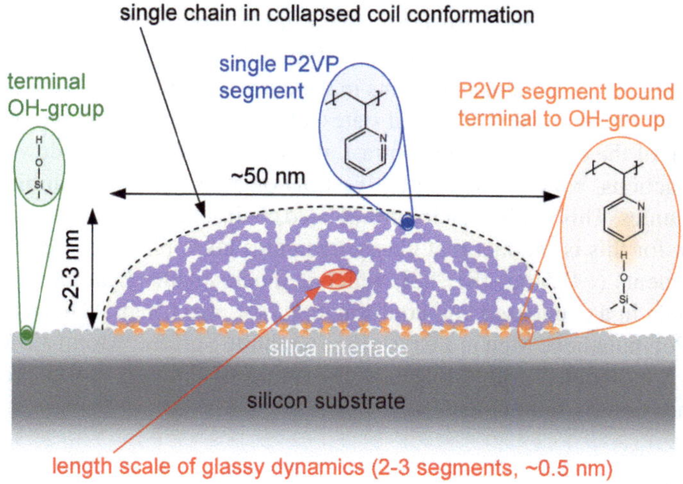

Fig. 9 Scheme of a single chain in a collapsed coil conformation supported on a solid substrate which establishes bonds to the segments as found for the P2VP-silica system (vertical and lateral extension not to scale)

contact with the surface is slowed down two possibilities are envisaged. Either the bound segments account for the slower relaxation modes while any other segment has bulk-like relaxation, or these segments are completely immobilized due to the bond and the nonbound segments in their immediate neighborhood exhibit slower dynamics. In both cases, about 15 % of slowed segments are expected to be found by BDS.

According to studies on the adsorption of polymer chains at solid interfaces [18, 35], the latter of the two possibilities seems more plausible. In these studies, Napolitano et al. found that an increase of the thickness of the adsorbed layer is accompanied by a reduction in the dielectric strength of the segmental relaxation of P2VP and PS on aluminum surfaces.[8] This means that adsorbed chains do not contribute to the dielectric relaxation process because they are immobilized. The fact, that Napolitano et al. find adsorbed (and therefore immobile) layers of ~5 nm thickness in P2VP with a M_W of 159 kg/mol [18] while in the present study nanodroplets of lesser height exhibit segmental mobility is attributed to the difference in molecular weight. This is inferred from a recent study of PS by the same authors which examines the kinetics of the adsorption [35]. Therein, an adsorbed layer thickness of 8 nm has been found to fully develop after annealing at $\sim 50\,K + T_g$ for more than 10 h for a M_W of 97 kg/mol. In contrast, an increase in M_W by less than a factor of 2 ($M_W = 160$ kg/mol) resulted in an increase of the time scale of adsorption kinetics by

[8] Despite the fact that the aluminum surfaces are different from the silica substrates in the present study, especially in the case of the upper electrode which is evaporated onto the polymer layer, the discussion of these results will show that there is no contradiction between them and the herein reported dynamics of isolated chains.

a factor of almost 100 (in this case, the authors could not observe the full development of the adsorbed layer in their experiment). Extrapolating this to the extremely high molecular weights of the isolated chains in the current study leads to the conclusion that the development of such adsorbed states takes orders of magnitudes longer than the duration of the experiment. In contrast, the segments not slowed down by interfacial interactions, which comprise the vast majority in the coils, exhibit bulk-like glassy dynamics. This holds even for the isolated chains, i.e., single macromolecules. The reason for this is the fact that glassy dynamics takes place on the length scale of 2–3 segments (~0.5 nm) [81, 82], which is much smaller than the extension of a single chain in a condensed coil conformation (see also Chaps. "Dielectric Relaxation of a Polybutadiene Melt at a Crystalline Graphite Surface: Atomistic Molecular Dynamics Simulations" and "Molecular Dynamics of Poly(*cis*-1,4-Isoprene) in 1- and 2-Dimensional Confinement").

4 Summary

Investigating the properties of isolated molecules is of major interest for new developments, especially miniaturization, in several branches of research, be it sensor technology, catalysis, nanomotors and machines, etc. This applies also to polymers as a class of materials which is in wide range of use in modern technology. On the cutting edge of what is experimentally possible, BDS in combination with a nanostructured electrode arrangement has been employed to study glassy dynamics in isolated P2VP polymer chains, which are deposited on a solid substrate in a condensed coil conformation. Remarkably, their dynamics is found to be mainly bulk-like; only a small fraction of segments, namely those in direct contact with the solid surface, are slowed down. Complementary IR measurements reveal that about half of them are bound to the surface due to the specific interactions between the pyridine rings and the terminal OH groups of the silica surface. These bound segments are highly likely to be immobilized while the remaining segments in this first layer at the interface are assigned to the slightly slowed fraction seen in the dielectric spectra.

As demonstrated for the system of isolated condensed P2VP chains on a silica surface, the sample capacitor particularly developed for these measurements paves the way for further investigations of the dynamics of isolated molecules. This comprises but is not limited to more complex molecules from copolymers to large proteins and may include variation of the substrate chemistry as well.

Acknowledgments We are grateful to DFG for funding within the research projects "Polymer-Solid Contacts: Interfaces and Interphases" SPP 1369 and "Polymers under multiple constraints: restricted and controlled molecular order and mobility" SFB TRR 102 as well as the graduate school Build MoNa. Further, we thank M. Wübbenhorst and J.L. Barrat for fruitful discussions regarding absorption of polymer chains at solid substrates and internal pressure in nanodroplets, respectively.

Appendix

A.1 Model Considerations

The nanostructured electrode arrangement is, on the one hand, superior to conventionally evaporated electrodes since first, it avoids an exhibition of the sample material to hot metal atoms and second, it enables the measurement of samples without a complete surface coverage, like the presented isolated polymer coils, in the first place. On the other hand, in this arrangement several different components are, besides the actual sample, located within the probe volume of the capacitor. The contributions of all these components superimpose in an intricate way yielding the total complex capacity. In order to analyze data obtained with this setup in more detail, it is necessary to employ equivalent circuit representations. Several equivalent circuits of complex capacities which model different possible descriptions are considered in the following.

A.1.1 Equivalent Circuits

A schematic representation of the sample cross-section is given in Fig. 10a together with four different equivalent circuits. For the latter, the schematic cross-section is divided into regions containing only one kind of material, each of which is then described by a single capacitor with the complex capacity $C^* = C' - i\,C''$ (C' and C'' denote the real and imaginary part of the capacity, respectively).

$$C^* = \varepsilon_0 \varepsilon^* A/d \tag{A.1}$$

C^* is defined by the respective complex dielectric function $\varepsilon^* = \varepsilon' - i\,\varepsilon''$ (with the real and imaginary parts ε' and ε'', respectively which will be referred to as permittivity and loss) as well as the area and the separation of the electrodes A and d, respectively. $\varepsilon_0 = 8.8548\cdots \times 10^{-12}$ As/Vm denotes the permittivity of vacuum. The specific choice of how these capacitors are connected to a each other resembles the presence or absence of paths of electrical conduction at boundaries and thereby affects the total capacitance of the whole arrangement. Since it is not clear which boundaries are sufficiently conductive to justify a connection, several different circuits (as shown in Fig. 10b–e) reflecting different degrees of complexity or surface conductivity are modeled in the following.

A.1.1.1 Model 1: Simple Network

In the simplest network, all regions which are—in parallel to the electrode—made of the same material are described by a single capacitor each (Fig. 10b). Thereby,

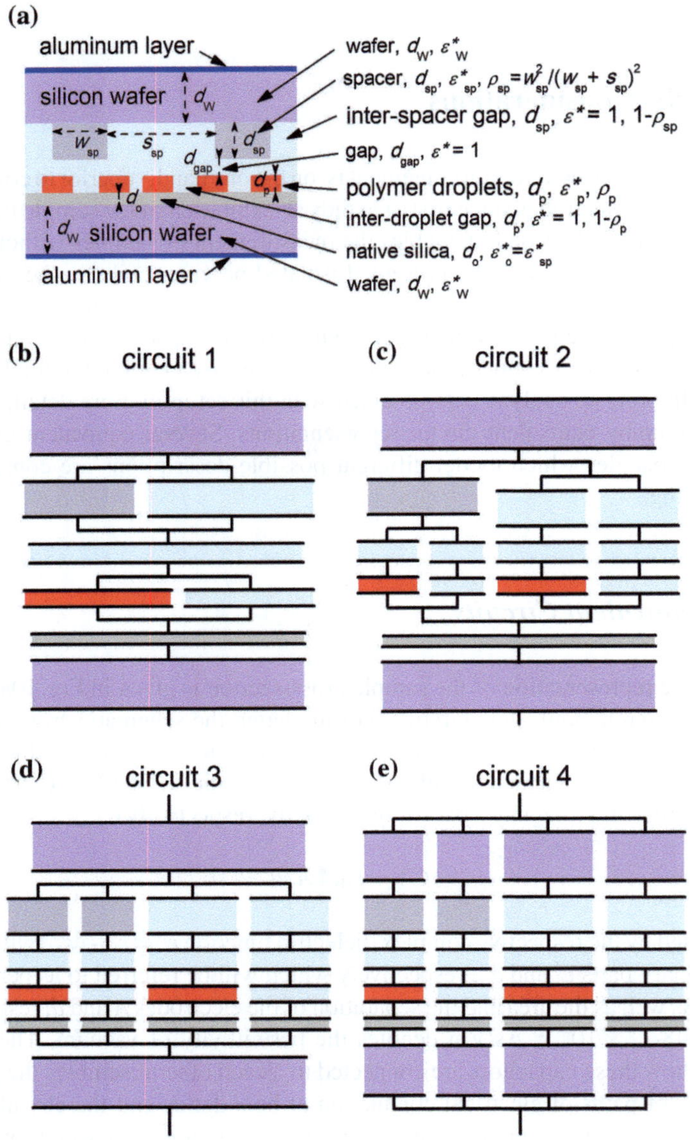

Fig. 10 **a** Simplified scheme of the sample cross section to match the description by a network of parallel plate capacitors (measures not to scale). **b–e** Different equivalent circuits for modeling the schemed cross section as discussed in the text. The same color code as in **a** is used for the different materials: *red*–polymer, *light purple*–silicon, *gray*–silica, *light blue*–air/nitrogen, *blue*–aluminum (since aluminum is considered an ideal conductor compared to the other components, it is not included in the equivalent circuits). For definition of the parameters depicted in **a** see Sect. A.1.1.1

the capacitor representing the spacers is combined with that of the empty space in-between them in a parallel arrangement which results in an effective complex capacitance $C^*{}_{sp}$, eff

$$C^*{}_{sp}, \mathrm{eff} = \frac{\varepsilon_0 \left(\rho_{sp}\varepsilon^*_{sp} + 1 - \rho_{sp}\right) A}{d_{sp}} \tag{A.2}$$

where ε^*_{sp} denotes the complex dielectric function of the spacers, d_{sp} their height and ρ_{sp} their density of coverage. Since both the base area of the structures and their planar spacing to nearest neighbors are quadratic, ρ_{sp} is easily deduced from the width w_{sp} and the distance in-between s_{sp} as

$$\rho_{sp} = \left(\frac{w_{sp}}{s_{sp} + w_{sp}}\right)^2 \tag{A.3}$$

In a similar way the effective complex capacity of the polymer nanodroplets and the space in-between $C^*{}_p$, eff is obtained

$$C^*{}_p, \mathrm{eff} = \frac{\varepsilon_0 \left(\rho_p\varepsilon^*_p + 1 - \rho_p\right) A}{d_p} \tag{A.4}$$

with the complex dielectric function of the polymer coils ε^*_p, their height d_{sp} and the density of coverage ρ_p. In this case the latter must be determined from AFM scans of the sample surface (for compact layers $\rho_p = 1$). Further, due to the simplification of the shape of the coils (this aspect will be addressed in Sect. A.1.2), the height reflects a certain average which must yield, in combination with the density of coverage, a volume that matches the experimentally determined value.

Together with the remaining capacitors, $C^*{}_W$, $C^*{}_{\mathrm{gap}}$, $C^*{}_o$ representing the wafer, an interfacial oxide layer and an additional gap, respectively, these two parallel circuits are connected in a serial arrangement and the total complex capacitance $C^*{}_{\mathrm{total}}$ is given by

$$\frac{1}{C^*{}_{\mathrm{total}}} = \frac{2}{C^*{}_W} + \frac{1}{C^*{}_{sp}, \mathrm{eff}} + \frac{1}{C^*{}_{\mathrm{gap}}} + \frac{1}{C^*{}_p, \mathrm{eff}} + \frac{1}{C^*{}_o} \tag{A.5}$$

The gap takes into account the fact that in most cases the separation between the electrodes is somewhat larger than the spacers. This can have different origins: While deviations by more than ~50 nm are very likely to be caused by particle contamination which is circumvented by a severe cleaning as described in Sect. 2.1, differences on the order of ~5–10 nm are presumably induced by surface irregularities and hardly avoidable. Since there is no experimental way to determine the height of this gap, it must be treated as a fit parameter in the analysis.

After inserting Eqs. A.2 and A.4 as well as equations of the type of Eq. A.1, into Eq. A.5 the permittivity of the vacuum ε_0 and the base area of the capacitor A cancel out and the total complex dielectric function $\varepsilon^*_{\text{total},1}$ of circuit 1 is obtained as

$$\frac{D}{\varepsilon^*_{\text{total},1}} = 2\frac{d_W}{\varepsilon^*_W} + \frac{d_{sp}}{\rho_{sp}\varepsilon^*_{sp} + 1 - \rho_{sp}} + \frac{d_{\text{gap}}}{\varepsilon^*_{\text{gap}}} + \frac{d_p}{\rho_p\varepsilon^*_p + 1 - \rho_p} + \frac{d_o}{\varepsilon^*_o} \tag{A.6}$$

where thicknesses and complex dielectric functions of the silicon wafers, the native or thermally oxidized silica layer, the gap, the polymeric layer and the silica spacers are given by d_W, ε^*_W, d_o, ε^*_o, d_{gap}, $\varepsilon^*_{\text{gap}} = 1$, d_p, ε^*_p, d_{sp}, $\varepsilon^*_{sp} = \varepsilon^*_o$, respectively. Technically, D is the distance between the electrodes

$$D = d_{sp} + d_{\text{gap}} + d_p + d_o \tag{A.7}$$

Note that D does not comprise the thickness of the silicon electrodes because this component contributes only in case it would resemble a dielectric rather than a conductor. This is the case at high frequencies ($f > 10^4$ Hz) causing an additional contribution which increases with frequency. It has to be considered in the analysis of weak signals like obtained from the isolated chains but it is not present in the following example measurement of a thin layer. At lower frequencies though, the wafers' conductivity is still sufficiently high to resemble a conductor, hence the negligence of its thickness is justified. Further, if Eq. A.6, rearranged with respect to $\varepsilon'_{\text{total},1}$ and/or $\varepsilon''_{\text{total},1}$, is applied as a fit function to measured permittivity and/or loss spectra, a correct analysis requires D to take the thickness value which has been used to extract this data from the measured capacity, irrespective of fulfilling Eq. A.7. This action traces back to the capacity, the actually measured quantity, and therefore prevents the propagation of any errors possibly arising from an estimation for D.

Admittedly, circuit 1 reflects a very simplified scenario since it comprises paths of charge transport which are physically implausible. For example, the path from the lower surface through the polymer nanodroplets (instead of the empty space in-between) and then, after bridging the gap, through the silica spacers is in the model open to all charges. In contrast, if in a real sample a polymer nanodroplet is located such that there is no spacer atop, the distance for charges to pass through the latter would be much larger than bridging just the space to the top electrode. Consequently, more refined circuits have to be considered (Fig. 10c–e). The rising degree of complexity is interpreted in terms of physical considerations.

A.1.1.2 Model 2: Conductive Silica Surface

Circuit 2 pays attention to the fact that the parts of the polymeric layer which have spacers on top are not electrically connected to those without. A parallel arrangement of these two paths is taking this into account. Physically, this scenario corresponds to a high surface conductivity of the silica components and its total dielectric function $\varepsilon^*_{\text{total},2}$ is given by

$$\frac{D}{\varepsilon^*_{\text{total},2}} = 2\frac{d_W}{\varepsilon^*_W} + \left(\frac{\rho_{sp}}{K^*} + \frac{(1-\rho_{sp})\rho_p}{d_{sp}+d_{\text{gap}}+\frac{d_p}{\varepsilon^*_p}} + \frac{(1-\rho_{sp})(1-\rho_p)}{d_{sp}+d_{\text{gap}}+d_p}\right)^{-1} + \frac{d_o}{\varepsilon^*_o}, \text{ where}$$

$$K^* = \frac{d_{sp}}{\varepsilon^*_{sp}} + \left(\frac{d_{\text{gap}}+d_p}{d_{\text{gap}}+d_p/\varepsilon^*_p}\right) \tag{A.8}$$

A.1.1.3 Model 3: Conductive Silicon Surface

In circuit 3, a similar way of reasoning is made so that the silica components, namely spacers and the native oxide layer, are divided and merged to the parallel paths in addition to the polymer. This reflects a low surface conductivity of the silica components while that one of the silicon electrodes remains high. The respective equation reads

$$c\frac{D}{\varepsilon^*_{\text{total},3}} = 2\frac{d_W}{\varepsilon^*_W} + \left(L^*_1 + L^*_2 + L^*_3 + L^*_4\right)^{-1}, \text{ where}$$

$$L^*_1 = \frac{\rho_{sp}\rho_p}{\frac{d_{sp}}{\varepsilon^*_{sp}} + d_{\text{gap}} + \frac{d_p}{\varepsilon^*_p} + \frac{d_o}{\varepsilon^*_o}}$$

$$L^*_2 = \frac{\left(1-\rho_{sp}\right)\rho_p}{d_{sp} + d_{\text{gap}} + \frac{d_p}{\varepsilon^*_p} + \frac{d_o}{\varepsilon^*_o}}$$

$$L^*_3 = \frac{\rho_{sp}\left(1-\rho_p\right)}{\frac{d_{sp}}{\varepsilon^*_{sp}} + d_{\text{gap}} + d_p + \frac{d_o}{\varepsilon^*_o}}$$

$$L^*_4 = \frac{\left(1-\rho_{sp}\right)\left(1-\rho_p\right)}{d_{sp} + d_{\text{gap}} + d_p + \frac{d_o}{\varepsilon^*_o}} \tag{A.9}$$

A.1.1.4 Model 4: No Surface Conductivity

The last considered circuit also lumps parts of the divided silicon electrodes into the parallel paths, hence this arrangement represents a low surface conductivity of this semi-conducting material. That may possibly occur at high frequencies and is described as follows

$$\frac{D}{\varepsilon^*_{\text{total},4}} = \left(\frac{\rho_{sp}\rho_p}{M^*_1} + \frac{(1-\rho_{sp})\rho_p}{M^*_2} + \frac{\rho_{sp}(1-\rho_p)}{M^*_3} + \frac{(1-\rho_{sp})(1-\rho_p)}{M^*_4}\right)^{-1}, \text{ where}$$

$$M_1^* = \frac{d_{sp}}{\varepsilon_{sp}^*} + d_{\text{gap}} + \frac{d_p}{\varepsilon_p^*} + \frac{d_o}{\varepsilon_o^*} + 2\frac{d_W}{\varepsilon_W^*}$$

$$M_2^* = d_{sp} + d_{\text{gap}} + \frac{d_p}{\varepsilon_p^*} + \frac{d_o}{\varepsilon_o^*} + 2\frac{d_W}{\varepsilon_W^*}$$

$$M_3^* = \frac{d_{sp}}{\varepsilon_{sp}^*} + d_{\text{gap}} + d_p + \frac{d_o}{\varepsilon_o^*} + 2\frac{d_W}{\varepsilon_W^*}$$

$$M_4^* = d_{sp} + d_{\text{gap}} + d_p + \frac{d_o}{\varepsilon_o^*} + 2\frac{d_W}{\varepsilon_W^*} \tag{A.10}$$

A.1.1.5 Model and Data: What Does It Tell About the Sample?

The introduced models are employed as fit functions, with a certain number of parameters, to the measured data of the whole sample. Therefore, Eqs. A.6, A.8, A.9 and A.10 have been rearranged to separate their real and imaginary parts, and then are used to fit ε' and ε'' at the same time with one set of parameters. As dielectric function of the polymer ε_p^* two Havriliak-Negami functions $\varepsilon_{HN,\alpha}^*$ and $\varepsilon_{HN,\beta}^*$ (Eq. 3) describing the α- and β-relaxation, respectively as well as a conductivity term are combined

$$\varepsilon_p^*(\omega) = \varepsilon_{HN,\alpha}^*(\omega) + \varepsilon_{HN,\beta}^*(\omega) - i\frac{a\sigma_0}{\varepsilon_0\omega^s} \tag{A.11}$$

where σ_0 denotes the DC-conductivity and s is an exponent describing the slope of the conductivity contribution in the spectra (for purely electronic conductivity $s = 1$ while for other types of charge transport $s < 1$ is observed [39]). The parameter a has the dimension (rad Hz)^s(Hz)$^{-1}$. Note that in Eq. A.11, a rigorous replacement of $\varepsilon_{HN,\alpha}^*$ and $\varepsilon_{HN,\beta}^*$ by Eq. 3 yields two constants ε_∞ (one for each process) which can be expressed by a single constant. The ε_∞ given in Table 2 represents such a combined constant for the whole ε_p^*. Additionally to the inclusion of Eq. A.11 in the four equivalent circuits, for comparison it is employed directly to analyze a bulk measurement of P2VP. The parameters describing the β-relaxation obtained from this analysis ($\Delta\varepsilon = 1.28$, $\alpha = 0.37$, $\beta = 1$ and $\tau_{HN} = 4.6 \times 10^{-8}$ s) were included as constants in the fit functions for two reasons. First, to reduce the number of free parameters in the model and second, the used example spectra do not cover the peak maximum of the β-relaxation hence there is not enough data to properly fit this feature. This approach is justified by the fact that the β-relaxation is assigned to local dynamics which is not expected to change in the present sample (~30 nm layer thickness).

For the dielectric functions of the silica components as well as the silicon electrodes, fixed values derived from measurements of the respective materials are used as stated above. As for the geometrical dimensions, the thicknesses of polymer layer and the gap are allowed to vary while the thicknesses of the interfacial silica, the

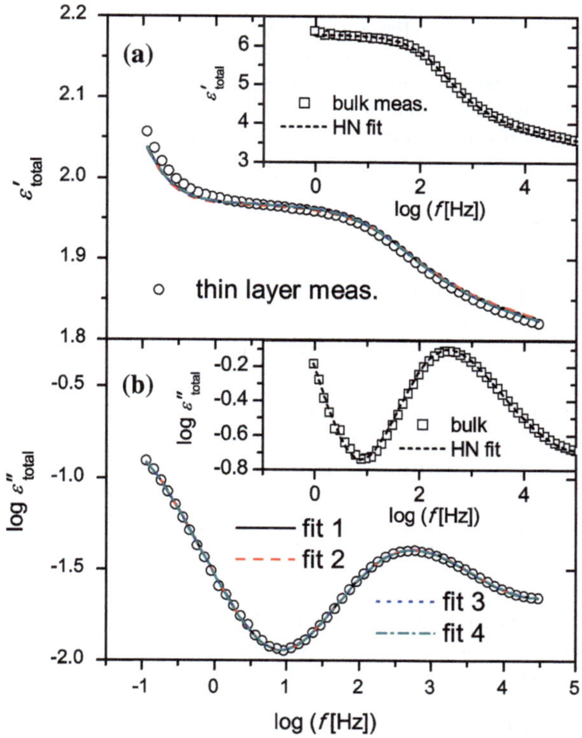

Fig. 11 **a** Permittivity and **b** loss spectra of a supported P2VP layer of ~30 nm thickness as measured in the nanostructured electrode arrangement at a temperature of 400 K and fits as indicated according to the four different equivalent circuits described by Eqs. A.6, A.8, A.9 and A.10 as depicted in Fig. 10. (*Insets*) Permittivity and loss spectra of bulk P2VP at a temperature of 400 K and a fit of Eq. A.11 to the data. The fit parameters are shown in Tables 2 and 3

silicon electrodes as well as the spacer height and their density of coverage are kept constant.[9] This classification of quantities into two categories—those subject to rather large uncertainties, and the ones which are very well defined—is supposed to reduce the number of free parameters.

The fit curves shown in Fig. 11 indicate that any of the modeled circuits can technically describe the data. This is confirmed by statistics revealing adjusted coefficients of determination (adjusted R^2) very close to 1 for all conducted fits. Furthermore, the values of most fit parameters coincide within the limits of uncertainty (Tables 2 and 3). Consequently, the specific choice of one of the tested models has only minor impact, hence none of them can be rejected as inappropriate.

[9] In detail, in a first step the parameters for thicknesses of the polymer layer and the gap were fixed to the experimentally deduced value while the other free parameters have been varied to achieve a best fit to the data. Then, in a second step, in addition to the latter, also the former parameter was allowed to change to find those minima in the parameter space being closest to the known geometrical dimensions.

Table 2 Fit parameters according to Eqs. A.6, A.8, A.9, A.10 and A.11 for a thin layer of P2VP at a temperature of 400 K as shown in Fig. 11: The permittivity at the high frequency limit ε_∞, dielectric strength $\Delta\varepsilon$, parameters for symmetric and asymmetric broadening α and β, respectively, the relaxation time of the Havriliak-Negami function τ_{HN}, DC-conductivity σ_0 and the slope parameter s of the conductivity contribution. For comparison the respective values of bulk P2VP according to Eq. A.11 are given

	α-relaxation					Conductivity	
	ε_∞	$\Delta\varepsilon$	α	β	τ_{HN} (10^{-4} s)	σ_0 (10^{-11} S/cm)	s
bulk	2.47	2.55	0.82	0.61	7.7	3.3	0.99
	(±1.27)	(±0.11)	(±0.01)	(±0.03)	(±0.5)	(±0.2)	(±0.03)
fit 1	1.71	1.76	0.74	0.71	8.6	2.1	0.80
	(±0.17)	(±0.12)	(±0.01)	(±0.03)	(±0.5)	(±0.2)	(±0.01)
fit 2	1.55	1.70	0.74	0.74	8.3	2.1	0.80
	(±0.17)	(±0.13)	(±0.01)	(±0.04)	(±0.6)	(±0.2)	(±0.01)
fit 3	1.69	1.75	0.74	0.72	8.6	2.1	0.80
	(±0.17)	(±0.12)	(±0.01)	(±0.03)	(±0.5)	(±0.2)	(±0.01)
fit 4	1.70	1.76	0.74	0.71	8.6	2.1	0.80
	(±0.17)	(±0.12)	(±0.01)	(±0.03)	(±0.5)	(±0.2)	(±0.01)

Table 3 Dimensions of the components, as measured by AFM* or ellipsometry**, of the thin P2VP layer from which the data shown in Fig. 11 has been obtained as well as the corresponding fit parameters and constants deduced from the fits according to Eqs. A.6, A.8, A.9 and A.10: Thicknesses of the polymer layer d_p, the gap atop the polymer layer d_g, the interfacial silica d_o, as well as the height of the spacers d_{sp}, the thickness of the silicon electrodes d_W, and the surface coverage of the spacers and the polymer ρ_{sp} and ρ_p, respectively

	Parameters					Constants	
	d_p (nm)	d_g (nm)	d_o (nm)	d_{sp} (nm)	d_W (μm)	ρ_{sp}	ρ_p
sample	30*	–	30**	37*	599-650	0.006	1
	(±5)		(±5)	(±1)			
fit 1	33.3	5.0					
	(±2.3)	(±0.2)					
fit 2	31.6	5.3					
	(±2.3)	(±0.2)	30	37	600	0.006	1
fit 3	32.8	4.7					
	(±2.3)	(±0.2)					
fit 4	33.0	4.7					
	(±2.3)	(±0.2)					

A comparison with the bulk values shows that the relaxation time τ_{HN} of the α-relaxation is nearly unchanged while the broadening parameters α and β, the dielectric strength $\Delta\varepsilon$, the permittivity in the high frequency limit ε_∞, the DC-conductivity σ_0 and its slope parameter s deviate (Table 2). The discrepancies in $\Delta\varepsilon$ and ε_∞ may be affected by the specific choice of the dielectric properties of the silica components; hence, these parameters are subject to much larger uncertainties then pretended by the fit result. Consequently, any further interpretation is not meaningful. Regarding the broadening parameters, it is more appropriate to compare the slopes of the wings of the relaxation process on the low and high frequency side, respectively. While the former is expressed by the symmetric broadening parameter α, the latter is given by the product of both parameters $\alpha\beta$. In fact this product is similar for the thin layer fits and the bulk. In contrast, the slope at the low frequency side differs which can not be explained so far. Furthermore, the values of the DC-conductivity σ_0 and its slope parameter a are slightly diminished compared to the bulk, where the latter indicates a less steep slope. An interpretation regarding the type (and potential change) of charge transport is however difficult and outside the scope of this work. Besides seeking for a physical interpretation, the change in slope affects the low frequency wing of the α-relaxation which means that herein may be the cause for the change in the symmetric broadening parameter α.

A.1.2 Impact of Height Distributions

Apart from the complicated geometry of the sample capacitor as a whole, the particular shape of the polymer nanodroplets (Sects. 3.1.1 and 3.1.3) is expected to affect the total dielectric response. Again, equivalent circuits are a versatile tool to study the impact of any arbitrary height distribution. For reasons of simplicity, a model capacitor containing only a single droplet and an empty space above it is considered (Fig. 12). The resulting system is approximated by several sets of two capacitances, each: one represents a certain fraction of the polymer, while the other models the empty space atop. These sets are connected parallel to each other, while their constituent components are serially linked (Fig. 12). The accuracy of this approach can be increased with the number of subsets included in the calculation. To keep the calculation as simple as possible, the empty space laterally in-between the droplets has been neglected as well since its contribution is simply superimposed to the overall signal. As it contains no loss, it introduces a purely linear down-scaling of the intensity which is constant in the whole spectrum and identical for any chosen height distribution of the polymer nanodroplet.

For a gradual monitoring of possible effects, the number of subsets is varied between 1 and 8. The case of only 1 subset represents just two layers of uniform thicknesses. To ensure comparability, the average height of the layers of all subsets in any of the following circuits is fixed to the same value. In particular, Fig. 13 shows modeled spectra for such circuits with a total height of 36 nm which comprise in the simplest case a 4 nm thick capacitor with dielectric properties of bulk P2VP. Every addition of a further subset or branch is accompanied by a change of the capacitor thicknesses in a way that the polymer capacitors of the subsets differ by 1 nm in height but their mean height remains fixed 4 nm. The thickness of the empty space capacitors is adjusted accordingly to give each subset a total height of 36 nm. For example, in the 2-branch circuit, the polymer capacitors are 3.5 and 4.5 nm high while their empty space counterparts have heights of 32.5 and 31.5 nm, respectively.

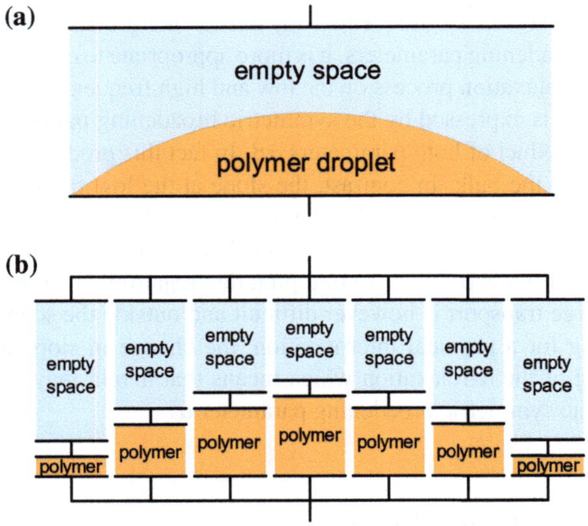

Fig. 12 **a** Sketch of a capacitor partially filled with a single polymer nanodroplet. **b** Schematic approach of an equivalent circuit for calculation of the total response

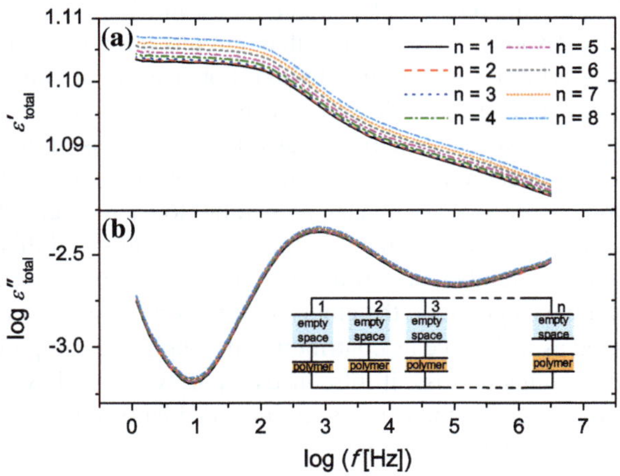

Fig. 13 **a** Permittivity and **b** loss spectra of several equivalent circuits as schemed reflecting different height distributions of a polymeric sample (for the depicted calculations P2VP bulk data is used) in an incompletely filled capacitor. Adapted from [38], supplementary materials. Reprinted with permission from AAAS

Comparing the dielectric functions of these circuits as depicted in Fig. 13, starting from a single subset, which corresponds to a uniform layer structure, to eight subsets, following a gradual distribution of heights, reveals only minor differences. The permittivity ε' increases by only ~0.5 % in the case of eight branches, which means that the thinnest and thickest polymer capacitors in the corresponding subsets have thicknesses of 0.5 nm and 7.5 nm, respectively. Similarly, such a slight increase in overall signal intensity is found for the dielectric loss ε''. Moreover, neither the position nor the shape of the relaxation peak is affected by this variation of the thickness distribution.

A.1.3 Implication

The previous Sects. A.1.1 and A.1.3 examine equivalent circuits which model the complicated composition of the whole sample including nanostructured electrodes, as well as the nonuniform height distribution of the actual sample material in a parallel plate capacitor, respectively. In both cases, neither the position nor the shape of the relaxation peak allotted to the polymeric component is affected. Only its intensity and thereby its apparent dielectric strength is reduced due to the very small filling fraction in the probe volume. Therefore, it is justified to compare the peak position and the peak width of the measured dielectric function of the whole sample. Especially for the latter, it is helpful to normalize the curves with respect to the peak maximum. However, this way of analysis excludes a quantitative extraction of the dielectric strength as shown in Sect. A.1.1.5. Consequently, this property can be used only as a relative measure of the amount of mobile segments, but it cannot be related accurately to the volume of sample material.

References

1. Keddie JL, Jones RAL, Cory RA (1994) Size-dependent depression of the glass transition temperature in polymer films. Europhys Lett 27:59
2. Keddie JL, Jones RAL, Cory RA (1994) Interface and surface effects on the glass-transition temperature in thin polymer films. Faraday Discuss Chem Soc 98:219
3. Forrest JA, Dalnoki-Veress K, Stevens JR, Dutcher JR (1996) Effect of free surfaces on the glass transition temperature of thin polymer films. Phys Rev Lett 77:2002
4. van Zanten JH, Wallace WE, Wu W (1996) Effect of strongly favorable substrate interactions on the thermal properties of ultrathin polymer films. Phys Rev E 53:R2053
5. Fukao K, Miyamoto Y (2000) Glass transitions and dynamics in thin polymer films: dielectric relaxation of thin films of polystyrene. Phys Rev E 61:1743
6. Fryer DS, Peters RD, Kim EJ et al (2001) Dependence of the glass transition temperature of polymer films on interfacial energy and thickness. Macromolecules 34:5627–5634. doi:10.1021/ma001932q
7. Tsui OKC, Zhang HF (2001) Effects of chain ends and chain entanglement on the glass transition temperature of polymer thin films. Macromolecules 34:9139–9142
8. Dalnoki-Veress K, Forrest JA, Murray C et al (2001) Molecular weight dependence of reductions in the glass transition temperature of thin, freely standing polymer films. Phys Rev E 63:031801
9. Ellison CJ, Kim SD, Hall DB, Torkelson JM (2002) Confinement and processing effects on glass transition temperature and physical aging in ultrathin polymer films: novel fluorescence measurements. Eur Phys J E 8:155

10. Serghei A, Kremer F (2003) Confinement-induced relaxation process in thin films of cis-polyisoprene. Phys Rev Lett 91:165702
11. Efremov MY, Olson EA, Zhang M et al (2004) Probing glass transition of ultrathin polymer films at a time scale of seconds using fast differential scanning calorimetry. Macromolecules 37:4607
12. Fakhraai Z, Sharp JS, Forrest JA (2004) Effect of sample preparation on the glass-transition of thin polystyrene films. J Polym Sci Part B Polym Phys 42:4503–4507
13. Lupaşcu V, Huth H, Schick C, Wübbenhorst M (2005) Specific heat and dielectric relaxations in ultra-thin polystyrene layers. Thermochim Acta 432:222
14. Serghei A, Mikhailova Y, Huth H et al (2005) Molecular dynamics of hyperbranched polyesters in the confinement of thin films. Eur Phys J E 17:199
15. Huth H, Minakov AA, Schick C (2006) Differential ac-chip calorimeter for glass transition measurements in ultrathin films. J Polym Sci Part B Polym Phys 44:2996
16. O'Connell PA, McKenna GB (2006) Dramatic stiffening of ultrathin polymer films in the rubbery regime. Eur Phys J E 20:143–150
17. Serghei A, Huth H, Schick C, Kremer F (2008) Glassy dynamics in thin polymer layers having a free upper interface. Macromolecules 41:3636
18. Napolitano S, Lupaşcu V, Wübbenhorst M (2008) Temperature dependence of the deviations from bulk behavior in ultrathin polymer films. Macromolecules 41:1061
19. Raegen AN, Massa MV, Forrest JA, Dalnoki-Veress K (2008) Effect of atmosphere on reductions in the glass transition of thin polystyrene films. Eur Phys J E 27:375
20. Lu H, Chen W, Russell TP (2009) Relaxation of thin films of polystyrene floating on ionic liquid surface. Macromolecules 42:9111–9117
21. Inoue R, Kanaya T, Nishida K et al (2009) Glass transition and molecular mobility in polymer thin films. Phys Rev E 80:031802. doi:10.1103/PhysRevE.80.031802
22. Erber M, Tress M, Mapesa E et al (2010) Glassy dynamics and glass transition in thin polymer layers of pmma deposited on different substrates. Macromolecules 43:7729
23. Mapesa EU, Erber M, Tress M et al (2010) Glassy dynamics in nanometer thin layers of polystyrene. Eur Phys J Spec Top 189:173
24. Tress M, Erber M, Mapesa EU et al (2010) Glassy dynamics and glass transition in nanometric thin layers of polystyrene. Macromolecules 43:9937–9944
25. Glynos E, Frieberg B, Oh H et al (2011) Role of molecular architecture on the vitrification of polymer thin films. Phys Rev Lett 106:128301. doi:10.1103/PhysRevLett.106.128301
26. Paeng K, Richert R, Ediger MD (2012) Molecular mobility in supported thin films of polystyrene, poly(methyl methacrylate), and poly(2-vinyl pyridine) probed by dye reorientation. Soft Matter 8:819–826. doi:10.1039/C1SM06501G
27. Efremov MY, Thode C, Nealey PF (2013) Demonstration of glass transition temperature shift in thin supported polystyrene films by internal reference method. Rev Sci Instrum 84:023905. doi:10.1063/1.4793226
28. Mapesa EU, Tress M, Schulz G et al (2013) Segmental and chain dynamics in nanometric layers of poly(cis-1,4-isoprene) as studied by broadband dielectric spectroscopy and temperature-modulated calorimetry. Soft Matter 9:10592–10598. doi:10.1039/C3SM51311D
29. Kipnusu WK, Elmahdy MM, Tress M, et al (2013) Molecular order and dynamics of nanometric thin layers of poly(styrene-b-1,4-isoprene) diblock copolymers. Macromolecules 0:null. doi:10.1021/ma4019334.
30. Napolitano S, Capponi S, Vanroy B (2013) Glassy dynamics of soft matter under 1d confinement: how irreversible adsorption affects molecular packing, mobility gradients and orientational polarization in thin films. Eur Phys J E 36:1–37. doi:10.1140/epje/i2013-13061-8
31. Boucher VM, Cangialosi D, Yin H et al (2012) T g depression and invariant segmental dynamics in polystyrene thin films. Soft Matter 8:5119–5122. doi:10.1039/C2SM25419K
32. Boucher VM, Cangialosi D, Alegría A, Colmenero J (2014) Accounting for the thickness dependence of the tg in supported ps films via the volume holes diffusion model. Thermochim Acta 575:233–237. doi:10.1016/j.tca.2013.10.023

33. Napolitano S, Cangialosi D (2013) Interfacial free volume and vitrification: reduction in tg in proximity of an adsorbing interface explained by the free volume holes diffusion model. Macromolecules 46:8051–8053. doi:10.1021/ma401368p
34. Serghei A, Tress M, Kremer F (2009) The glass transition of thin polymer films in relation to the interfacial dynamics. J Chem Phys 131:154904
35. Napolitano S, Wübbenhorst M (2011) The lifetime of the deviations from bulk behaviour in polymers confined at the nanoscale. Nat Commun 2:260
36. Forrest JA (2013) What can we learn about a dynamical length scale in glasses from measurements of surface mobility? J Chem Phys 139:084702. doi:10.1063/1.4818989
37. Ediger MD, Forrest JA (2014) Dynamics near free surfaces and the glass transition in thin polymer films: A view to the, future. Macromolecules. doi:10.1021/ma4017696
38. Tress M, Mapesa EU, Kossack W et al (2013) Glassy dynamics in condensed isolated polymer chains. Science 341:1371–1374. doi:10.1126/science.1238950
39. Kremer F, Schönhals A (2003) Broadband dielectric spectroscopy. Springer, Berlin Heidelberg
40. Kumaki J, Nishikawa Y, Hashimoto T (1996) Visualization of single-chain conformations of a synthetic polymer with atomic force microscopy. J Am Chem Soc 118:3321–3322. doi:10.1021/ja9538593
41. Minko S, Kiriy A, Gorodyska G, Stamm M (2002) Single flexible hydrophobic polyelectrolyte molecules adsorbed on solid substrate: transition between a stretched chain, necklace-like conformation and a globule. J Am Chem Soc 124:3218–3219. doi:10.1021/ja017767r
42. Kiriy A, Gorodyska G, Minko S et al (2002) Cascade of coil-globule conformational transitions of single flexible polyelectrolyte molecules in poor solvent. J Am Chem Soc 124:13454–13462. doi:10.1021/ja0261168
43. Kumaki J, Hashimoto T (2003) Conformational change in an isolated single synthetic polymer chain on a mica surface observed by atomic force microscopy. J Am Chem Soc 125:4907–4917. doi:10.1021/ja0290429
44. Kiriy A, Gorodyska G, Minko S et al (2003) Chemical contrasting in a single polymer molecule afm experiment. J Am Chem Soc 125:11202–11203
45. Kiriy A, Gorodyska G, Minko S et al (2003) Single molecules and associates of heteroarm star copolymer visualized by atomic force microscopy. Macromolecules 36:8704–8711. doi:10.1021/ma0348270
46. Gallyamov MO, Khokhlov AR, Möller M (2005) Real-time imaging of the coil-globule transition of single adsorbed poly(2-vinylpyridine) molecules. Macromol Rapid Commun 26: 456–460. doi:10.1002/marc.200400602
47. Roiter Y, Minko S (2005) Afm single molecule experiments at the solid-liquid interface. in situ conformation of adsorbed flexible polyelectrolyte chains. J Am Chem Soc 127:15688–15689. doi:10.1021/ja0558239
48. Fukao K, Miyamoto Y (1999) Glass transition temperature and dynamics of alpha-process in thin polymer films. Europhys Lett 46:649
49. Labahn D, Mix R, Schönhals A (2009) Dielectric relaxation of ultrathin films of supported polysulfone. Phys Rev E 79:011801
50. Yin H, Napolitano S, Schönhals A (2012) Molecular mobility and glass transition of thin films of poly(bisphenol a carbonate). Macromolecules 45:1652–1662. doi:10.1021/ma202127p
51. Serghei A, Kremer F (2008) Broadband dielectric studies on the interfacial dynamics enabled by use of nanostructured electrodes. Rev Sci Instrum 79:026101
52. García-Turiel J, Jérôme B (2007) Solvent retention in thin polymer films studied by gas chromatography. Colloid Polym Sci 285:1617–1623
53. Perlich J, Körstgens V, Metwalli E et al (2009) Solvent content in thin spin-coated polystyrene homopolymer films. Macromolecules 42:337
54. Serghei A, Huth H, Schellenberger M et al (2005) Pattern formation in thin polystyrene films induced by an enhanced mobility in ambient air. Phys Rev E 71:061801
55. Serghei A, Kremer F (2008) Metastable states of glassy dynamics, possibly mimicking confinement-effects in thin polymer films. Macromol Chem Phys 209:810

56. Parsonage E, Tirrell M, Watanabe H, Nuzzo RG (1991) Adsorption of poly(2-vinylpyridine)-poly (styrene) block copolymers from toluene solutions. Macromolecules 24:1987–1995. doi:10.1021/ma00008a041
57. Huguenard C, Pefferkorn E (1994) Rate of adsorption of diblock copolymers from micellar solutions on solid-liquid interfaces. Macromolecules 27:5271–5276. doi:10.1021/ma00097a005
58. Priestley RD, Ellison CJ, Broadbelt LJ, Torkelson JM (2005) Structural relaxation of polymer glasses at surfaces, interfaces, and in between. Science 309:456–459. doi:10.1126/science.1112217
59. Rittigstein P, Priestley RD, Broadbelt LJ, Torkelson JM (2007) Model polymer nanocomposites provide an understanding of confinement effects in real nanocomposites. Nat Mater 6:278. doi:10.1038/nmat1870
60. Iacob C, Sangoro JR, Papadopoulos P et al (2010) Charge transport and diffusion of ionic liquids in nanoporous silica membranes. Phys Chem Chem Phys 12:13798–13803. doi:10.1039/C004546B
61. Iacob C (2012) Rotational and translational diffusion in glass-forming ionic liquids confined in nanoporous silica. Universität Leipzig
62. Lehmann V, Stengl R, Luigart A (2000) On the morphology and the electrochemical formation mechanism of mesoporous silicon. Mater Sci Eng B 69–70:11–22. doi:10.1016/S0921-5107(99)00286-X
63. Pap AE, Kordás K, Tóth G et al (2005) Thermal oxidation of porous silicon: study on structure. Appl Phys Lett 86:041501. doi:10.1063/1.1853519
64. Kumar P, Huber P (2007) Effect of etching parameter on pore size and porosity of electrochemically formed nanoporous silicon. J Nanomater 2007:89718. doi:10.1155/2007/89718
65. Dondos A, Benoit H (1971) The influence of solvents on unperturbed dimensions of polymer in solution. Macromolecules 4:279–283. doi:10.1021/ma60021a003
66. Puterman M, Koenig JL, Lando JB (1979) Conformational transitions of poly-2-vinylpyridine in aqueous solutions as a function of neutralization. i. raman and infrared studies. J Macromol Sci Part B 16:89–116. doi:10.1080/00222347908212284
67. Wang S, Zhao J (2007) First-order conformation transition of single poly(2-vinylpyridine) molecules in aqueous solutions. J Chem Phys 126. doi:10.1063/1.2711804
68. Minko S, Usov D, Goreshnik E, Stamm M (2001) Environment-adopting surfaces with reversibly switchable morphology. Macromol Rapid Commun 22:206–211. doi:10.1002/1521-3927
69. Mark JE (2007) Physical properties of polymers handbook. Springer, New York
70. Sauer BB, Dee GT (2002) Surface tension and melt cohesive energy density of polymer melts including high melting and high glass transition polymers. Macromolecules 35:7024–7030. doi:10.1021/ma0202437
71. Havriliak S, Negami S (1967) A complex plane representation of dielectric and mechanical relaxation processes in some polymers. Polymer 8:161
72. Vogel H (1921) Das temperaturabhängigkeitsgesetz der viskosität von flüssigkeiten. Phys Z 22:645
73. Fulcher GS (1925) Analysis of recent measurements of the viscosity of glasses. J Am Ceram Soc 8:339
74. Tammann G, Hesse G (1926) Die abhängigkeit der viscosität von der temperatur bei unterkühlten flüssigkeiten. Z Für Anorg Allg Chem 156:245
75. Griffiths PR, de Haseth JA (1986) Fourier transform infrared spectroscopy. Wiley-Interscience
76. Akiyama Y, Fujita S, Senboku H et al (2008) An in situ high pressure ftir study on molecular interactions of ketones, esters, and amides with dense phase carbon dioxide. J Supercrit Fluids 46:197–205. doi:10.1016/j.supflu.2008.03.009
77. Nakano H, Kuwahara Y, Oue S et al (2005) Formation of poly(2-vinylpyridine) films on zn by galvanostatic electropolymerization. Mater Trans 46:281–286
78. Panov VP, Kazarin LA, Dubrovin VI et al (1974) Infrared spectra of atactic poly-4-vinylpyridine. J Appl Spectrosc 21:1504–1510. doi:10.1007/BF00604430

79. Socrates G (2001) Infrared and raman characteristic group frequencies. Tables and charts, 3rd edn. Wiley, New York
80. Paolantoni M, Sassi P, Morresi A, Cataliotti RS (2005) Infrared study of 1-octanol liquid structure. Chem Phys 310:169–178. doi:10.1016/j.chemphys.2004.10.027
81. Bahar I, Erman B, Kremer F, Fischer EW (1992) Segmental motions of cis-polyisoprene in the bulk state: interpretation of dielectric relaxation data. Macromolecules 25:816
82. Berthier L, Biroli G, Bouchaud J-P et al (2005) Direct experimental evidence of a growing length scale accompanying the glass transition. Science 310:1797–1800. doi:10.1126/science.1120714

Molecular Dynamics of Poly(*cis*-1,4-Isoprene) in 1- and 2-Dimensional Confinement

Emmanuel Urandu Mapesa, Martin Tress, Manfred Reiche and Friedrich Kremer

Abstract Broadband Dielectric Spectroscopy (BDS)—in combination with a nanostructured electrode arrangement—is employed to study thin layers of poly(*cis*-1,4-isoprene) (PI). PI is further probed in the 2D confining space of Anodic Aluminum Oxide (AAO) nanopores. Being a *type A* polymer, PI presents an unrivaled opportunity to investigate two distinct relaxation modes taking place at two different length scales: the segmental motion (corresponding to the dynamic glass transition) which involves structures of about one nanometer in size, and the so-called normal mode which represents the global dynamics of the chain. We report that while the structural relaxation shows no dependence on either layer thickness or molecular weight, the normal mode—actually the fluctuation of the end-to-end vector of the unperturbed chain—is dramatically influenced by confinement: (i) its relaxation strength is layer-thickness-dependent; (ii) for PI having a molecular weight M_w comparable to M_c (i.e. the critical molecular weight below which Rouse dynamics dominate), the mean spectral position does not shift with layer thickness, (iii) in contrast, when $M_w > M_c$, the relaxation strength and rate of the normal mode respond to the confinement; (iv) it is demonstrated—for the first time—that the concentration of the mother solution from which the thin layers are spin-cast has an impact on the chain dynamics; and (v) the extent by which the normal mode is affected depends on the dimensionality of confinement. Overall, these results show that while the chain dynamics are altered in a manifold of ways (due, for instance, to interactions with the confining surface), the structural relaxation retains most of its bulk-like nature. The latter observation is due to the fact that the length scale underlying the dynamic glass transition is less than a nanometer.

E. U. Mapesa (✉) · M. Tress · F. Kremer
Instutute for Experimental Physics I, University of Leipzig, Linnestraße 5,
04103 Leipzig, Germany
e-mail: urandu@physik.uni-leipzig.de

M. Reiche
Max-Planck Institute for Microstructure Physics, Weinberg 2, 06120 Halle (Saale), Germany

F. Kremer (ed.), *Dynamics in Geometrical Confinement*, Advances in Dielectrics,
DOI: 10.1007/978-3-319-06100-9_4, © Springer International Publishing Switzerland 2014

Keywords Poly(*cis*-1,4-isoprene) · Dynamic glass transition · Broadband dielectric spectroscopy (BDS) · Segmental mode · Normal mode · Nanostructured electrodes

Abbreviations

AAO	Anodic aluminum oxide
ACC	Alternating current calorimetry
AFM	Atomic force microscopy
BDS	Broadband dielectric spectroscopy
BLS	Brillouin light scattering
DMA	Dynamic mechanical analysis
DR	Dye-reorientation
DSC	Differential scanning calorimetry
FS	Fluorescence spectroscopy
h	Hour(s)
HN	Havriliak–Negami
Hz	Hertz
M_c	Critical molecular weight
M_n	Number-averaged molecular weight
mol	Mole(s)
M_w	Weight-averaged molecular weight
NBI	Nano-bubble inflation
NM	Normal mode
NS	Neutron scattering
PI	Poly(*cis*-1,4-isoprene)
PLS	Positron lifetime spectroscopy
PTFE	Polytetrafluoroethylene
rms	Root-mean-square
R_g	Radius of gyration
s	Second(s)
SM	Segmental mode
TES	Thermal expansion spectroscopy
TSC	Terminal subchains
VFT	Vogel–Fulcher–Tammann
WLF	Williams–Landel–Ferry

1 Introduction

The nature of the glass transition remains a subject of intense studies in soft matter science even after several decades of devoted investigations. This arises from the fact that while several theoretical models have been advanced to explain this phenomenon, none vigorously explains all aspects of glass formation. In an attempt to make a break-through, considerable research effort is now focussed on understanding glass formers in confinement. The cooperative concepts of Adam and Gibbs [1] inform this approach since it is reasoned that for a sufficiently supercooled liquid, structural relaxation can only occur when a number of constituent molecules rearrange collectively. This cooperativity therefore points at so-called finite size effects, that is, changes in the dynamics and glass transition temperature when the material is confined to length scales near the cooperative size. Such confinement can be achieved and studied using, *inter alia*, nanoporous media, entangled polymeric systems, or thin films. The latter provide an ideal sample geometry since the extent of confinement can be remarkably diminished simply by reducing the thickness of the films.

Focus on glass formers in confinement is also due to their current and potential technological applications. In many of these applications (e.g., photoresists, batteries, sensors for smart drug delivery, nanolubrication, nanoadhesion, etc), the materials are used in miniaturized dimensions for purposes of increasing performance and convenience. Consequently, they have—due to an increased surface area to volume ratio—a large fraction of segments near an interface. How aspects of this miniaturization—confinement effects, interfacial interactions, and finite size effects—influence overall properties of the system must be understood for appropriate innovations. To this end, several experimental techniques (e.g., Ellipsometry [2–30], X-ray Reflectivity [25, 31–36], Differential Scanning Calorimetry [6, 8, 37–44], Positron Lifetime Spectroscopy [28, 45, 46], AC-Chip Calorimetry [25, 47–50], Fluorescence Spectroscopy [22, 51–63], Neutron Scattering [21, 36, 64–66], Brillouin Light Scattering [4, 67–70], Rheology [71–73], Atomic Force Spectroscopy [74, 75], Nano-bubble inflation [76–80], dewetting [81–83], Optical Waveguide Spectroscopy [84], and Broadband Dielectric Spectroscopy [24–26, 38, 44, 45, 47, 85–105]) have been employed to study key parameters (such as molecular weight [3, 7, 19, 25, 45, 86], tacticity [5, 11, 106, 107], measurement ambience [20, 88, 90, 95, 108, 109], nature of polymer–substrate interaction [3, 5, 6, 14, 26, 31, 67, 84, 109, 110], concentration of mother solutions [104], and even the type of measuring technique [14, 29, 101, 111]) that may influence the observed dynamics in confinement.

Most of the attention has been paid to the effect of confinement on short time-scale motions (α- and β-relaxations) of the glass formers [112, 113]. These studies have shown, for example, that the segmental relaxation is largely unaffected, in its rate, by the geometrical restriction of thin films [24–26, 39, 45, 50, 63, 74, 95, 106, 114–117]. Longer time-scale motions (so-called normal modes) have not received as much attention [118–120]. The reason for this—at least with respect to BDS—is the fact that there is just a handful of so-called *Type A* [121] polymers which can be synthesized with narrow molecular weight distributions. Poly(*cis*-1,4-isoprene) (PI) is one such polymer which, therefore, presents a unique opportunity to simultaneously probe the segmental and normal mode processes. Because of the lack of symmetry in its chemical structure, it has nonzero components of the dipole moment perpendicular and parallel to the chain contour; the former causes a dielectric relaxation due to segmental motions, while the latter is responsible for the normal mode process [121, 122].

In this chapter, we investigate the effects of confinement, for a series of molecular weights, M_w, by systematically varying the layer thicknesses, d, from bulk to dimensions comparable to the respective radius of gyration, R_g. Further, we explore a hitherto presumed parameter: the concentration, c, of the spin-cast solution. In the final part, we present inaugural results for PI confined in Anodic Aluminum Oxide (AAO) membranes and compare this with the case of thin layers. By this we demonstrate that molecular dynamics exhibit a dependence on the dimensionality of confinement.

2 Theory

The one-sided Fourier or pure imaginary Laplace transform of the time derivative of the normalized response function, $\Phi(t)$, for the dielectric polarization of a macroscopic system delivers the system's complex dielectric permittivity, $\varepsilon^*(\omega)$ [123–125]:

$$\varepsilon^*(\omega) = \varepsilon_\infty + (\varepsilon_s - \varepsilon_\infty) \int_0^\infty exp[-i\omega t] \left\{ -\frac{\mathrm{d}\Phi(t)}{\mathrm{d}t} \right\} \mathrm{d}t \tag{1}$$

where ε_s and ε_∞ are, respectively, the limiting low- and high-frequency permittivities. The response function $\Phi(t)$ comprises contributions of the dipole moment components parallel, $\mu^{||}$, and perpendicular, $\mu^{\perp}$, to the chain backbone [126]. The former components correspond to the end-to-end vector motion, defined as:

$$\sum_j \sum_l \langle \mu_{ij}^{||}(0).\mu_{ij}(t) \rangle = \mu^2 \langle \boldsymbol{r}_i(0).\boldsymbol{r}_i(t) \rangle \tag{2}$$

Hence $\Phi(t)$ is written as

$$\Phi(t) = \frac{\mu^2 \langle \mathbf{r}_t(0).\mathbf{r}_i(t) \rangle + \sum_j \sum_l \langle \mu_{ij}^{\perp}(0) \mu_{ij}^{\perp}(t) \rangle}{\mu^2 \langle \mathbf{r}_i(0).\mathbf{r}_i(0) \rangle + \sum_j \sum_l \langle \mu_{ij}^{\perp}(0) \mu_{ij}^{\perp}(0) \rangle} \tag{3}$$

where $\langle \boldsymbol{r}_i(0).\boldsymbol{r}_i(t) \rangle$ is the end-to-end vector autocorrelation function for the chain i at time t, and μ is the constant having the meaning of the parallel dipole moment per unit contour length. The second term of Eq. (3) is simply reduced to the autocorrelation function of the perpendicular dipole moment component representing segmental motions. Detailed analysis is available in references [125, 127].

To describe the chain dynamics of a polymer, the Rouse–Zimm [128, 129] bead-spring model is often invoked, and is successful from dilute to concentrated polymer solutions up to bulk, so long as the molecular weight is below the critical entanglement molecular weight, M_c. For higher molecular weights, the model does not hold although it forms the basis for other models which account for entanglements, that is, reptation dynamics [130, 131]. For unentangled chains, the Rouse–Zimm theory gives the correlation function for the end-to-end vector as

$$\langle \boldsymbol{r}(0).\boldsymbol{r}(t)\rangle = \frac{8\langle r^2\rangle}{\pi^2}\sum\nolimits_p \frac{1}{p^2}e^{(-t/\tau_p)} \tag{4}$$

$$\tau_p = \frac{\xi N^2 b^2}{3\pi^2 k_B T p^2} \tag{5}$$

where τ_p is the relaxation rate of the pth Rouse mode ($p = 1, 3, 5, \ldots$ and $p = 1$ corresponds to the longest Rouse mode), ξ is the monomeric friction coefficient, N is the degree of polymerization, and b is the effective bond length. Since N is proportional to the molecular weight M, this theory predicts for the longest relaxation time $\tau_1 \propto M^2$. It should be noted here that τ_1 dominates the dielectric response, and shows up as the peak maximum in the dielectric spectrum of the normal mode. For entangled polymer chains, the equations above are modified according to de Gennes' [130] tube model:

$$\langle \boldsymbol{r}(0).\boldsymbol{r}(t)\rangle = \frac{8\langle r^2\rangle}{\pi^2}\sum\nolimits_p \frac{1}{p^2}e^{(-p^2 t/\tau_d)} \tag{6}$$

$$\tau_p = \frac{\xi N^3 b^2}{\pi^2 k_B T}\left(\frac{b}{d}\right)^2 = 3\tau_R N\left(\frac{b}{d}\right)^2 \tag{7}$$

where d is the tube diameter, τ_d is the longest relaxation time for the tube disengagement, $\tau_R = (\xi N^2 b^2/3\pi^2 k_B T)$. This model predicts the M^3 dependence of τ_d.

For all thin polymer films prepared from samples with $M_w \gg M_c$, we observe a normal mode that is no longer bulk-like. We, therefore, designate it as terminal subchain (TSC) mode, and provide an explanation in the Discussion section.

3 Experimental

3.1 Thin Layers: Sample Preparation and Measurement

Highly doped silicon wafers with orientation (100) were purchased from MicroFab Bremen. These wafers have a root mean square (rms) roughness of 0.23 nm (as measured on a scan area of 1 μm^2 by Atomic Force Microscopy), specific resistivity less than 3 mΩ cm, and a 30-nm thermally oxidized surface layer. First, a layer of Al (~80 nm) is evaporated on the backside of the wafers in high vacuum (10^{-7} mbar) and a coating of photoresist is then spin-cast on the frontside. These two layers serve as electrical contacts and as protection during subsequent cutting, respectively. After cutting into small pieces, typically 4 mm × 8 mm, the wafer dice are rinsed in acetone to remove the photoresist; purged with a plasma cleaner and a snow-jet gun to eject any organic contaminants; sonicated successively in acetone (10 min)

Table 1 Characteristics of the PI samples

Sample[a]	$10^3 M_w$	M_w/M_n	$\langle R_{EE}^2\rangle^{0.5}$[b] (nm)	$\langle R_g^2\rangle^{0.5}$[c] (nm)	T_g^d $\pm 0.5K$
PI-12	11.6	1.06	8.9	3.5	208.3
PI-25	24.5	1.06	11.4	5.3	09.2
PI-45	44.5	1.06	17.4	6.9	208.8
PI-53	53	1.06	19	7.5	209.4
PI-75	75	1.08	22.6	9	210.4

[a]The microstructure (data provided by Polymer Source Inc.—Canada) is 80 % (*cis*-1,4), 15 % (*trans*-1,4) and 5.0 % (vinyl-3,4) for all samples
[b] Calculated [112] assuming a Gaussian chain, $\langle R_{EE}^2\rangle = Na^2$, where N is the degree of polymerization and a is the statistical segment length, $a = 0.68$ nm
[c] Evaluated [132] as $\langle R_g^2\rangle = 1.07 \times 10^3 M_w$
[d] Measured by differential scanning calorimetry at a cooling rate of 10 K/min

and dichloromethane (10 min); and, dried under nitrogen flow and on a hot plate (at 423 K for 5 min) before eventual spin-coating of polymer films.

All the poly(*cis*-1,4-isoprene) (Polymer Source Inc.) thin layers (for properties, see Table 1) were prepared by spin-casting from chloroform (Sigma-Aldrich, purity ≥99.9 %) solutions at a rate of 3,000 rpm. Always handled using glass syringes (neoLab GmbH), the solutions were filtered through PTFE (polytetrafuoroethylene) membranes with 200-nm pores. In one series of the experiments, for different molecular weights, the layer thicknesses were systematically decreased by reducing the concentration of the polymer in (semi-dilute) solution, and carrying out one spinning process. In another, several thin layers with nearly the same thickness (from PI-53) were prepared from solutions with varying concentrations (from semi-dilute to dilute regimes). In this case, with reducing concentration, several depositions had to be made to attain the required film thickness. All the films—supported on highly conductive silicon wafers—were annealed at 400 K for 24 h in an oil-free high vacuum (10^{-6} mbar). Details on the annealing and vacuum system are available in Chap. "Rotational Diffusion of Guest Molecules Confined in Uni-Directional Nanopores."

The absolute film thickness and the surface topography were determined by employing a Veeco Dimension 3000 Metrology Atomic Force Microscopy (AFM) (Digital Instruments, Veeco Metrology Group) operated in tapping mode by using silicon tips (nominal radius of curvature <10 nm) with a force constant of about 40 N/m and a resonance frequency in the range 200–400 kHz. The film thickness was measured with 1–2 nm precision by analyzing AFM images of scratches applied to the polymer film with a sharp steel blade. The surface topography was probed both for the as-prepared film, and the same film after all dielectric measurements. The latter step is mandatory in order to exclude dewetting effects. The topography was analyzed in terms of the root mean square roughness, R_{rms}, as calculated for at least seven images taken from different spots on the sample. Images were scanned with a resolution of 512 × 512 pixels at room temperature and processed and analyzed

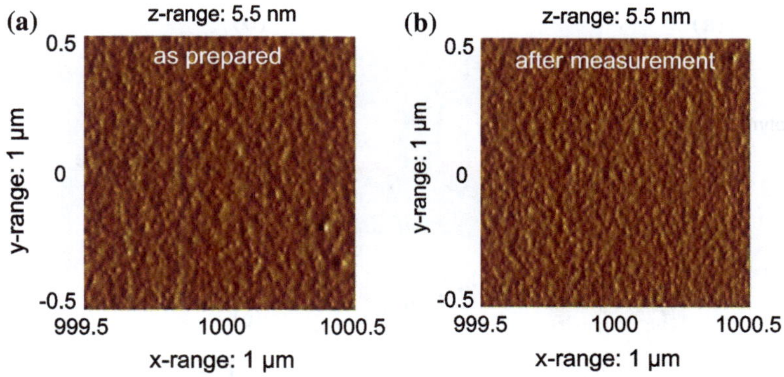

Fig. 1 AFM images of the surface of a thin polyisoprene (from PI-12) layer (thickness, 81 nm) as prepared (**a**) and after annealing and dielectric measurements (**b**). The rms roughness values are 1.7 ± 0.1 and 1.6 ± 0.1 nm, respectively

using the Gwyddion software 2.25 [133]. The procedure for calculation of R_{rms} was computer implemented as follows. First, a slope correction of the image is carried out using a second- or third-order polynomial plane fit. R_{rms}, for each image, is then calculated as

$$R_{\text{rms}} = \sqrt{\frac{1}{MN} \sum_{k=0}^{M-1} \sum_{l=0}^{N-1} [Z(x_k, y_l) - \mu]^2} \tag{8}$$

where μ is the mean height defined as, $\mu = \frac{1}{MN} \sum_{k=0}^{M-1} \sum_{l=0}^{N-1} Z(x_k, y_l)$, M and N are the dimensions of the image ($M = N$, for a quadrangular image), and z is the height of a given pixel in the image defined by its coordinates. Notably, μ is not subtracted from the height values before squaring, and hence a proper plane correction is mandatory. It was found that R_{rms} ranges between 1 and 2 nm (on a 10 μm^2 scan area) with a standard deviation less than 0.2 nm, while the variability in the film thickness is in the range 10–20 %. Figure 1 shows the typical surface of a sample at two instances: just after preparation, and later, at the end of experiments (i.e., 24 h of annealing, and 2 days of dielectric measurements). An insignificant variation (~6 %) in the R_{rms} is observed, implying a stable film topography.

Broadband Dielectric Spectroscopy (besides Nuclear Magnetic Resonance) is *the* experimental tool to study molecular dynamics in a broad frequency (10^{-6} Hz to 10^{+12} Hz) and temperature range [134]. It has the additional advantage that the sensitivity of the measurements increases with decreasing separation of the capacitor electrodes and hence with decreasing amount of sample material. For the current studies, the sample capacitors were assembled using a novel technique (Fig. 2a–e) where a regular matrix of highly insulating silica nanopillars serves as spacers. Ultraflat and highly conductive silicon wafers are used as electrodes (rms roughness of 0.6 nm on a micrometric scale, resistivity <5 mΩ cm). Detailed information on the use of this capacitor arrangement is available in Ref. [135], and a

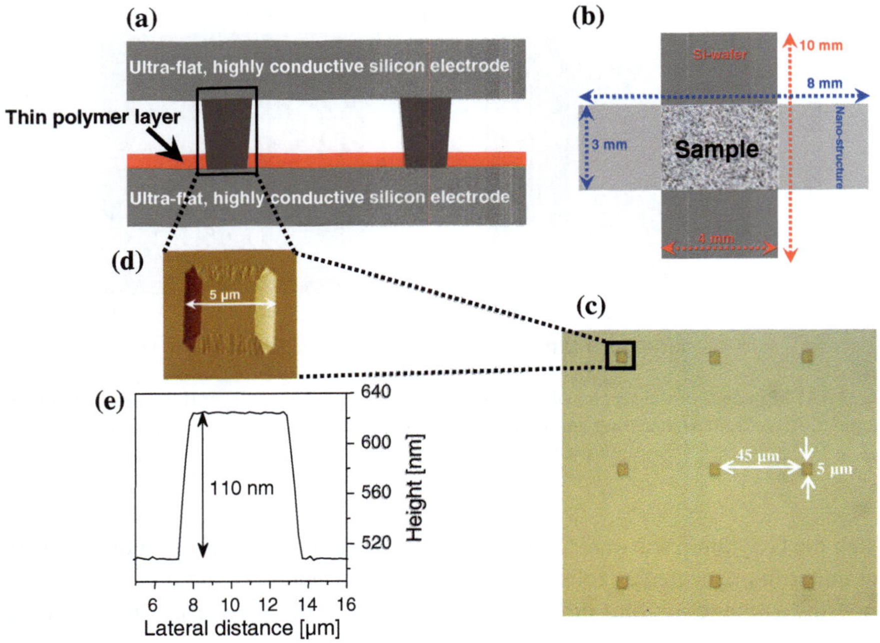

Fig. 2 **a** A side-view scheme of the sample capacitor assembled using ultraflat highly conductive silicon wafers as electrodes and insulating silica nanostructures as spacers; **b** top-view of the same sample capacitor; **c** optical microscope image of a matrix of quadratic silica nanospacers embedded on a conductive silicon wafer; **d** tapping mode AFM-height image of a single spacer of 110 nm in height and lateral dimensions of 5 μm; and **e** the corresponding profile of the AFM-height image. [Adopted from Ref. [104]. Reproduced by permission of the Royal Society of Chemistry

further analysis using equivalent circuits is given in the appendix of Chap. "Molecular Dynamics of Condensed (Semi-) Isolated Polymer Chains". It is important to note that, at least up to now, the absolute value of the dielectric strength, $\Delta\varepsilon$, of a given material probed in this geometry cannot be obtained. But this does not impair the current studies since the relaxation strength of the segmental mode can be used as a reference. The imaginary part of the (total) measured permittivity is denoted as $\varepsilon''_{\text{total}}$ and used in the analysis. Previous evaluations (see [105, 136]) have shown that the mean relaxation rate and the shape of the dielectric response (of the material under study) are not altered by the presence of the air gap.

The dielectric measurements of the bulk samples were done using a high-resolution Alpha Analyzer (Novocontrol), while for all the thin layers, the measurements were performed using an Andeen-Hagerling impedance bridge which has an accuracy of $\leq 10^{-5}$ in $\tan(\delta)$, corresponding to an error in the dielectric loss ε'' smaller than the symbol size in all presentations. Temperature regulation for both systems was executed by a Quattro System (Novocontrol) using a jet of dry nitrogen, thereby ensuring relative and absolute errors better than 0.1 and 2 K, respectively.

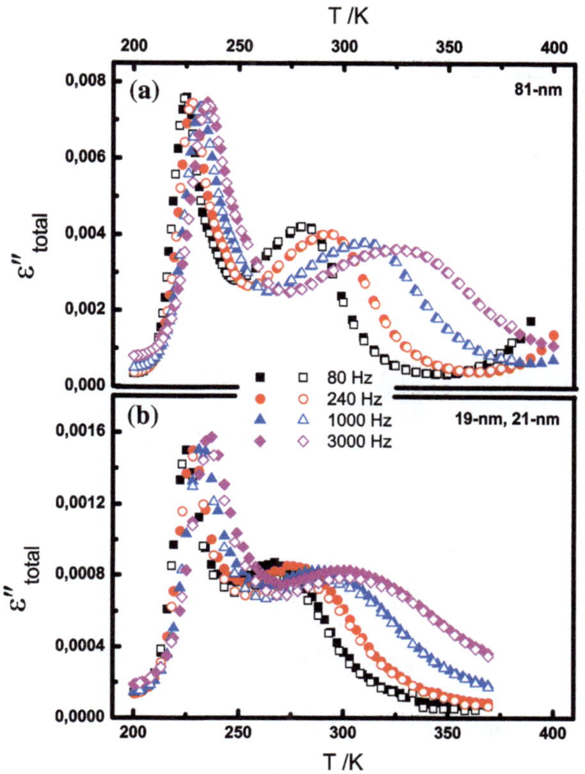

Fig. 3 Dielectric loss as a function of temperature, at different frequencies (as indicated), for **a** a polyisoprene (PI-12) sample of thickness 81 ± 2 nm measured after 24 h of annealing at 400 K (filled symbols), and after a subsequent annealing for 48 h (empty symbols), and **b** two polyisoprene (PI-12) samples prepared from the same solution, and hence having nearly the same thickness, i.e., 19 ± 2 nm (empty symbols) and 21 ± 2 nm (filled symbols). Unless otherwise indicated, the experimental uncertainty is less than the size of the symbols

Figure 3 shows representative results from an experiment carried to check the sample stability in terms of its dielectric response, as well as the reproducibility of the sample preparation. Here, a thin PI film with a thickness of 81 nm was annealed for 24 h at 400 K in high vacuum (10^{-6} mbar) before the first dielectric measurements. Thereafter, this sample was kept at 400 K in dry nitrogen flow for 48 h and remeasured. No changes are observed (in the temperature dependence of the dielectric loss) for the two relaxation processes as accessed by BDS (Fig. 3a). In Fig. 3b, data is displayed for two identically prepared and measured PI film samples of thickness ~20 nm; it is evident that the dielectric data thus obtained is replicate, ascertaining the reproducibility of the sample preparation and measurement procedure.

Table 2 Characteristics of the AAO membranes used for these studies

	I	II
Diameter of membrane ($\pm$0.2 mm)	13	13
Thickness ($\pm 2\,\mu$m)	48	50
Porosity (%)	11	10
Pore diameter (nm)	55 ± 6	18 ± 3
Pore density (cm^{-2})	2×10^9	5×10^{10}

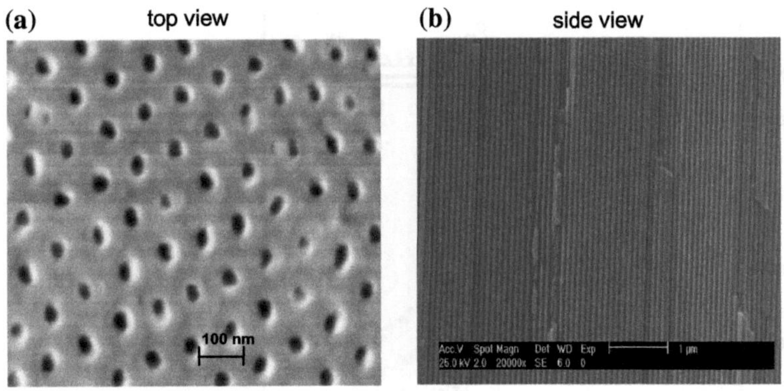

Fig. 4 Scanning Electron Microscope images of an AAO membrane with pores having a mean diameter of 55 nm, viewed from the top (**a**), and from the side after cleaving along the pore direction (**b**). The first image displays the well-known hexagonal array of cylindrical pores [139], while the second demonstrates the unidirectional nature of the channels

3.1.1 PI in Anodic Aluminum Oxide Nanopore Media

Aluminum can be oxidized by means of an electrochemical procedure which changes the surface chemistry of the metal to produce an anodic oxide layer. During this anodization process, a self-organized, highly ordered array of cylindrical-shaped pores can be produced (in dependence on oxidation parameters) with controllable pore diameters, periodicity, and porosity [137, 138]. AAO membranes are, therefore, used as templates in a variety of nanotechnology applications because the need for expensive lithographical techniques is circumvented.

AAO membranes with unidirectional pore channels were purchased from Synkera Technologies Inc., USA, (see Table 2 and Fig. 4) and infiltrated with poly(*cis*-1,4-isoprene) for dielectric studies. The experimental procedure was carried out as follows: (i) the empty membrane was weighed before being annealed at 573 K for 24 h in oil-free high vacuum (10^{-6} mbar); (ii) the system was cooled down to 300 K to allow for injection of 0.5 ml of a PI/chloroform (23 vol % PI) onto the membrane (still in high vacuum); (iii) after 12 h of infiltration at 300 K, the membrane was annealed at 374 K for 1 h to remove solvent, and then (iv) cooled down again to 300 K for the

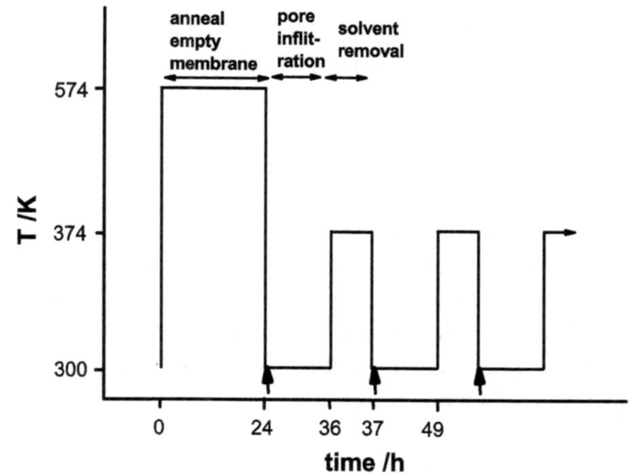

Fig. 5 A scheme describing the fractionated pore-infiltration procedure applied in this work. The thick arrows show the instances when injection of the solution is made

next injection of solution. This was repeated over several days, and a filling factor (as determined from weight measurements) of ≥70 % is achieved in 6 days (for PI-53 in 55-nm pores). Figure 5 is a schematic representation of the pore-filling procedure used in this work. A scheme of the main components of the annealing and pore-filling chamber is provided in Fig. 5.4.

3.2 Data Analysis

The empirical Havriliak–Negami (HN) equation [140] is commonly used to quantify the main features of a dielectric relaxation process, i.e., the relaxation strength, relaxation time, and the shape parameters of the relaxation peak. It is a generalization of the well-known Cole–Cole [141] and Cole–Davidson [142] relations, and defines the complex dielectric function, $\varepsilon^*(\omega)$, as

$$\varepsilon^*(\omega) = \varepsilon_\infty + \Delta\varepsilon/(1 + (i\omega\tau_{HN})^\beta)^\gamma \tag{9}$$

where $\Delta\varepsilon$ is the dielectric strength ($\Delta\varepsilon = \varepsilon_s - \varepsilon_\infty$), τ_{HN} denotes the characteristic relaxation time, β is a parameter characterizing a symmetrical broadening of the distribution of relaxation times, and γ accounts for asymmetrical broadening ($0 < \beta, \gamma \leq 1$). $\omega = 2\pi f$ is the angular frequency of the applied external electric field. The relaxation time at maximum loss, $\tau_{\max}$, is analytically obtained as $\tau_{\max} = \tau_{HN}\left[(\sin(\beta\gamma\pi/2 + 2\gamma))/(\sin(\beta\pi/2 + 2\gamma))\right]^{1/\beta}$. $\tau_{\max}$ not τ_{HN} , is used to generate activation plots (Fig. 13) since the latter is a function of shape parameters β and γ. When $\tau_{\max}$ has a non-Arrhenius temperature dependence, as is characteristic of the structural relaxation for all glass formers, it is described by the Vogel–Fulcher–Tammann (VFT) equation [143–145]:

$$\tau_{\max}(T) = \tau_0 exp\left(-\frac{DT_v}{T - T_V}\right) \tag{10}$$

where τ_0 and D are fit parameters (the latter is referred to as the fragility parameter, and is oft used to quantify the extent by which the temperature dependence of the relaxation times deviates from Arrhenius behavior) and T_V is the so-called Vogel temperature. It has been shown that the normal mode in bulk PI is also VFT-like [122, 146]. The VFT equation is mathematically equivalent to the Williams–Landel–Ferry (WLF) equation [147] which reads

$$log\frac{\tau_{\max}}{\tau_0} = \frac{C_1(T - T_0)}{C_2 + T - T_0} \tag{11}$$

where T_0 is a reference temperature, τ_0 is the corresponding relaxation time, and C_1 and C_2 are fitting parameters. The parameters of these two equations are related [122] by $DT_V = C_1C_2$ and $T_V = T_0 - C_2$.

Combining Eqs. (9) and (10), and assuming a temperature independence of the dielectric strength and shape parameters, delivers a temperature-dependent description of the complex dielectric function:

$$\varepsilon^*(\omega) = \varepsilon_\infty + \frac{\Delta\varepsilon}{\left[1 + \left(i\omega\tau_0 exp\left\{-\frac{DT_V}{T-T_V}\right\}\right)^\beta\right]^\gamma} \tag{12}$$

To extract the distribution of relaxation times for the two dynamic processes in PI, each was modeled, as illustrated in Fig. 6, in a step-by-step manner using Eq. (12). First, a function which best describes the segmental mode was calculated and subtracted from the measured (raw) data (Fig. 6a). This delivers a separated normal mode peak (Fig. 6b), which is then easily modeled as next step (Fig. 6c). Figure 7 is a representative display of data (PI-45) described using this procedure. Notice that fitting Eq. (12) directly to the data would involve at least eight free parameters, a procedure which would not deliver unique values.

With τ_{HN} and the shape parameters β and γ, the distribution of the relaxation times, $L(\tau)$, is computed employing an analytically derived equation [134, 148] of the form

$$L(\tau) = \frac{1}{\pi}x^{\beta\gamma}\left\{\sin\left(\gamma\Theta(x)\right)\Omega(x)\right\} \tag{13}$$

where, $x = \frac{\tau}{\tau_{HN}}\Theta(x) = arctan\left[\frac{\sin(\pi\beta)}{x^\beta+\cos(\pi\beta)}\right]$, and $\Omega(x) = [1 + 2x^\beta\cos(\pi\beta) + x^{2\beta}]^{-x/2}$

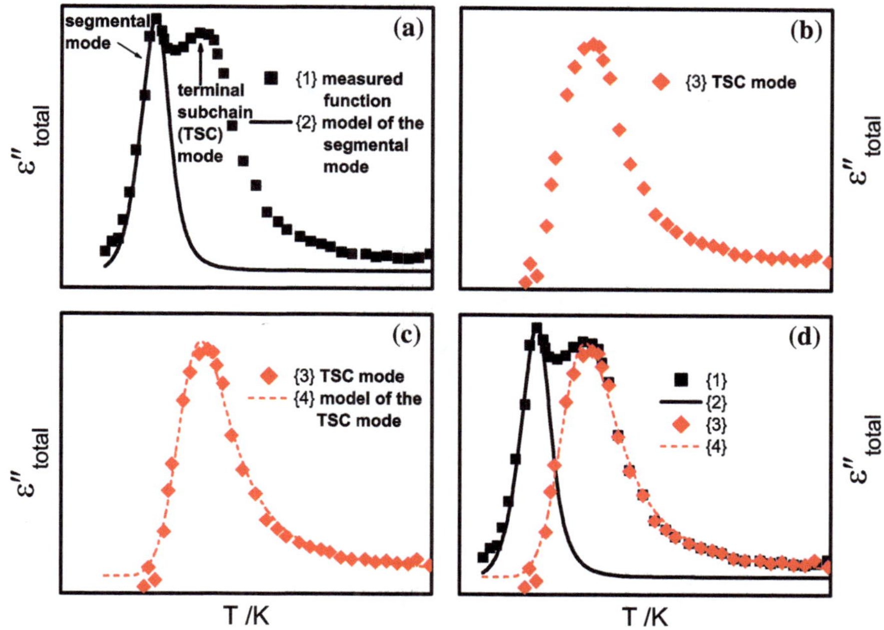

Fig. 6 An illustration of how dielectric loss data in the temperature representation is described step-by-step: **a** the raw data {*1*} showing the two relaxation processes present in a thin spin-cast PI sample, and a model function {*2*} calculated using Eq. (12) to fit the segmental mode; **b** a separated terminal subchain (TSC) mode {*3*} obtained by subtracting {*2*} from {*1*}; **c** a model function {*4*} evaluated to fit the TSC mode; and **d** a plot showing all the functions {*1*}, {*2*}, {*3*}, and {*4*}

3.3 Results

It is conventional in this kind of studies to vary the concentration of the polymer in solution in order to get films with desired thicknesses. Figure 8 presents the results for such a study for PI-12 and PI-75 where all samples were prepared from semi-dilute solutions by a single deposition procedure. Here, the layer thicknesses were systematically adjusted from bulk to dimensions comparable to the radius of gyration, R_g, of the respective chain in bulk. It is evident, for both molecular weights, that the segmental peak remains unaffected—in its spectral position and shape—within the limits of experimental uncertainty. Concerning chain dynamics, for $M_w \cong M_c$(PI-12), the normal mode remains bulk-like in its relaxation rate down to the thinnest layer ($d = 6\,\text{nm}$) studied, that is, $d \sim 1.7R_g$ (Fig. 8a, c, e). For $M_w = 7.5M_c$ (PI-75), the normal mode for all the thin layers is not bulk-like in its relaxation rate and shape. We, therefore, refer to this mode as the terminal subchain (TSC) mode (and explain it later). We observe that the TSC modes become faster and reduce in their dielectric strength upon increasing confinement (Fig. 8b, d, f).

In order to check how these results would evolve with a gradual variation of molecular weights, three more molecular weights (lying between PI-12 and PI-75) were investigated. Figure 9 is a summary of the results at a representative frequency of

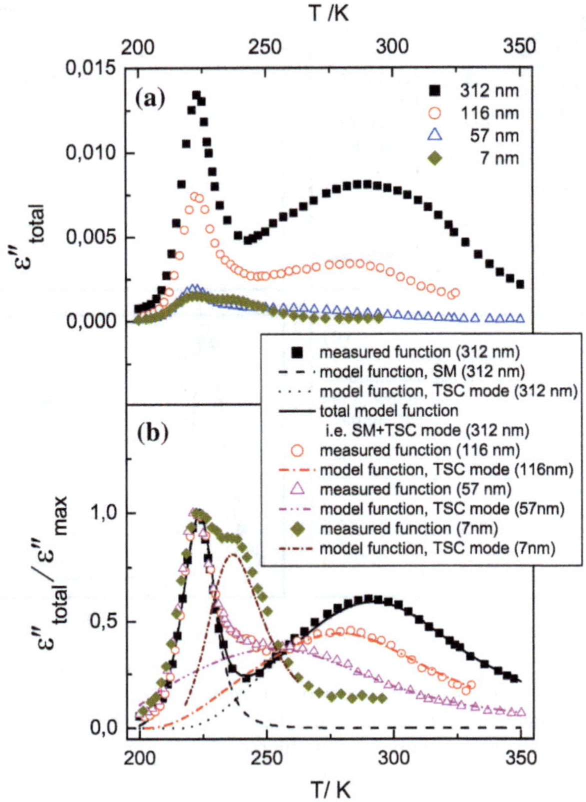

Fig. 7 **a** The measured dielectric loss, $\varepsilon''_{\text{total}}$, plotted as a function of temperature (at 80 Hz) for thin layers (thicknesses indicated) prepared from PI-45. **b** The same data in **a** normalized w.r.t. the maximum loss value of the segmental mode (SM). The lines represent functions (shown in the normalized data for graphical clarity) calculated to describe the data as explained in Fig. 6. The parameters used to construct these model functions are recorded in Table 3

80 Hz. It is observed that (i) irrespective of the molecular weight, the segmental mode is not shifted by a reduction in film thickness; (ii) its shape parameters (see Table 3) remain invariant with thickness, even when $d \approx R_g$, implying an unbroadened peak at this level of confinement; and (iii) TSC modes are present, vary in their shape (Table 3), and approach the segmental mode with decreasing film thickness (see also Fig. 10).

It is significant to note that the results so far presented involve a variation of the concentration of the mother solution used to prepare the thin layers. However, the findings for PI-12 point to the fact that entanglements play a role in the chain dynamics. To probe this intuition further, we carried out experiments where the film thickness and molecular weight are kept constant, but the solution concentration, c, systematically varied. Figure 11 is a plot of the results thus obtained (PI-53) and shows the dielectric loss (as a function of temperature) in its dependence on the

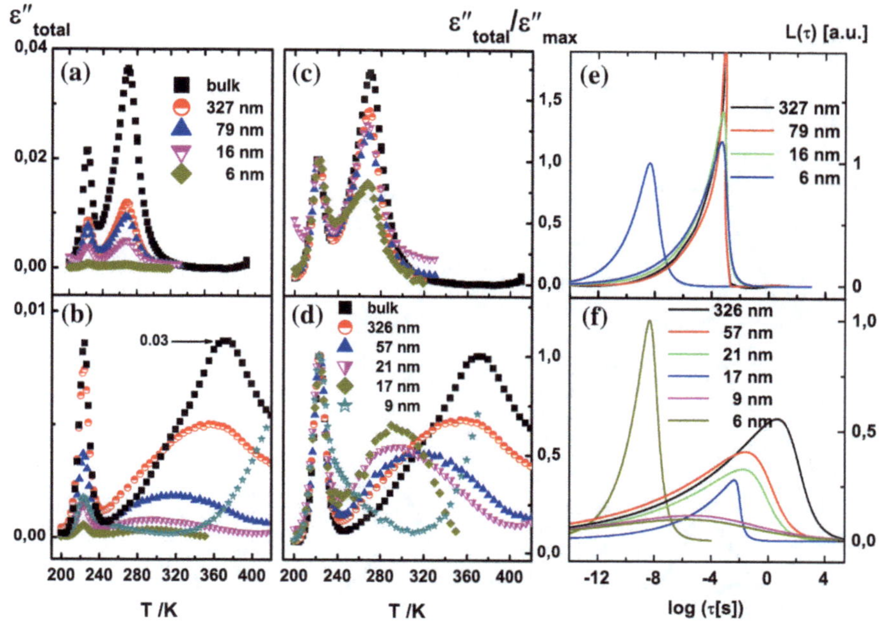

Fig. 8 Temperature dependence of the imaginary part, ε''_{total}, of the measured complex permittivity at 80 Hz for **a** PI-12 and **b** PI-75. In **c** and **d**, respectively, the same dielectric data is shown normalized w.r.t. the maximum loss value of the segmental mode. **e** and **f** display the corresponding relaxation time distribution functions at $T = 312$ K. Panels **a**, **c**, and **b**, **d** share common legends. The bulk data in **b** does not share the same y-axis scale with the rest of the curves. [Adapted from Ref. [104]. Reproduced by permission of the Royal Society of Chemistry]

concentration, *c*, of the mother solution used to deposit thin layers. *c* was varied from 4.15 to 0.32 % vol. of PI, that is, from semi-dilute to dilute regimes, and several samples with thickness $d \approx 7$ nm prepared. In order to attain this same thickness, the number of depositions required (at the same spinning rate) had to be increased with reducing concentration. It is evident for all samples studied, when compared to bulk, that the temperature position of the segmental relaxation peak remains unchanged. Additionally, the TSC mode—shifted by about 95 K to lower temperatures—is present, and its strength reduces with decreasing concentration of the mother solution (Fig. 11a, b). The corresponding distributions of the relaxation times for both the segmental and TSC modes are shown in Fig. 11c, and confirm the unaffected nature of the former mode, together with the *c*-dependence of the latter.

Having determined the effect of *c* on the resulting dynamics, we now proceed to keep *c* fixed (in the semi-dilute regime), and vary the molecular, M_w, in order to unravel the influence of the size of the chain on the molecular dynamics in confinement. The results of this experiment (for PI-45, PI-53 and PI-75) are displayed in Fig. 12 where films with thicknesses, *d*, ~10, ~15 and ~60 nm were prepared from solutions with *c* equal to 0.05, 0.25, and 1.25 mg/ml, respectively. It is consistently

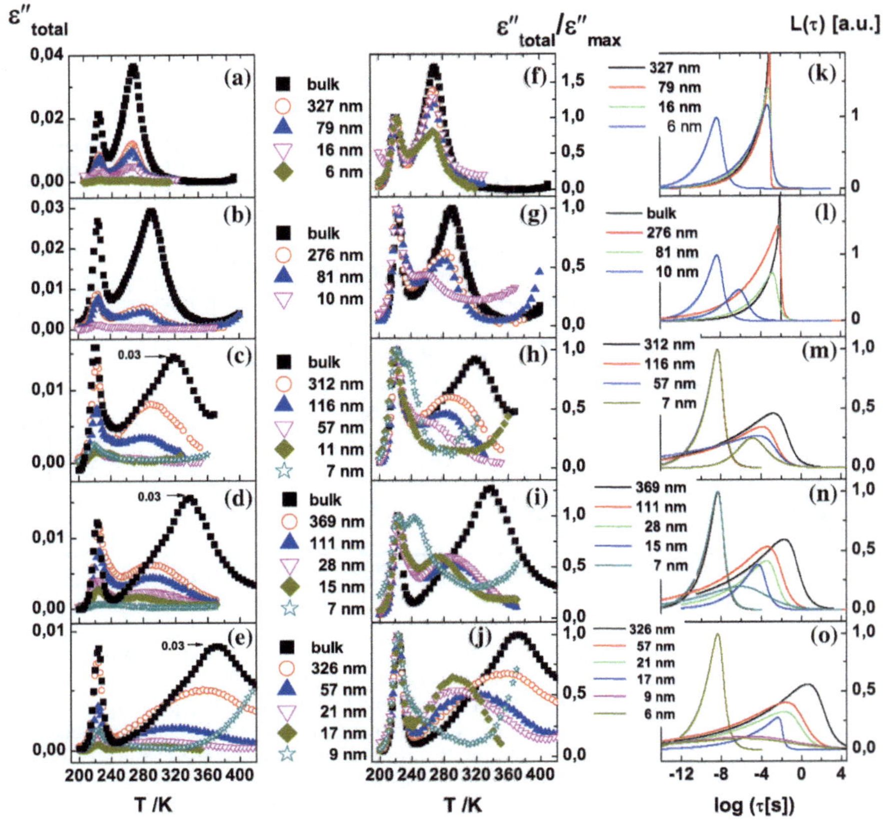

Fig. 9 Temperature dependence of the imaginary part, ε''_{total}, of the complex permittivity at 80 Hz for **a** PI-12, **b** PI-25, **c** PI-45, **d** PI-53, and **e** PI-75. In **f–j**, respectively, the same dielectric data is shown normalized w.r.t. the maximum loss value of the segmental mode. **k–o** display the corresponding relaxation time distribution functions at $T = 312$ K. The bulk data in **c**, **d**, **e** does not share the same y-axis scale with the rest of the curves. [Adapted from Ref. [104]. Reproduced by permission of the Royal Society of Chemistry]

clear that while the structural relaxation remains unchanged, the TSC mode becomes faster with decreasing chain length.

Figure 13 provides an overview of the activation map for the all the samples studied. It shows, for the segmental mode, that the temperature dependence of the relaxation rates is independent of both the layer thickness and molecular weight. Furthermore, a VFT fit to the data, extrapolated to a relaxation time of 100 s, delivers a dielectrically determined T_g value which is in quantitative agreement with what is obtained (for bulk) by differential scanning calorimetry (DSC). Similar results have been obtained for other polymer systems, e.g., in [25, 26, 95, 105].

Table 3 Havriliak–Negami (HN) shape parameters

Sample	Thickness (nm)	Segmental mode		Normal mode	
		β	γ	β	γ
PI-12	Bulk	0.74	0.46	0.99	0.84
	327	0.74	0.43	0.95	0.85
	79	0.72	0.42	0.97	0.82
	16	0.71	0.42	0.96	0.79
	6	0.73	0.41	0.92	0.75
PI-25	Bulk	0.72	0.72	0.69	0.52
	276	0.71	0.71	0.54	0.45
	81	0.72	0.72	0.49	0.47
	10	0.69	0.69	0.50	0.38
PI-45	Bulk	0.74	0.45	0.99	0.40
	312	0.72	0.40	0.55	0.14
	116	0.72	0.40	0.55	0.14
	57	0.71	0.41	0.51	0.10
	11	0.70	0.43	0.55	0.84
	7	0.71	0.42	0.39	0.90
PI-53	Bulk	0.72	0.49	0.71	0.35
	369	0.71	0.45	0.54	0.15
	111	0.70	0.42	0.54	0.15
	28	0.69	0.43	0.54	0.21
	15	0.70	0.41	0.65	0.30
	7	0.70	0.42	0.28	0.35
PI-75	Bulk	0.73	0.51	0.63	0.36
	326	0.70	0.40	0.50	0.14
	57	0.70	0.40	0.45	0.14
	21	0.61	0.40	0.47	0.15
	17	0.69	0.40	0.85	0.15
	9	0.70	0.40	0.17	0.35
	6	0.70	0.40	0.17	0.34

[Adapted from Ref. [104]. Reproduced by permission of the Royal Society of Chemistry]

4 Discussion

The preceding experimental findings can be summed up as follows: (i) the segmental relaxation, i.e., the dynamic glass transition in poly(*cis*-1,4-isoprene) is not shifted or broadened independent of all the studied parameters; in contrast (ii) the chain dynamics corresponding to the fluctuation of the end-to-end vector of the polymer chain or of the free tails of tethered chains (terminal subchains, TSC), are strongly affected by the concentration from which the polymer layers are deposited, the thickness of the layers, and the molecular weight of the sample. An interpretation of these observations is mainly of a qualitative nature because atomistic simulations of the dynamic glass transition, and more importantly of entangled systems, are not yet available.

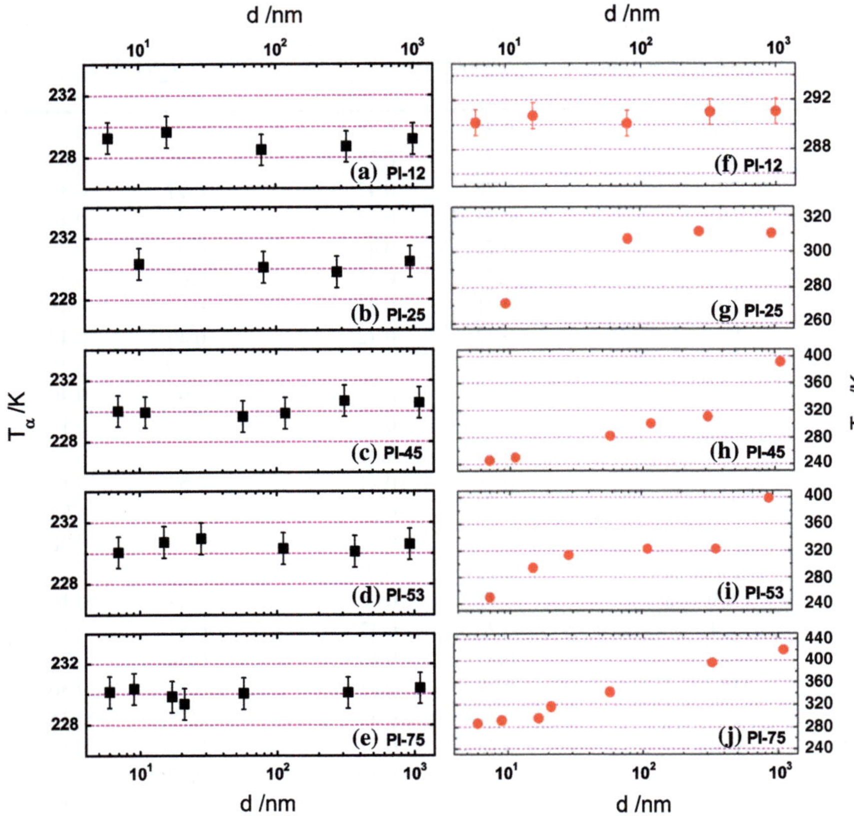

Fig. 10 Thickness (d)-dependence of the characteristic temperature T_{α} (*left panels*) and T_{NM} (*right panels*) for the segmental and normal modes, respectively, as measured at 1 kHz for different molecular weights (as indicated) of PI. Unless indicated, the experimental uncertainty is smaller than the size of the symbols

It has been generally [149], and specifically for PI [150], shown that the segmental mode is a fluctuation occurring at a length scale of about 2–3 chain segments, corresponding to about 7 Å. This is a size far below the external, one-dimensional confinement of $d \geq 6$ nm, given by the layer thickness (Fig. 14). At this extent of confinement, no changes in the dynamic glass transition are expected, at least for measurements carried out in the linear-response regime [101]. As has been recently proven for poly(vinylpyridine) on silica [105], certainly some polymer segments at the polymer-substrate interface (for the case of predominantly attractive interactions) will be immobilized [100], with respect to the time-scale of the experiment. This explains the reduction in the dielectric strength of the segmental mode as, for instance, observed in Fig. 9. Unfortunately, we cannot carry out a quantitative analysis of the dielectric strength because of the geometry applied in these measurements—there are

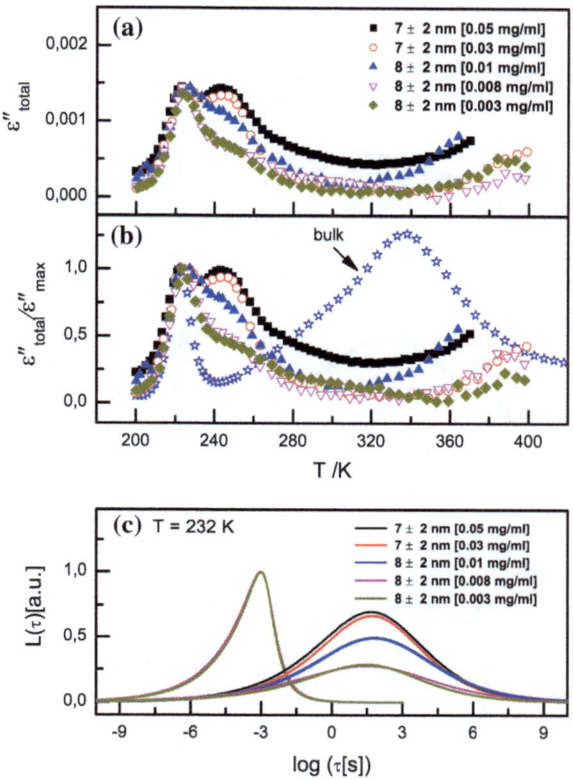

Fig. 11 **a** Dielectric loss $\varepsilon''_{\text{total}}$ *versus* temperature at 80 Hz (PI-53) for thin layers of thickness ~7 nm, each prepared from a different solution (in terms of concentration, *c*), with *c* varied from dilute (0.008 mg/ml) to semi-dilute (0.05 mg/ml) regimes. In **b**, the same data is shown normalized w.r.t. the maximum loss value of the segmental mode. For comparison, data for a bulk sample is included. Panels **a** and **b** share a common legend. **c** The relaxation time distributions at $T = 232$ K for the various thin layers. Notice that all curves for the segmental mode superimpose. [Adapted from Ref. [104]. Reproduced by permission of the Royal Society of Chemistry]

inaccuracies in defining the macroscopic evenness of the nanostructures at extensions of 4 mm

We provide here an elaborate discussion of the chain dynamics. In the bulk state, the main parameters of a chain depend on the macromolecular nature of the material, with the characteristic dimension being the radius of gyration, R_g, which is proportional to the molecular weight. The chain has a Gaussian distribution of the distances of one chain end from the other [151]. As first predicted by Flory [152], a given test chain occupies ~1 % of the volume within the nearly spherical region encompassed by the chain itself, while the rest is occupied by many other entangled chains and free volume. The length scale characterizing the chain dynamics is therefore that of the end-to-end distance (Fig. 14a). However, this conformation of the chain must be modified when it comes in contact with a solid substrate (Fig. 14b), or when spin-cast from solution into a thin geometrically constrained layer (Fig. 14c, d). In the spin-coating process, the polymer solution is transferred—within a time much shorter than the time required for the whole chain to relax—to a molten, nonequilibrium state [153], in which most of the solvent is removed. As observed in the current experiments, this results in a suppression of the normal mode relaxation (see Figs. 7, 8, 9, 12 and 13) in nanometric PI layers and is explained by several factors which add

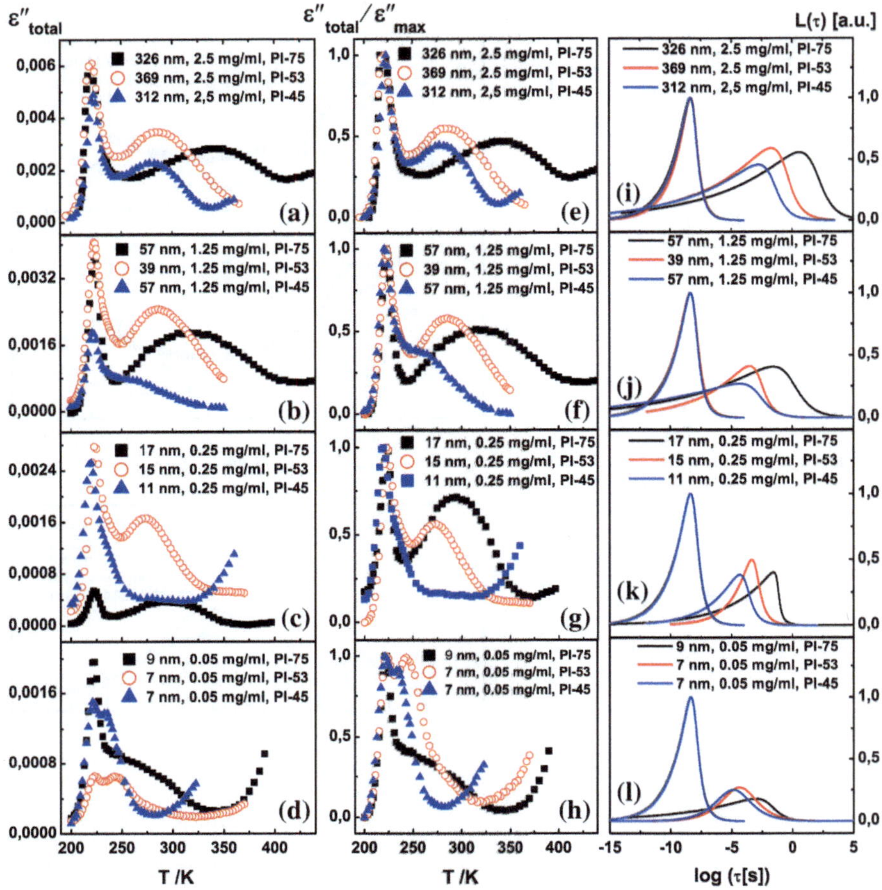

Fig. 12 Temperature dependence (at 80 Hz) of the measured dielectric loss, $\varepsilon''_{\text{total}}$, for thin layers (prepared from different molecular weights of PI, but the same solution concentrations) with comparable thicknesses: **a** 312–369 nm; **b** 39–57 nm; **c** 11–17 nm; and 7–9 nm. The same data normalized w.r.t. the maximum loss value of the segmental mode is plotted in **e**–**h**. Panels **i**–**l** show the respective distribution of relaxation times at 312 K. [Adapted from Ref. [104]. Reproduced by permission of the Royal Society of Chemistry]

up: (i) chain segments in the immediate vicinity of, and attractively interacting with the substrate become immobilized, leading to a suppression of the fluctuations of the end-to-end vector; (ii) the end-to-end vector of a chain prepared in a nanometric layer preferentially orients perpendicular to the $\boldsymbol{E}$ vector of the external electric field in the sample capacitor, and hence its contribution to the relaxation process is no longer dielectrically observable; (iii) topological constraints imposed by the spin-coating procedure might as well interrupt the normal mode process.

For polymer chains prepared from solutions in the dilute regime (c = 0.003 mg/ml), a weak relaxation is observed close to the segmental mode. It is assigned to

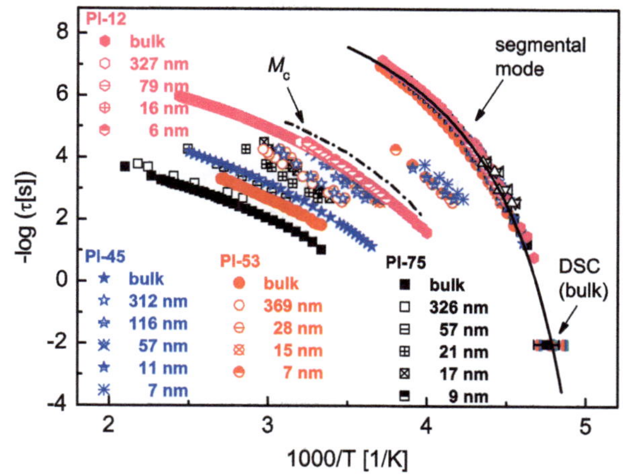

Fig. 13 Activation plot for the segmental, normal, and terminal subchain modes of PI samples prepared from different molecular weights, as indicated. Complementary data obtained from AC Calorimetry for PI-53 is presented by open triangles: bulk (*up triangles*), 100 nm (*down triangles*) and 15 nm (*side triangles*). T_g data determined for bulk samples by DSC is included as well. The *solid line* is a VFT fit to the segmental mode, with parameters: $\log\tau_0 = 10.6$, $DT_v = 299$ K and $T_v = 185$ K, while the *dash-dotted line* is calculated for the normal mode as expected for $M_w = M_c = 10^4$ g/mol. [Adopted from Ref. [104]. Reproduced by permission of the Royal Society of Chemistry]

the relaxation of terminal subchains (TSC) as depicted in Fig. 14d. The length of the TSC can be estimated from the frequency position of the respective peak to be about 90 segments. Increasing the polymer concentration in the spin-cast solution leads to an overlap of the polymer chains in the semi-dilute regime ($c > 0.008$ mg/ml), and hence to an increase in the strength of the TSC relaxations (Fig. 11a, b). It is also noteworthy that as the concentration of the solution is increased (and hence the number of spin-casting processes reduced), the angle between the TSC and the external field $\boldsymbol{E}$ increases (Fig. 14c, d); this should also account for the increase in the strength of the dielectric response. Remarkably, the relaxation time distribution of the TSC relaxation shows very minimal dependence on the polymer concentration. The TSC relaxation, however, shows pronounced molecular weight dependence, as demonstrated for samples prepared from semi-dilute solutions (Fig. 12). With increasing molecular weight, the peak of the TSC relaxation shifts to higher temperatures corresponding to lower relaxation times.

The fact that entanglements matter for the TSC relaxation is clearly reflected in the M_w-dependence of the chain dynamics in confinement (Fig. 7): for $M_w = 11.6$ kg/mol, which is close to $M_c = 10$ kg/mol, the cross-over molecular weight from Rouse- to reptation dynamics, the normal mode is not changed in its spectral position and relaxation time distribution (Fig. 8a, c, e)—only the strength of the relaxation is decreased. This is in pronounced contrast to all measurements on samples having $M_w \gg M_c$, where TSC relaxations are observed.

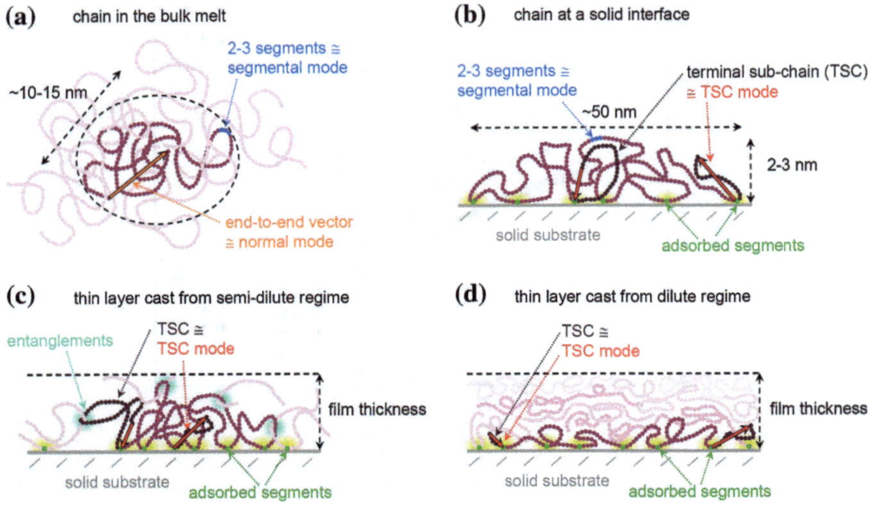

Fig. 14 Schematic representation of different polymer chain conformations induced by preparation. In the bulk melt, the chain has its equilibrium conformation and the molecular relaxations detected by BDS correspond to the segmental motions as well as the fluctuations of the end-to-end vector (**a**). In thin layers, the interaction with the solid substrate leads to adsorption of chain segments directly at the interface which are therefore immobilized (**b**). This interferes with the bulk normal mode: the dipole moment along the chain between pinned segments cannot perform fluctuations anymore; hence, only the free chain ends after the first and/or last adsorbed segment, the so-called terminal subchains (TSC), perform relaxation motions which we refer to as TSC modes. Spin-casting a thin layer from a semi-dilute solution (where chains still have overlap/entanglements) presumably results in an entangled network (**c**) with adsorption of segments causing TSC. For preparations from dilute concentrations (no overlap/entanglements of chains), several depositions are necessary which suggest that chains individually wet the substrate and are strongly stretched in parallel to the substrate (**d**). Consequently, the dipole moment of the TSC has only a minor component in the direction of the applied electrical field, causing a reduction in the strength of the detected signal

As reviewed in the theoretical section, the Rouse and reptation theories, respectively, predict M^2 and M^3 dependencies of the mean relaxation times of the normal mode, below and above $M_c = 10^4$ g/mol. In Fig. 15, a plot of the mean relaxation times *versus* molecular weight is presented as a function of PI layer thickness. It must be clarified here that, for the thin layers, the molecular weights presented are those of the mother (starting) sample before dissolution in chloroform. We find, for the bulk samples, the relation $\tau \propto M^{3.5\mp0.1}$, which is in quantitative agreement with reported results in the literature, e.g., in [113, 122] (see inset Fig. 15). For the case of spin-cast samples, we observe a gradual reduction in the gradient of the $\log\tau$ *versus* $\log M_w$ plots as the layer thickness is decreased. It is thus inferred that the dynamics of the TSC modes approach Rouse dynamics with reducing layer thickness, even though the mother chains are entangled.

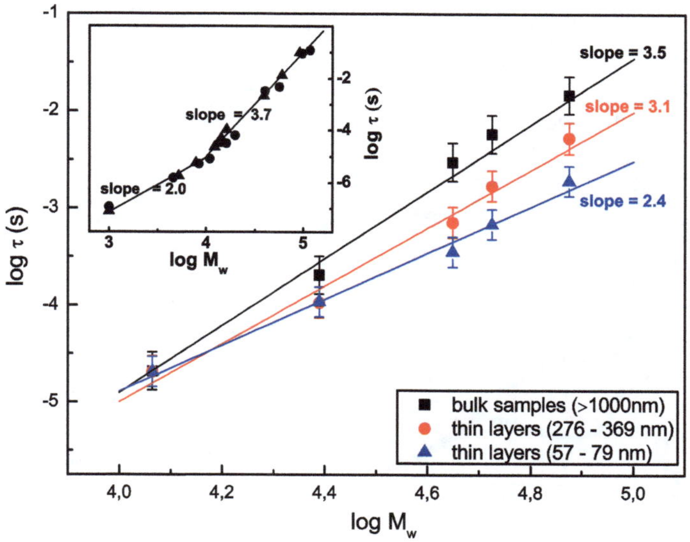

Fig. 15 Dependence of the mean dielectric relaxation time on molecular weight, M_w, for the normal mode (bulk samples) and terminal subchain mode (spin-cast layers) of poly(*cis*-1,4-isoprene) at 320 K. *Inset* data for bulk linear PI—Reprinted with permission from [122]. Copyright 1990 American Chemical Society

4.1 1-D Versus 2-D Confinement of PI

We consider thin films (with one free interface) as being one-dimensional (1-D) confinement, since a given chain experiences constraints only in one direction, while, with the same reasoning, unidirectional nanopores offer a two-dimensional (2-D) confining space.

Figure 16 shows data obtained (after subtraction of the alumina background) for PI measured in AAO nanopores (pore diameter, 55 nm (Fig. 16a) and 18 nm (Fig. 16b)) and the comparison with the case of PI in thin layers (Fig. 16c, d). In addition to the fact that the 2-D confinement measurements are reproducible, given the coincidence between the heating and cooling runs, it is also observed that (i) the mean spectral position of the segmental mode remains unchanged, when 1- and 2D confined systems are compared to bulk data; (ii) the segmental mode is broadened in the AAO nanopore matrix, a phenomenon not observed for 1D confinement; (iii) the broadening of the segmental mode is pore-size dependent (Fig. 16e); and (iv) the (remaining) normal mode (also referred to as TSC mode) becomes faster with reducing pore diameter (inset Fig. 16).

The reason for the first observation (concerning the dynamic glass transition) is the same as already provided for the thin films in the Discussion section. Fundamentally, the 2D confining space $\geq$18 nm (i.e., the pore diameter) is still large enough to allow for fluctuation of sufficient basic units (2–3 segments) that underlie the (dynamic) glass transition. On the other hand, the broadening in the structural relaxation process,

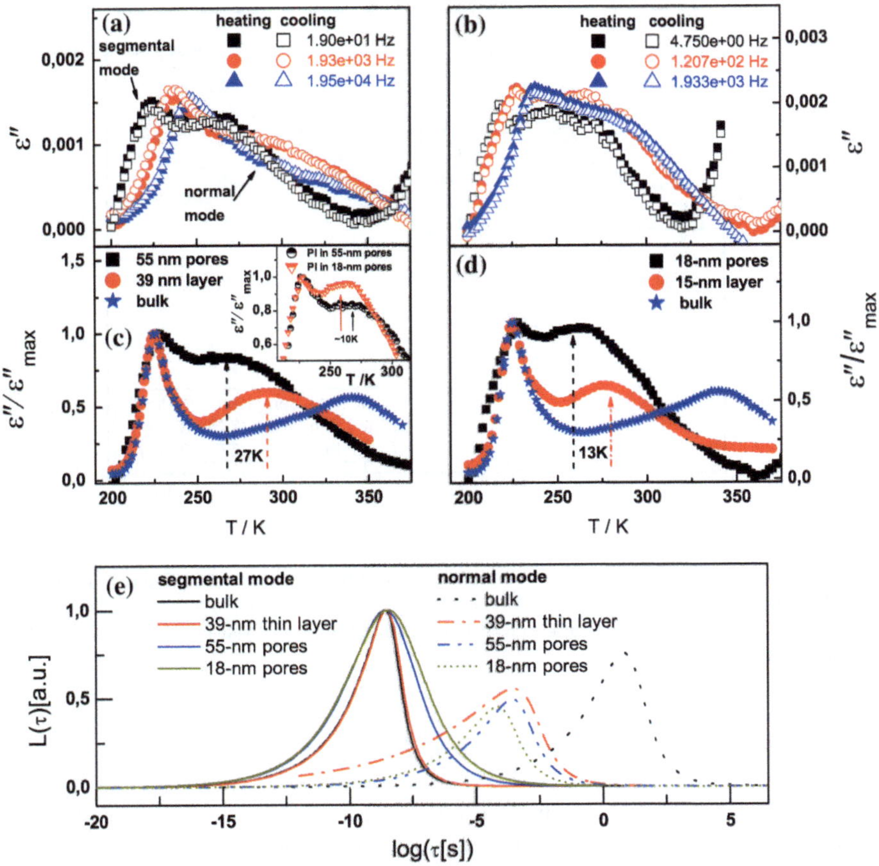

Fig. 16 Temperature dependence of the dielectric loss, ε'', (heating and cooling runs) for PI-53 measured in AAO templates with pores of diameter **a** 55 nm and **b** 18 nm. In **c** and **d**, a comparison of normalized data is made for PI measured (at 120 Hz) in pores and in thin films of comparable size to the pore diameter, as indicated. Bulk data is included as well for completeness. Panel **e** shows the respective distribution functions of the relaxation times at $T = 312$ K. *Inset* Normalized loss data (measured at 120 Hz) is plotted as function of temperature for PI in 55- and 18-nm pores

in pores as compared to thin films, is a consequence of the increase in the number of segments per unit volume that interact with the solid surface. This is because the surface area to volume ratio increases by at least three orders of magnitude from thin films to nanopores. A further reduction in the pore diameter would increase the chain–surface interaction, and hence the pore-size dependent broadening as observed. The

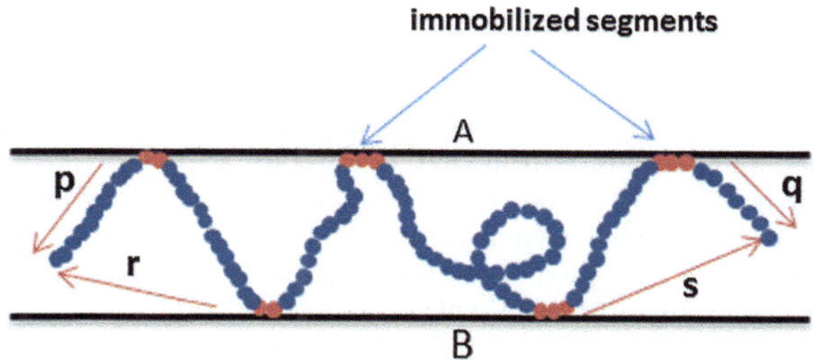

Fig. 17 A schematic view of a chain (viewed in one plane inside a nanopore) constrained by the solid surfaces A and B. Segments of the chain in (attractive) interaction with the surfaces are adsorbed, and hence become immobilized. The tail ends (terminal subchains) are free to fluctuate with resultant vectors **p** and **q** or **r** and **s** in the presence or absence, respectively, of wall A

scheme in Fig. 17 presents a possible explanation for why the TSC mode in nanopores is faster than that in a thin film of comparable size. For the pore system, taking a snapshot view in one plane, the chain interacts with the two surfaces A and B so that the resultant terminal subchains have vectors **p** and **q**. If surface A is removed—producing a system equivalent to an uncapped thin film—the TSCs have vectors **r** and **s**, which are, intuitively, longer than **p** and **q**; the fluctuation of the latter should therefore be faster.

5 Summary

Being a *Type A* polymer, poly(*cis*-1,4-isoprene) is an ideal candidate for studying both segmental and chain dynamics in confinement. The use of the nanostructured electrode arrangement, on the other hand, avails the experimental possibility of probing ultrathin layers by BDS. It has been shown in this chapter—for several molecular weights of PI lying in the range $10 < M_w < 80\,\mathrm{kg/mol}$—that while the structural relaxation remains largely unaffected by changes in layer thickness (down to 7 nm), molecular weight, and the polymer concentration in the spin-cast solution, the chain dynamics are drastically altered in dependence on these parameters. It has also been shown, courtesy of inaugural experiments, that chain dynamics in confinement depend on the dimensionality of the confining geometry. For a full understanding of this dependence, there is need for more experiments devoted to comparing 1D with 2D confining systems, as well as the corresponding molecular simulations.

Acknowledgments E.U.M. and M.T. gratefully appreciate financial support from the Deutsche Forschungsgemeinschaft in a project (DFG SPP 1369) devoted to "Polymer-Solid Contacts: Interfaces and Interphases," and the Graduate School BuildMona, respectively. Funding to F.K. by DFG in the framework of the Collaborative Research Centre SFB-TRR 102 is highly acknowledged. We also wish to thank Roxana Ene-Figuli for carrying out the DSC measurements at the Karlsruhe Institute of Technology (KIT).

References

1. Adam G, Gibbs JH (1965) On the temperature dependence of cooperative relaxation properties in glass-forming liquids. J Chem Phys 43:139–146
2. Keddie JL, Jones RAL, Cory RA (1994) Size-dependent depression of the glass transition temperature in polymer films. Europhys Lett 27:59–64
3. Keddie JL, Jones RAL, Cory RA (1994) Interface and surface effects on the glass-transition temperature in thin polymer films. Faraday Discuss 98:219–230
4. Forrest JA, Dalnoki-Veress K, Dutcher JR (1997) Interface and chain confinement effects on the glass transition temperature of thin polymer films. Phys Rev E 56:5705–5716
5. Grohens Y, Brogly M, Labbe C, David M-O, Schultz J (1998) Glass transition of stereoregular poly (methyl methacrylate) at interfaces. Langmuir 14:2929–2932
6. Fryer DS, Nealey PF, de Pablo JJ (2000) Thermal probe measurements of the glass transition temperature for ultrathin polymer films as a function of thickness. Macromolecules 33:6439–6447
7. Dalnoki-Veress K, Forrest J, Murray C, Gigault C, Dutcher J (2001) Molecular weight dependence of reductions in the glass transition temperature of thin, freely standing polymer films. Phys Rev E 63:031801
8. Fryer DS, Peters RD, Kim EJ, Tomaszewski JE, de Pablo JJ, Nealey PF et al (2001) Dependence of the glass transition temperature of polymer films on interfacial energy and thickness. Macromolecules 34:5627–5634
9. Kawana S, Jones R (2001) Character of the glass transition in thin supported polymer films. Phys Rev E 63:0215011–0215016
10. Tsui OKC, Zhang HF (2001) Effects of chain ends and chain entanglement on the glass transition temperature of polymer thin films. Macromolecules 34:9139–9142
11. Grohens Y, Hamon L, Reiter G, Soldera A, Holl Y (2002) Some relevant parameters affecting the glass transition of supported ultra-thin polymer films. Eur Phys J E: Soft Matter 8:217–224
12. Xie F, Zhang HF, Lee FK, Du B, Tsui OKC, Yokoe Y et al (2002) Effect of low surface energy chain ends on the glass transition temperature of polymer thin films. Macromolecules 35:1491–1492
13. Sharp J, Forrest J (2003) Free surfaces cause reductions in the glass transition temperature of thin polystyrene films. Phys Rev Lett 91:2357011–2357014
14. Sharp J, Forrest J (2003) Dielectric and ellipsometric studies of the dynamics in thin films of isotactic poly(methylmethacrylate) with one free surface. Phys Rev E 67:0318051–0318059
15. D'Amour J (2004) Influence of substrate chemistry on the properties of ultrathin polymer films. Microelectron Eng 73–74:209–217
16. Fakhraai Z, Sharp JS, Forrest JA (2004) Effect of sample preparation on the glass-transition of thin polystyrene films. J Polym Sci Part B: Polym Phys 42:4503–4507
17. Singh L, Ludovice PJ, Henderson CL (2004) Influence of molecular weight and film thickness on the glass transition temperature and coefficient of thermal expansion of supported ultrathin polymer films. Thin Solid Films 449:231–241
18. Fakhraai Z, Forrest J (2005) Probing slow dynamics in supported thin polymer films. Phys Rev Lett 95:025701
19. Roth CB, Pound A, Kamp SW, Murray CA, Dutcher JR (2006) Molecular-weight dependence of the glass transition temperature of freely-standing poly(methyl methacrylate) films. Eur Phys J E 20:441–448
20. Raegen AN, Massa MV, Forrest JA, Dalnoki-Veress K (2008) Effect of atmosphere on reductions in the glass transition of thin polystyrene films. Eur Phys J E 27:375–377
21. Inoue R, Kanaya T, Nishida K, Tsukushi I, Telling M, Gabrys B et al (2009) Glass transition and molecular mobility in polymer thin films. Phys Rev E 80:031802
22. Kim S, Hewlett SA, Roth CB, Torkelson JM (2009) Confinement effects on glass transition temperature, transition breadth, and expansivity: comparison of ellipsometry and fluorescence measurements on polystyrene films. Eur Phys J E 30:83–92

23. Lu H, Chen W, Russell TP (2009) Relaxation of thin films of polystyrene floating on ionic liquid surface. Macromolecules 42:9111–9117
24. Mapesa EU, Erber M, Tress M, Eichhorn K-J, Serghei A, Voit B et al (2010) Glassy dynamics in nanometer thin layers of polystyrene. Eur Phys J Spec Top 189:173–180
25. Tress M, Erber M, Mapesa EU, Huth H, Müller J, Serghei A et al (2010) Glassy dynamics and glass transition in nanometric thin layers of polystyrene. Macromolecules 43:9937–9944
26. Erber M, Tress M, Mapesa EU, Serghei A, Eichhorn K-J, Voit B et al (2010) Glassy dynamics and glass transition in thin polymer layers of pmma deposited on different substrates. Macromolecules 43:7729–7733
27. Clough A, Peng D, Yang Z, Tsui OKC (2011) Glass transition temperature of polymer films that slip. Macromolecules 44:1649–1653
28. Glynos E, Frieberg B, Oh H, Liu M, Gidley DW, Green PF (2011) Role of molecular architecture on the vitrification of polymer thin films. Phys Rev Lett 106:128301
29. Bäumchen O, McGraw JD, Forrest JA, Dalnoki-Veress K (2012) Reduced glass transition temperatures in thin polymer films: surface effect or artifact? Phys Rev Lett 109:055701
30. Pye JE, Roth CB (2011) Two simultaneous mechanisms causing glass transition temperature reductions in high molecular weight freestanding polymer films as measured by transmission ellipsometry. Phys Rev Lett 107:235701
31. Reiter G (1994) Dewetting as a probe of polymer mobility in thin films. Macromolecules 27:3046–3052
32. Wallace WE, Van Zanten JH, Wu WL (1995) Influence of an impenetrable interface on a polymer glass-transition temperature. Phys Rev E 52:R3329–R3332
33. Tsui OKC, Russell TP, Hawker CJ (2001) Effect of interfacial interactions on the glass transition of polymer thin films. Macromolecules 34:5535–5539
34. Inoue R, Kanaya T, Miyazaki T, Nishida K, Tsukushi I, Shibata K (2006) Glass transition and thermal expansivity of polystyrene thin films. Mater Sci Eng: A 442:367–370
35. Miyazaki T, Inoue R, Nishida K, Kanaya T (2007) X-ray reflectivity studies on glass transition of free standing polystyrene thin films. Eur Phys J Spec Top 141:203–206
36. Kanaya T, Inoue R, Kawashima K, Miyazaki T, Tsukushi I, Shibata K et al (2009) Glassy dynamics and heterogeneity of polymer thin films. J Phys Soc Jpn 78:041004
37. Tate RS, Fryer DS, Pasqualini S, Montague MF, de Pablo JJ, Nealey PF (2001) Extraordinary elevation of the glass transition temperature of thin polymer films grafted to silicon oxide substrates. J Chem Phys 115:9982
38. Fukao K, Miyamoto Y (2001) Slow dynamics near glass transitions in thin polymer films. Phys Rev E 64:0118031–01180319
39. Efremov MY, Olson E, Zhang M, Zhang Z, Allen L (2003) Glass transition in ultrathin polymer films: calorimetric study. Phys Rev Lett 91:0857031–0857034
40. Efremov MY, Olson EA, Zhang M, Zhang Z, Allen LH (2004) Probing glass transition of ultrathin polymer films at a time scale of seconds using fast differential scanning calorimetry. Macromolecules 37:4607–4616
41. Koh YP, McKenna GB, Simon SL (2006) Calorimetric glass transition temperature and absolute heat capacity of polystyrene ultrathin films. J Polym Sci Part B: Polym Phys 44:3518–3527
42. Koh YP, Simon SL (2008) Structural relaxation of stacked ultrathin polystyrene films. J Polym Sci Part B: Polym Phys 46:2741–2753
43. Bernazzani P, Sanchez RF (2009) Structural and thermal behavior of polystyrene thin films using ATR-FTIR-NanoDSC measurements. J Therm Anal Calorim 96:727–732
44. Fukao K, Terasawa T, Oda Y, Nakamura K, Tahara D (2011) Glass transition dynamics of stacked thin polymer films. Phys Rev E 84:041808
45. DeMaggio GB, Frieze WE, Gidley DW, Zhu M, Hristov HA, Yee AF (1997) Interface and surface effects on the glass transition in thin polystyrene films. Phys Rev Lett 78:1524–1527
46. Jean YC, Zhang R, Cao H, Yuan J-P, Huang C-M, Nielsen B et al (1997) Glass transition of polystyrene near the surface studied by slow-positron-annihilation spectroscopy. Phys Rev B 56:R8459

47. Lupaşcu V, Huth H, Schick C, Wübbenhorst M (2005) Specific heat and dielectric relaxations in ultra-thin polystyrene layers. Thermochim Acta 432:222–228
48. Huth H, Minakov AA, Schick C (2006) Differential AC-chip calorimeter for glass transition measurements in ultrathin films. J Polym Sci Part B: Polym Phys 44:2996–3005
49. Huth H, Minakov AA, Serghei A, Kremer F, Schick C (2007) Differential AC-chip calorimeter for glass transition measurements in ultra thin polymeric films. Eur Phys J Spec Top 141:153–160
50. Schick C (2010) Glass transition under confinement-what can be learned from calorimetry. Eur Phys J Spec Top 189:3–36
51. Ellison CJ, Torkelson JM (2002) Sensing the glass transition in thin and ultrathin polymer films via fluorescence probes and labels. J Polym Sci Part B: Polym Phys 40:2745–2758
52. Ellison CJ, Kim SD, Hall DB, Torkelson JM (2002) Confinement and processing effects on glass transition temperature and physical aging in ultrathin polymer films: Novel fluorescence measurements. Eur Phys J E: Soft Matter 8:155–166
53. Ellison CJ, Torkelson JM (2003) The distribution of glass-transition temperatures in nanoscopically confined glass formers. Nat Mater 2:695–700
54. Ellison C, Ruszkowski R, Fredin N, Torkelson J (2004) Dramatic reduction of the effect of nanoconfinement on the glass transition of polymer films via addition of small-molecule diluent. Phys Rev Lett 92:095702
55. Ellison CJ, Mundra MK, Torkelson JM (2005) Impacts of polystyrene molecular weight and modification to the repeat unit structure on the glass transition-nanoconfinement effect and the cooperativity length scale. Macromolecules 38:1767–1778
56. Mundra MK, Ellison CJ, Behling RE, Torkelson JM (2006) Confinement, composition, and spin-coating effects on the glass transition and stress relaxation of thin films of polystyrene and styrene-containing random copolymers: Sensing by intrinsic fluorescence. Polymer 47:7747–7759
57. Mundra MK, Ellison CJ, Rittigstein P, Torkelson JM (2007) Fluorescence studies of confinement in polymer films and nanocomposites: glass transition temperature, plasticizer effects, and sensitivity to stress relaxation and local polarity. Eur Phys J Spec Top 141:143–151
58. Kim S, Roth CB, Torkelson JM (2008) Effect of nanoscale confinement on the glass transition temperature of free-standing polymer films: novel, self-referencing fluorescence method. J Polym Sci Part B: Polym Phys 46:2754–2764
59. Kim S, Torkelson JM (2011) Distribution of glass transition temperatures in free-standing. Nanoconfined polystyrene films: a test of de Gennes' sliding motion mechanism. Macromolecules 44:4546–4553
60. Paeng K, Swallen SF, Ediger MD (2011) Direct measurement of molecular motion in free-standing polystyrene thin films. J Am Chem Soc 133:8444–8447
61. Paeng K, Ediger MD (2011) Molecular motion in free-standing thin films of poly(methyl methacrylate), poly(4- tert -butylstyrene), poly(α-methylstyrene), and poly(2-vinylpyridine). Macromolecules 44:7034–7042
62. Paeng K, Lee H-N, Swallen SF, Ediger MD (2011) Temperature-ramping measurement of dye reorientation to probe molecular motion in polymer glasses. J Chem Phys 134:024901
63. Paeng K, Richert R, Ediger MD (2012) Molecular mobility in supported thin films of polystyrene, poly(methyl methacrylate), and poly(2-vinyl pyridine) probed by dye reorientation. Soft Matter 8:819–826
64. Kanaya T, Miyazaki T, Watanabe H, Nishida K, Yamano H, Tasaki S et al (2003) Annealing effects on thickness of polystyrene thin films as studied by neutron reflectivity. Polymer 44:3769–3773
65. Inoue R, Kawashima K, Matsui K, Kanaya T, Nishida K, Matsuba G et al (2011) Distributions of glass-transition temperature and thermal expansivity in multilayered polystyrene thin films studied by neutron reflectivity. Phys Rev E 83:021801
66. Inoue R, Kawashima K, Matsui K, Nakamura M, Nishida K, Kanaya T et al (2011) Interfacial properties of polystyrene thin films as revealed by neutron reflectivity. Phys Rev E 84:031802

67. Forrest JA, Dalnoki-Veress K, Stevens JR, Dutcher JR (1996) Effect of free surfaces on the glass transition temperature of thin polymer films. Phys Rev Lett 77:2002–2005
68. Forrest JA, Mattsson J (2000) Reductions of the glass transition temperature in thin polymer films: probing the length scale of cooperative dynamics. Phys Rev E 61:R53–R56
69. Mattsson J, Forrest JA, Börjesson L (2000) Quantifying glass transition behavior in ultrathin free-standing polymer films. Phys Rev E 62:5187
70. Cheng W, Sainidou R, Burgardt P, Stefanou N, Kiyanova A, Efremov M et al (2007) Elastic properties and glass transition of supported polymer thin films. Macromolecules 40:7283–7290
71. Akabori K, Tanaka K, Kajiyama T, Takahara A (2003) Anomalous surface relaxation process in polystyrene ultrathin films. Macromolecules 36:4937–4943
72. Akabori K, Tanaka K, Nagamura T, Takahara A, Kajiyama T (2005) Molecular motion in ultrathin polystyrene films: dynamic mechanical analysis of surface and interfacial effects. Macromolecules 38:9735–9741
73. Akabori K-I, Tanaka K, Takahara A, Kajiyama T, Nagamura T (2007) Substrate effect on mechanical relaxation of polystyrene in ultrathin films. Eur Phys J Spec Top 141:173–180
74. Ge S, Pu Y, Zhang W, Rafailovich M, Sokolov J, Buenviaje C et al (2000) Shear modulation force microscopy study of near surface glass transition temperatures. Phys Rev Lett 85:2340–2343
75. Dinelli F, Ricci A, Sgrilli T, Baschieri P, Pingue P, Puttaswamy M et al (2011) Nanoscale viscoelastic behavior of the surface of thick polystyrene films as a function of temperature. Macromolecules 44:987–992
76. O'Connell PA (2005) Rheological measurements of the thermoviscoelastic response of ultrathin polymer films. Science 307:1760–1763
77. O'Connell PA, McKenna GB (2006) Dramatic stiffening of ultrathin polymer films in the rubbery regime. Eur Phys J E 20:143–150
78. O'Connell PA, McKenna GB (2007) Novel nanobubble inflation method for determining the viscoelastic properties of ultrathin polymer films. Rev Sci Instrum 78:013901
79. O'Connell PA, Hutcheson SA, McKenna GB (2008) Creep behavior of ultra-thin polymer films. J Polym Sci Part B: Polym Phys 46:1952–1965
80. O'Connell PA, McKenna GB (2008) A novel nano-bubble inflation method for determining the viscoelastic properties of ultrathin polymer films. Scanning 30:184–196
81. Bodiguel H, Fretigny C (2007) Viscoelastic properties of ultrathin polystyrene films. Macromolecules 40:7291–7298
82. Yang Z, Peng D, Clough A, Lam C-H, Tsui OKC (2010) Is the dynamics of polystyrene films consistent with their glass transition temperature? Eur Phys J Spec Top 189:155–164
83. Wang J, McKenna GB (2013) Viscoelastic and glass transition properties of ultrathin polystyrene films by dewetting from liquid glycerol. Macromolecules 46:2485–2495
84. Prucker O, Christian S, Bock H, Rühe J, Frank WC, Knoll W (1998) On the glass transition in ultrathin polymer films of different molecular architecture. Macromol Chem Phys 199:1435–1444
85. Fukao K, Miyamoto Y (1999) Glass transition temperature and dynamics of α-process in thin polymer films. Europhys Lett 46:649–654
86. Fukao K, Miyamoto Y (2000) Glass transitions and dynamics in thin polymer films: dielectric relaxation of thin films of polystyrene. Phys Rev E 61:1743–1754
87. Fukao K, Uno S, Miyamoto Y, Hoshino A, Miyaji H (2002) Relaxation dynamics in thin supported polymer films. J Non-Cryst Solids 307–310:517–523
88. Serghei A, Huth H, Schellenberger M, Schick C, Kremer F (2005) Pattern formation in thin polystyrene films induced by an enhanced mobility in ambient air. Phys Rev E 71:0618011–0618014
89. Lupaşcu V, Picken SJ, Wübbenhorst M (2006) Cooperative and non-cooperative dynamics in ultra-thin films of polystyrene studied by dielectric spectroscopy and capacitive dilatometry. J Non-Cryst Solids 352:5594–5600

90. Serghei A, Kremer F (2006) Unexpected preparative effects on the properties of thin polymer films. Prog Colloid Polym Sci 132:33–40
91. Napolitano S, Wübbenhorst M (2007) Dielectric signature of a dead layer in ultrathin films of a nonpolar polymer. J Phys Chem B 111:9197–9199
92. Priestley R, Broadbelt L, Torkelson J, Fukao K (2007) Glass transition and α-relaxation dynamics of thin films of labeled polystyrene. Phys Rev E 75:061806
93. Svanberg C (2007) Glass transition relaxations in thin suspended polymer films. Macromolecules 40:312–315
94. Serghei A, Huth H, Schick C, Kremer F (2008) Glassy dynamics in thin polymer layers having a free upper interface. Macromolecules 41:3636–3639
95. Serghei A, Kremer F (2008) Metastable states of glassy dynamics, possibly mimicking confinement-effects in thin polymer films. Macromol Chem Phys 209:810–817
96. Rotella C, Napolitano S, Wübbenhorst M (2009) Segmental mobility and glass transition temperature of freely suspended ultrathin polymer membranes. Macromolecules 42:1415–1417
97. Napolitano S, Wübbenhorst M (2010) Structural relaxation and dynamic fragility of freely standing polymer films. Polymer 51:5309–5312
98. Rotella C, Napolitano S, De Cremer L, Koeckelberghs G, Wübbenhorst M (2010) Distribution of segmental mobility in ultrathin polymer films. Macromolecules 43:8686–8691
99. Napolitano S, Rotella C, Wübbenhorst M (2011) Is the reduction in tracer diffusivity under nanoscopic confinement related to a frustrated segmental mobility? Macromol Rapid Commun 32:844–848
100. Napolitano S, Wübbenhorst M (2011) The lifetime of the deviations from bulk behaviour in polymers confined at the nanoscale. Nat Commun 2:260–267
101. Boucher VM, Cangialosi D, Yin H, Schönhals A, Alegría A, Colmenero J (2012) Tg depression and invariant segmental dynamics in polystyrene thin films. Soft Matter 8:5119–5122
102. Boucher VM, Cangialosi D, Alegría A, Colmenero J (2012) Time dependence of the segmental relaxation time of poly(vinyl acetate)-silica nanocomposites. Phys Rev E 86:041501
103. Kremer F, Hartmann L, Serghei A, Pouret P, Leger L (2003) Molecular dynamics in thin grafted and spin-coated polymer layers. Eur Phys J E: Soft Matter 12:139–142
104. Mapesa EU, Tress M, Schulz G, Huth H, Schick C, Reiche M et al (2013) Segmental and chain dynamics in nanometric layers of poly(cis-1,4-isoprene) as studied by broadband dielectric spectroscopy and temperature-modulated calorimetry. Soft Matter 9:10592
105. Tress M, Mapesa EU, Kossack W, Kipnusu WK, Reiche M, Kremer F (2013) Glassy dynamics in condensed isolated polymer coils. Science 341:1371–1374
106. Fukao K (2003) Dynamics in thin polymer films by dielectric spectroscopy. Eur Phys J E: Soft Matter 12:119–125
107. Karasz FE, MacKnight WJ (1968) The influence of stereoregularity on the glass transition temperatures of vinyl polymers. Macromolecules 1:537–540
108. Nguyen HK, Labardi M, Capaccioli S, Lucchesi M, Rolla P, Prevosto D (2012) Interfacial and annealing effects on primary α-relaxation of ultrathin polymer films investigated at nanoscale. Macromolecules 45:2138–2144
109. Nguyen HK, Labardi M, Lucchesi M, Rolla P, Prevosto D (2013) Plasticization in ultrathin polymer films: the role of supporting substrate and annealing. Macromolecules 46:555–561
110. van Zanten JH, Wallace WE, Wu W (1996) Effect of strongly favorable substrate interactions on the thermal properties of ultrathin polymer films. Phys Rev E 53:R2053
111. Serghei A (2008) Challenges in glassy dynamics of polymers. Macromol Chem Phys 209:1415–1423
112. Petychakis L, Floudas G, Fleischer G (1997) Chain dynamics of polyisoprene confined in porous media. A dielectric spectroscopy study. Europhys Lett 40:685–690
113. Adachi K, Kotaka T (1985) Dielectric normal mode process in undiluted cis-polyisoprene. Macromolecules 18:466–472
114. Liu Y, Russell TP, Samant MG, Stöhr J, Brown HR, Cossy-Favre A et al (1997) Surface relaxations in polymers. Macromolecules 30:7768–7771

115. Serghei A, Kremer F, Kob W (2003) Chain conformation in thin polymer layers as revealed by simulations of ideal random walks. Eur Phys J E: Soft Matter 12:143–146
116. Serghei A, Kremer F (2003) Confinement-induced relaxation process in thin films of cis-polyisoprene. Phys Rev Lett 91:1657021–1657024
117. Labahn D, Mix R, Schönhals A (2009) Dielectric relaxation of ultrathin films of supported polysulfone. Phys Rev E 79:01108011–01108018
118. Kojio K, Jeon S, Granick S (2002) Confinement-induced differences between dielectric normal modes and segmental modes of an ion-conducting polymer. Eur Phys J E: Soft Matter 8:167–173
119. Jeon S, Granick S (2001) A polymer's dielectric normal modes depend on its film thickness when confined between nonwetting surfaces. Macromolecules 34:8490–8495
120. Holt AP, Sangoro JR, Wang Y, Agapov AL, Sokolov AP (2013) Chain and segmental dynamics of poly(2-vinylpyridine) nanocomposites. Macromolecules 46:4168–4173
121. Stockmayer WH (1967) Dielectric dispersion in solutions of flexible polymers. Pure Appl Chem 15:539–554
122. Boese D, Kremer F (1990) Molecular dynamics in bulk cis-polyisoprene as studied by dielectric spectroscopy. Macromolecules 23:829–835
123. Cole RH (1965) Correlation function theory of dielectric relaxation. J Chem Phys 42:637
124. Williams G, Watts DC (1970) Non-symmetrical dielectric relaxation behaviour arising from a simple empirical decay function. Trans Faraday Soc 66:80–85
125. Williams G (1972) Use of the dipole correlation function in dielectric relaxation. Chem Rev 72:55–69
126. Adachi K, Kotaka T (1988) Dielectric normal mode process in semidilute and concentrated solutions of cis-polyisoprene. Macromolecules 21:157–164
127. Williams G, Cook M, Hains PJ (1972) Molecular motion in amorphous polymers. Consideration of the mechanism for α, β and ($\alpha\beta$) dielectric relaxations. J Chem Soc Faraday Trans 2(68):1045–1050
128. Rouse PE (1953) A theory of the linear viscoelastic properties of dilute solutions of coiling polymers. J Chem Phys 21:1272
129. Zimm BH (1956) Dynamics of polymer molecules in dilute solution: viscoelasticity. Flow birefringence and dielectric Loss. J Chem Phys 24:269
130. de Gennes PG (1971) Reptation of a polymer chain in the presence of fixed obstacles. J Chem Phys 55:572
131. Doi M, Edwards SF (1978) Dynamics of concentrated polymer systems. Part 1.—Brownian motion in the equilibrium state. J Chem Soc Faraday Trans 2(74):1789–1801
132. Hadjichristidis N, Roovers JEL (1974) Synthesis and solution properties of linear, four-branched, and six-branched star polyisoprenes. J Polym Sci: Polym Phys Ed 12:2521–2533
133. Klapetek P, Necas D (2013) Data analysis software. Central European Institute of Technology, Kotlářská
134. Kremer F, Schönhals A (eds) (2003) Broadband dielectric spectroscopy. Springer, Berlin
135. Serghei A, Kremer F (2008) Broadband dielectric studies on the interfacial dynamics enabled by use of nanostructured electrodes. Rev Sci Instrum 79:0261011–0261012
136. Serghei A, Tress M, Kremer F (2009) The glass transition of thin polymer films in relation to the interfacial dynamics. J Chem Phys 131:154904
137. Schwirn K, Lee W, Hillebrand R, Steinhart M, Nielsch K, Gösele U (2008) Self-ordered anodic aluminum oxide formed by H_2 SO_4 hard anodization. ACS Nano 2:302–310
138. Zaraska L, Sulka GD, Jaskuła M (2011) Anodic alumina membranes with defined pore diameters and thicknesses obtained by adjusting the anodizing duration and pore opening/widening time. J Solid State Electrochem 15:2427–2436
139. Poinern GEJ, Ali N, Fawcett D (2011) Progress in nano-engineered anodic aluminum oxide membrane development. Materials 4:487–526
140. Havriliak S, Negami S (1967) A complex plane representation of dielectric and mechanical relaxation processes in some polymers. Polymer 8:161–210

141. Cole KS, Cole RH (1941) Dispersion and absorption in dielectrics I. Alternating current characteristics. J Chem Phys 9:341
142. Davidson DW, Cole RH (1951) Dielectric relaxation in glycerol, propylene glycol, and n-propanol. J Chem Phys 19:1484
143. Vogel H (1921) The law of the relationship between viscosity of liquids and the temperature. Physik Z 22:645–646
144. Fulcher GS (1925) Analysis of recent measurements of the viscosity of glasses. J Am Ceram Soc 8:339–355
145. Tammann G, Hesse G (1926) Die Abhängigkeit der Viscosität von der Temperatur bei Unterkühlten Flüssigkeiten. Z Anorg Allg Chem 156:245–257
146. Floudas G, Reisinger T (1999) Pressure dependence of the local and global dynamics of polyisoprene. J Chem Phys 111:5201
147. Williams ML, Landel RF, Ferry JD (1955) The temperature dependence of relaxation mechanisms in amorphous polymers and other glass-forming liquids. J Am Chem Soc 77:3701–3707
148. Bello A, Laredo E, Grimau M (1999) Distribution of relaxation times from dielectric spectroscopy using Monte Carlo simulated annealing: application to α-PVDF. Phys Rev B 60:12764–12774
149. Schönhals A, Goering H, Schick C, Frick B, Zorn R (2004) Glass transition of polymers confined to nanoporous glasses. Colloid Polym Sci 282:882–891
150. Bahar I, Erman B, Kremer F, Fischer EW (1992) Segmental motions of cis-polyisoprene in the bulk state: interpretation of dielectric relaxation data. Macromolecules 25:816–825
151. Frank CW, Rao V, Despotopoulou MM, Pease RFW, Hinsberg WD, Miller RD et al (1996) Structure in thin and ultra-thin spin-cast polymer films. Science 247:912–915
152. Flory PJ (1953) Principles of polymer chemistry. Cornell University Press, New York
153. García-Turiel J, Jérôme B (2007) Solvent retention in thin polymer films studied by gas chromatography. Colloid Polym Sci 285:1617–1623

Rotational Diffusion of Guest Molecules Confined in Uni-directional Nanopores

Wycliffe K. Kipnusu, Ciprian Iacob, Malgorzata Jasiurkowska-Delaporte, Wilhelm Kossack, Joshua R. Sangoro and Friedrich Kremer

Abstract Broadband dielectric spectroscopy (BDS) is employed to study the rotational diffusion of Tris(2-ethylhexyl)phosphate (TEHP), a glass-former, and 4-heptyl-4′-isothiocyanatobiphenyl (7BT), a liquid crystal, both confined in nanoporous silica membranes having *uni-directional* pores with diameters in the range 4–10.4 nm. It is observed that upon cooling, the glassy dynamics (α-process) of TEHP is enhanced near the calorimetric glass transition. This confinement effect is attributed to a slight reduction in density of the liquid in the nanopores. The secondary β-relaxation in TEHP is however unaffected by the geometrical confinement. Silanization of the inner pore surfaces has no measurable effect on the mobility of the guest molecules. For the case of liquid crystal 7BT, two relaxation processes originating from librations about the molecule's short (δ-process) and long axes (β_{LC}-process) are observed. The former becomes suppressed with decreasing pore diameter, while the latter is nearly unaffected with a tendency to become faster with decreasing pore diameter, an effect caused by orientational ordering due to geometrical constraints.

W. K. Kipnusu (✉) · W. Kossack · F. Kremer
Institute of Experimental Physics I, University of Leipzig, Linnéstraße 5,
04103 Leipzig, Germany
e-mail: Kipnusu@physik.uni-leipzig.de

C. Iacob
Department of Materials Science and Engineering, Penn State University, University Park, PA 16802, USA

M. Jasiurkowska-Delaporte
NanoBioMedical Centre, Adam Mickiewicz University, Umultowska 85,
61-614 Poznan, Poland

J. R. Sangoro
Department of Chemical and Biomolecular Engineering, University of Tennessee,
Knoxville, TN 37996-1600, USA

F. Kremer (ed.), *Dynamics in Geometrical Confinement*, Advances in Dielectrics,
DOI: 10.1007/978-3-319-06100-9_5,

Keywords Uni-directional nanopores · Rotational diffusion · Electrochemical etching · 4-heptyl-4′-isothiocyanatobiphenyl

Abbreviations

BDS	Broadband dielectric spectroscopy
TEHP	Tris(2-ethylhexyl)phosphate
7BT	4-heptyl-4′-isothiocyanatobiphenyl
NMR	Nuclear magnetic resonance
DSC	Differential scanning calorimetry
MWS	Maxwell–Wagner–Sillars
LCs	Liquid crystals
SmE	Smectic E
HF	Hydrofluoric acid
Si	Silicon
pSi	Porous silicon
$pSiO_2$	Porous silica
SEM	Scanning electron micrograph
h	Hour(s)
s	Second(s)
HMDS	Hexamethyldisilazane
FTIR	Fourier transform infrared
HP	Hewlett Packard
HN	Havriliak–Negami
I	Isotropic
N	Nematic
S	Smetic
RTD	Relaxation time distribution
VFT	Vogel–Fulcher–Tammann
T	Temperature
T_g	Glass transition temperature
DFT	Density functional theory
NCS	Isothiocyanate
Hz	Hertz

1 Introduction

Porous materials have a large surface area to volume ratio. When guest molecules are imbibed in these materials, their rotational diffusion/dynamics is affected by the type of guest/host interactions in addition to the effects of geometrical restrictions. The type of probe molecules (simple glass formers, liquid crystals, polymers) and confining matrix are also key influencing factors. Several experimental techniques for instance: Rayleigh Light Scattering [1, 2], X-ray Diffraction, [3] Neutron Scattering, [4] Raman Scattering, [5] Nuclear Magnetic Resonance (NMR) Spectroscopy, [6–8]

Differential Scanning Calorimetry (DSC), [9, 10] Optical Kerr Effect Spectroscopy, [11, 12] and Broadband Dielectric Spectroscopy (BDS), [13–22] have been applied with an attempt to unravel the impact of confinement on the dynamics of molecules having different sizes and architectures restricted in nanopores. In most of these studies, porous glasses (vycor, sol–gel processed silica, silicates), molecular sieves, anopore membranes, nucleopores, ZrO_2, Zeolites, and vermiculite clay have been utilized as confining matrices [13, 15, 23–32]. However, these studies are hampered by the lack of definite microstructures of the nanoporous materials used. In recent studies, [33–35] the current authors circumvented this problem by utilizing nanoporous silica having uni-directional nanopores. This chapter contains details of the preparation of these nanopores and the rotational diffusion (studied by BDS) of two classes of materials: (i) a simple glass former; Tris(2-ethylhexy)phosphate (TEHP) and (ii) liquid crystal; 4-heptyl-4′-isothiocyanatobiphenyl (7BT) confined therein.

Owing to its ability to probe molecular fluctuations over a wide frequency and temperature range, BDS is an ideal tool to study the dynamics of guest molecules restricted in nanopores. However, for quantitative analysis, it requires well-defined geometry of the confining matrix. If the sample is contained in porous host systems having intersecting pores or random cavities of indefinite morphologies, (as is the case for most porous system discussed above) then the permittivity of the matrix and the guest molecules become non-additive quantities. This complicates the analysis of the dielectric data because one has to clarify how much of the dielectric mixing affects the apparent confinement-induced effects. Maxwell and Wagner solved this problem of how to get the net permittivity of a mixture of two dielectrics. The approach was further refined by Sillars and is now referred to as Maxwell–Wagner–Sillars (MWS) polarization. In this framework, the permittivity of fillers (guest molecules) assuming that their shapes can be adjusted in terms of polarization factor p, is given by: [36]

$$\varepsilon_c(\omega) = \varepsilon_m(\omega)\frac{p\varepsilon_f(\omega) + (1-p)\varepsilon_m(\omega) + \phi(1-p)(\varepsilon_f(\omega) - \varepsilon_m(\omega))}{p\varepsilon_f(\omega) + (1-p)\varepsilon_m(\omega) - \phi p(\varepsilon_f(\omega) - \varepsilon_m(\omega))} \qquad (1)$$

where ε_m and ε_f are the permittivities of the confining matrix, and filler/guest molecules respectively, ε_c is the effective permittivity of the confined /filler material, and ϕ is the volume filling fraction. Richert [36] showed that the results based on this model are very sensitive to variation of p, which generally defines the geometry of the confining space. Besides the MWS, several other models such as Maxwell Garnett formula, Hanai–Bruggeman equation, and the Rayleigh's approach fail because of the assumption of the topology and the limiting case of $\phi < 0.2$ [36, 37]. It is therefore imperative to use nanoporous host systems having well-defined geometries.

By utilizing such host systems with uni-directional nanopores, the current authors show that the glassy dynamics of confined TEHP increases with decreasing pores sizes when approaching the calorimetric class transition temperature, while the static dielectric relaxation strength remains relatively bulk-like. Under this confinement, the rotational diffusion is determined by the interplay between surface and confinement effects [38]. The former is observed for attractive interactions between guest molecules and the pore walls resulting in a decrease in the relaxation rate of the dynamic glass transition. Silanization of the pore surfaces removes this effect. Confinement effects are observed for repulsive interactions between guest and host systems and are manifested by an increase of the molecular mobility closer to T_g [34]. With decreasing pore diameter, confinement effects become more pronounced. Silanization has no or negligible influence in this case. Considering liquid crystals (LCs), finite pore sizes and surface effects induce nontrivial alterations of the bulk-like properties. For instance, one can observe the induction of thermodynamic phases, the shift of phase transition temperatures, as well as orientational and translational order. Typically, as the pore diameter decreases, the phase transition temperature is reduced with respect to the bulk one [27, 37, 39, 40]. Below a critical value comparable to the correlation length of the liquid crystalline order parameter, the discontinuous phase transitions are replaced by the gradual occurrence of orientational order [41–45].

TEHP is mainly used in extraction of heavy metal ions, [46] as plasticizers/antiplasticizers [47] and in ion-selective membrane electrodes [48, 49]. It can also act as flame retardant in lithium ion batteries and polymer gel electrolytes [50, 51]. 7BT can be used as a component of liquid crystalline mixtures suitable for twist-type displays [52, 53]. In bulk, it shows smectic E (SmE) phase with orthorhombic arrangement within the molecular layers in the temperature range 330–345 K [54–56]. The current development of device miniaturization requires knowledge of the properties of materials confined in nanometric length scales such as nanopores.

2 Preparation of Nanoporous Silica Membranes

Electrochemical partial dissolution of silicon in the presence of hydrofluoric acid (HF) presents an easy and cheap method to obtain porous silicon (pSi). The defects, which result from non-uniform native oxide layer on the surface of the silicon wafers, serve as initial spots for the etching process. Several mechanisms leading to pore formation have been proposed. The presence of holes (carrier charges) in the bulk Si plays a key role during this process [57]. The diffusion of holes against energetic barriers is responsible for the rate (self) limiting nature of this process [58]. They are driven to the tip of the etched pits by the applied electric field resulting in the anisotropic dissolution of Si. The hole depleted regions are passivated [57]. Lehmann and Gösele [59] proposed a generally accepted mechanism of pore formation (Scheme 1).

When immersed in an HF electrolyte, the Si surface is terminated by hydrogen atoms. Since the induced polarization of Si–H bonds is very low, dissolution of Si does not take place at this stage until a hole reaches the hydrogen saturated surface. At this step, fluoride ions displace hydrogen atoms leading to the formation of Si–F (step 1 and 2). The structure is now generally polarized and an injection of an electron into the Si substrate takes place. Another hydrogen atom is ejected resulting

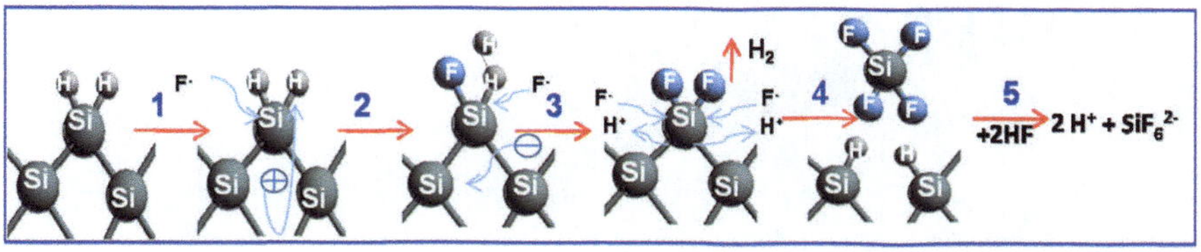

Scheme 1 Anodic dissolution mechanism of Si in the presence of HF electrolyte solution, associated with pSi formation according to Lehmann and Gösele [59]. The different steps in the reaction scheme are explained in the text

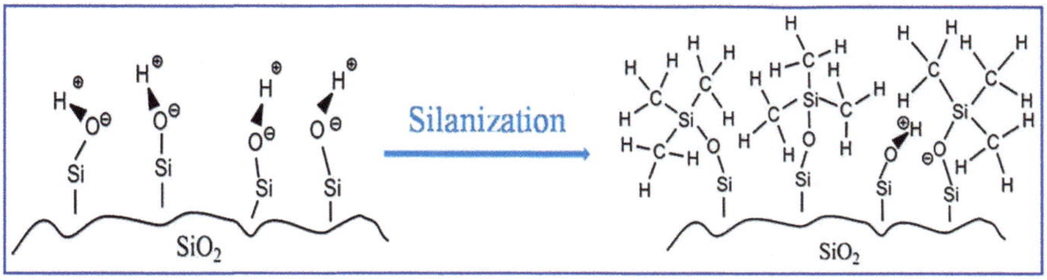

Scheme 2 Schematic outline of the difference between native (*left panel*) and silanized (*right panel*) of $pSiO_2$

in the release of hydrogen gas (step 3). A further attack on the two remaining Si–H by HF occurs in step 4 forming SiF_4, which spontaneously reacts with HF to form stable $H_2SiF_6{}^{2-}$(step 5). Zang [60] proposed an alternative model which involves competitive dissolution of Si through direct and indirect reaction paths. The former occurs as described above while the latter takes place via the chemical dissolution of the Si–O–Si formed in the presence of H_2O.

The desired pore sizes, morphology, porosity, and thickness of the pSi layer can be tuned by a careful selection of the resistivity (dopant and doping ratio) and crystallographic orientation of the Si wafers, and by controlling the HF concentration, current density, and etching time. Highly and moderately doped p-type Si wafers with crystal orientation of ⟨100⟩ are suitable for the formation of non-intersecting mesopores in the range of 2–10 nm. To achieve this, the current density and HF concentration must be tuned to the right levels. Pore formation takes place below a critical current density above which electro-polishing occurs. The details of the effects of these parameters are shown in Ref. [61].

The pSi membranes used in this work were prepared by electrochemical etching of highly doped ($\rho \leq 0.005\ \Omega$-cm) *p*-type ⟨100⟩ oriented mono-crystalline silicon wafers in a home built anodization cell. The electrolyte contained hydrofluoric acid (HF—48 %) and ethanol (C_2H_5OH—99 %) purchased from sigma Aldrich and mixed in the ratio of 1:1. Current densities (*j*) in the range of 20–120 mA cm^{-2} were applied to obtain pore diameters between 4 and 10 nm with porosity varying from 9 to 23 %. In each case, the etching time was adjusted to maintain the thickness of the pSi at around 50 μm. All these anodization parameters were controlled by computer

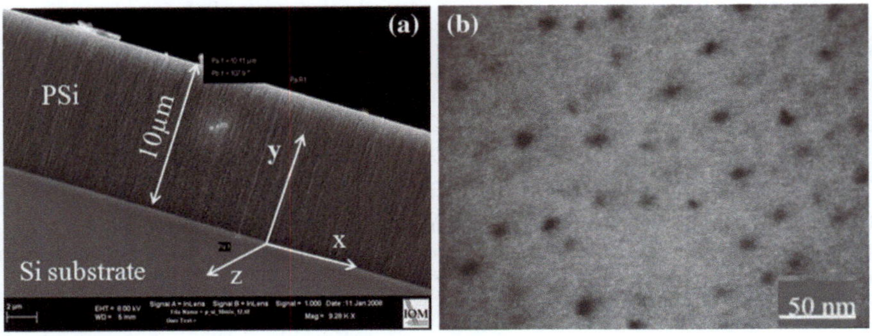

Fig. 1 **a** Scanning electron micrograph (SEM)-cross section of the edged silicon showing the nonintersecting nanopores. **b** SEM micrograph of the surface of a porous silica membrane

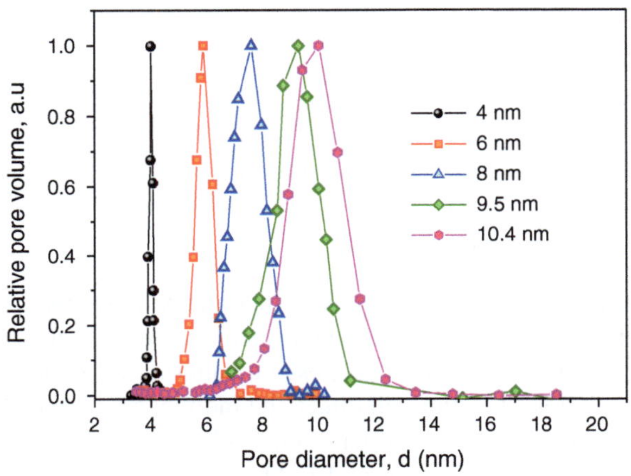

Fig. 2 The pore size distribution obtained by NMR cryoporometry. The mean pore diameters of different preparations are indicated

software. A final electro-polishing step ($j = 700\,\text{mA}\,\text{cm}^{-2}$) was applied for 3–4 s to lift the pSi from the Si substrate. The obtained pSi membranes were subsequently oxidized thermally in an oven (Vulcan 3–550, Neytech) at 1,100 K (heating rate 3 K/min) for 6 h to form completely transparent and insulating nanoporous silica ($pSiO_2$) with uni-directional pores. Scanning electron micrograph (SEM) image of the cross section of pSi (Fig. 1a) reveals non-intersecting pores. The SEM image of the top surface of $pSiO_2$ is shown in Fig. 1b. Pore size distribution as revealed by nuclear magnetic resonance (NMR) cryoporometry is depicted in Fig. 2. A detailed study of pore morphology and characterization is described elsewhere [62].

The surface of the prepared $pSiO_2$ was further modified by silanization. In this process, $pSiO_2$ were activated by annealing at 573 K in high vacuum (10^{-6} mbar). The details of the vacuum chamber are described in Sect. 3 of this chapter. Hexamethyldisilazane (HMDS), (purity $\geq 99.0\,\%$, purchased from Sigma-Aldrich)

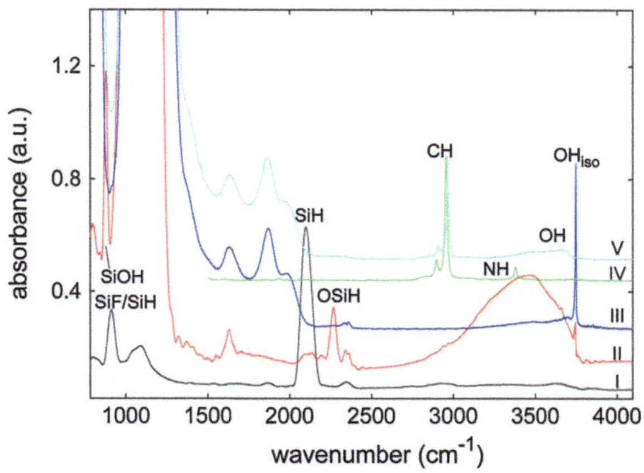

Fig. 3 FTIR spectra of (*I*) as prepared nanoporous silicon (pSi), (*II*) partially oxidized pSi, (*III*) fully oxidized pSi ($pSiO_2$), (*IV*) Hexamethyldisiloxane (HMDS), and (*V*) silanized $pSiO_2$

was injected into the vacuum chamber containing $pSiO_2$ at 350 K. Under these conditions, HMDS reacts with the silanol groups leaving trimethylsilyl as terminal groups on the silica surface (see Scheme 2). The chamber was then evacuated after 3 h at 350 K for 6 h to remove the unreacted HMDS. Fourier Transform infrared (FTIR) spectroscopy was used to characterize the as-prepared pSi, the partially and fully oxidized pSi ($pSiO_2$), and the surface modified $pSiO_2$ (Fig. 3). The freshly etched pSi (spectrum I) shows a strong peak around 2,200 cm^{-1} assigned to vibrations of Si–H bond and peaks at about 900 cm^{-1} are attributed to either Si–F or Si–H stretching vibrations. This proves that basically the surface of freshly prepared pSi is hydride terminated. Upon oxidation, Si–H is slowly replaced by stable Si–OH bonds. This is depicted in partially oxidized pSi (spectrum II), which shows diminishing intensity of Si–H and a growth of a broad peak at around 3,300 cm^{-1} due to H-bonds of adsorbed H_2O and another narrow peak at about 3,700 cm^{-1} assigned to isolated OH. The intensity of this band remarkably increases when the membrane is fully oxidized (spectrum III). This is reversed when the membrane is treated with HDMS during silanization. The FTIR of HDMS and fully silanized $pSiO_2$ membrane are shown in spectra (IV) and (V), respectively, proving the exchange of silanol with methyl groups. The C–H stretching bands in the region 2,800–3,000 cm^{-1} verify the presence of methyl groups.

3 Infiltration of the Guest Molecules into the Nanoporous Silica

Before filling the membranes with the material under study, they were annealed in a high vacuum chamber (10^{-6} mbar) at 573 K for 24 h to remove water and other volatile impurities. This was done in a home-built vacuum chamber (Fig. 4), which is evacuated in two steps. The first step involves a membrane pump that brings the pressure in the chamber down to $\sim 10^{-3}$ mbar before a turbo-molecular system pump is engaged in the second step to achieve a vacuum better than 10^{-5} mbar. In the

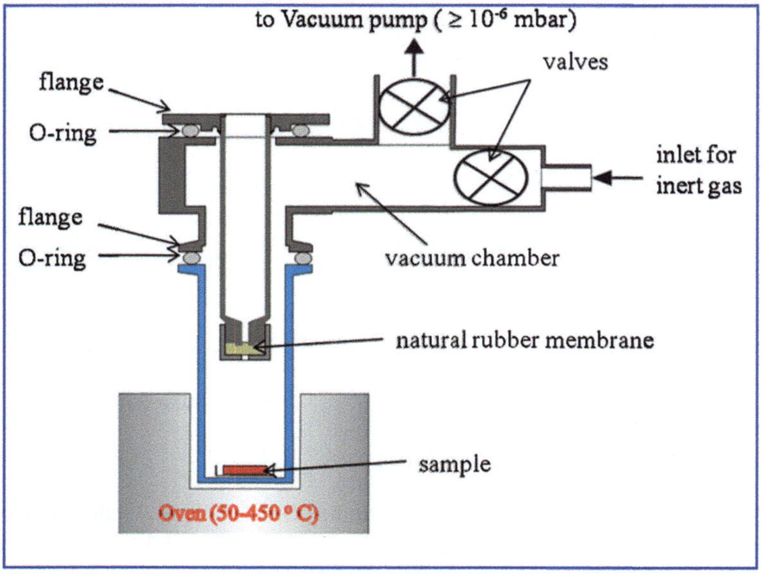

Fig. 4 A custom-made vacuum annealing chamber that is evacuated by a two-step pumping system that employs a turbo molecular pump and an oil-free membrane pump. The sample is placed inside a fused silica glass tube which is heated by an oven equipped with a temperature controller

process, nanoporous silica samples contained in fused silica glass tube were heated to 573 K by radiation using an oven equipped with a temperature controller. After 24 h, the temperature was then decreased to 300 K before the probe sample (liquid) was injected into the closed vacuum where the pores were filled by capillary wetting. The natural rubber (which is used only once) ensures that the vacuum chamber remains airtight even after the injection of the sample. A similar procedure was used during silanization with the exception that the chamber was evacuated to get rid of the ungrafted HDMS (as described above) before injecting the probe sample. By this method, the samples (pSi and the infiltrated material) are not exposed to ambient air nor to oil vapor. For the case of the liquid crystal (7BT), the powdered sample was placed on top of the annealed membranes in Argon atmospheres and then returned in the vacuum chamber where it was kept at the temperature corresponding to its isotropic phase (363 K).

3.1 Broadband Dielectric Spectroscopy

A thin aluminum foil (0.8 μm) was placed on both sides of the membranes to improve the electrical contacts. The samples were then placed between two platinum electrodes and measured by BDS within the temperature (120–353 K), which was controlled by a jet of pure nitrogen with stability better than 0.1 K using a Quatro controller. The frequency range of (10^{-3}–10^{7}) and (10^{7}–10^{9}) Hz was spanned using a Novocontrol high-resolution alpha analyzer and Hewlett Packard (HP) 4291A impedance analyzer respectively. In order to describe the dielectric spectra quantitatively, a superposition of model functions according to Havriliak–Negami (HN) and a conductivity contribution Eq. (2) was used to fit the isothermal dielectric loss data.

$$\varepsilon^*_{\mathrm{HN}}(\omega) = \varepsilon'(\omega) - \mathrm{i}\varepsilon''(\omega) = \varepsilon_\infty + \frac{\Delta\varepsilon}{\left[1 + (\mathrm{i}\omega\tau_{\mathrm{HN}})^\beta\right]^\gamma} + \frac{\sigma_0}{\mathrm{i}\omega\varepsilon_0} \tag{2}$$

where ε' and ε'' are real and imaginary (dielectric loss) parts of the complex dielectric function, $\Delta\varepsilon$ is the dielectric relaxation strength, τ_{HN} is the characteristic time constant that is related to relaxation time at the maximum loss ($\tau_{\max}$) by Eq. (3):

$$\tau_{\max} = \tau_{\mathrm{HN}} \left[\sin \frac{\pi\,\beta\gamma}{2 + 2\gamma}\right]^{1/\beta} \left[\sin \frac{\pi\,\beta}{2 + 2\gamma}\right]^{-1/\beta} \tag{3}$$

σ_0 is the dc-conductivity, ε_0 the permittivity in free space, β and $\gamma(0 < \beta, \beta\gamma \leq 1)$ represent symmetric and asymmetric broadening of the loss peaks respectively [37]. This approach implies a power law equal to β and $-\beta\gamma$ in the low and high frequency limits of the loss peak respectively. At low frequencies ($f < f_{\max}$), increase in the dielectric loss spectra with decreasing frequency is mainly due to a dc-conductivity contribution [63]. A superposition of Debye-functions with different relaxation times τ (Eq. 4) is considered to be a formal description of the broadened relaxation processes,

$$\varepsilon^*(\omega) = \varepsilon_\infty + \Delta\varepsilon \int \frac{L(\tau)}{1 + \mathrm{i}\omega\tau} \mathrm{d}\ln\tau \tag{4}$$

where the relaxation time distribution $L(\tau)$ fulfills the normalization condition shown in Eq. (5) below.

$$\int L(\tau)\mathrm{d}\ln\tau = 1 \tag{5}$$

The macroscopic dielectric response of LCs depends on the microscopic orientation of molecules in different phases. LCs have different thermotropic phases classified according to their degree of order. The isotropic (I) phase is found at higher temperatures followed by nematic (N) phase where the molecules are randomly positioned and oriented [64]. In this phase, the molecules have an average orientation along a common director denoted by a unit vector **n**. Upon further reduction in temperature, a transition to smectic (S) phases occurs where higher degree of order is found. In this phase, the mean orientation of molecules can be parallel (smetic A), or tilted (smectic C) with respect to the direction of the layer normal [37]. Smetic phases (B, E, F...) of highly organized molecules within a layer are also possible [37].

The electric permittivity of ordered LC is described by a second rank tensor $\varepsilon_{\alpha\beta}$ where the electric susceptibility is defined as [64]:

$$\chi_{\alpha\beta}^{\text{electric}} = \frac{p_\alpha}{\varepsilon_0 E_\beta} = \frac{P_\alpha}{\varepsilon_0(\varepsilon_{\alpha\beta} - \delta_{\alpha\beta})} \tag{6}$$

where *P* is the induced polarization and *E* the applied electric field. The symmetry of the LC phase determines the number of components of the permittivity tensor. According to Kramers–Kronig relation, both the real and imaginary parts of the permittivity are frequency dependent and can be expressed as [64]:

$$\varepsilon_i'(\omega) = \varepsilon_i'(\infty) + \frac{2}{\pi}\int_0^\infty \frac{\omega\varepsilon_i'(\omega)}{(\omega^2 - \omega_0^2)}\mathrm{d}\omega \tag{7}$$

$$\varepsilon_i''(\omega) = \frac{2\omega_0}{\pi}\int_0^\infty \frac{\omega\varepsilon_i''(\omega)}{(\omega^2 - \omega_0^2)}\mathrm{d}\omega \tag{8}$$

In LCs, Eqs. (7) and (8) are also tensor quantities. For uniaxial nematic phases, this tensor has components parallel and perpendicular to the nematic director **n** whose dielectric function $\varepsilon_{\parallel}^*(\omega)$ and $\varepsilon_{\perp}^*(\omega)$ respectively consists of weighted sum of the underlying relaxation modes.

$$\varepsilon_{\parallel}^*(\omega) = \varepsilon_{\infty,\parallel} + \frac{G}{3k_\mathrm{B}T}\left[(1+2S)\mu_l^2 C_{\parallel}^l(\omega) + (1+S)\mu_t^2 C_{\parallel}^t(\omega)\right] \tag{9}$$

$$\varepsilon_{\perp}^*(\omega) = \varepsilon_{\infty,\perp} + \frac{G}{3k_\mathrm{B}T}\left[(1-S)\mu_l^2 C_{\perp}^l(\omega) + (1+\frac{S}{2})\mu_t^2 C_{\perp}^t(\omega)\right] \tag{10}$$

where $\varepsilon_{\infty,\parallel}$ and $\varepsilon_{\infty,\perp}$ are the limiting high frequency permittivities parallel and perpendicular to the local director, *G* is a constant, and *S* is the order parameter. $C_j^i(\omega)(i = l, t; j = \parallel, \perp)$ denote the one-sided Fourier transformations of the correlation functions of the longitudinal and transverse components of the dipole moment vector of the mesogenic unit projected parallel and perpendicular to **n**.

The relaxation mode, which determines $\varepsilon_{\parallel}^*(\omega)$ is due to rotational fluctuations of the molecule around its short axis. It occurs at lower frequencies and is called δ-process. There are three other relaxation modes, which involve librations of the molecule about its long axis. These modes have nearly identical relaxation time and form a broad β_{LC}-relaxation process at higher frequencies that is related to $\varepsilon_{\perp}^*(\omega)$ [37]. The dielectric strength of the δ process relative to that of the tumbling mode depends on the order parameter *S*.

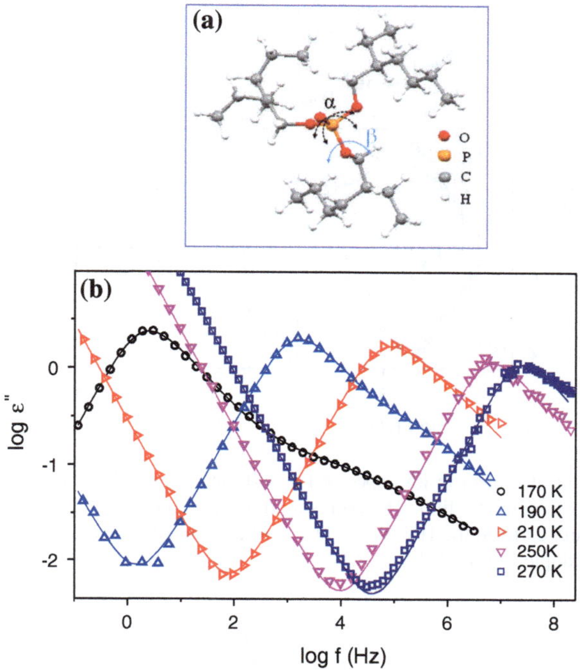

Fig. 5 **a** Molecular structure of TEHP with the dielectric relaxations (α and β) indicated. **b** Dielectric loss of bulk TEHP at selected temperatures in the frequency range of 10^{-1}–10^{9} Hz. *Lines* represent fits by the empirical Havriliak–Negami (HN) function. The *error bars* are smaller than the *symbols* if not explicitly stated otherwise

3.2 Results and Discussion

3.2.1 Rotational Diffusion of TEHP

A scheme of the molecular structure of TEHP and its raw dielectric spectra (imaginary part) in the bulk state for a few selected temperatures are shown in Fig. 5a, b respectively. The polar unit of P=O and P–O–C moieties at the core are surrounded by non-polar aliphatic chains. The dielectric loss spectra of molecules under confinement (Fig. 6) were corrected to eliminate the contribution from the silica membrane as discussed later. Intensity indicated by relaxation time distribution (RTD) shown in Fig. 6d–f reduces with decreasing pore sizes. The RTD of confined molecules shows negligible broadening compared to the bulk at higher temperature (Fig. 6f) which increase with decreasing temperature (Fig. 6d, e).The peak position shifts to higher frequencies, which implies a faster α-relaxation with decreasing pore diameters.

The relaxation rate ω_α for bulk TEHP is well described by the empirical Vogel–Fulcher–Tammann (VFT) equation given as:

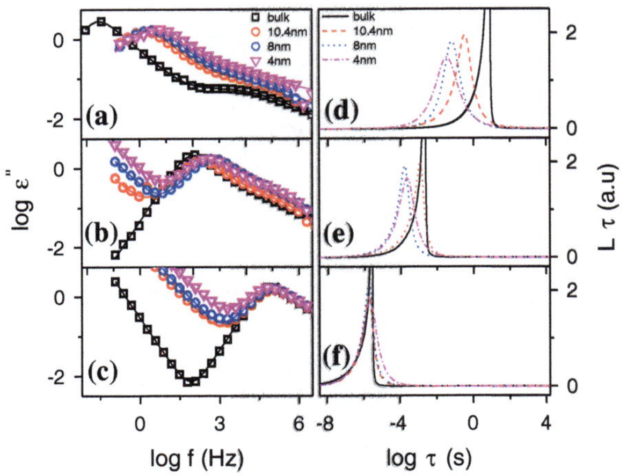

Fig. 6 Comparison of dielectric loss spectra for bulk and confined TEHP at **a** 162, **b** 180 and **c** 210 K (The *lines* denote HN fits). The corresponding relaxation time distribution (RTD) of the α-process as calculated from the HN fits for bulk and confined TEHP are shown in (**d–f**), (**e**) and (**f**) respectively. Adopted from [33]

$$\omega_\alpha(T) = \omega_\infty \exp\left(\frac{-DT_0}{T - T_0}\right) \tag{11}$$

where $\omega_\infty = \tau_\infty^{-1}$, the high temperature limit of the relaxation rate; D is a constant related to fragility—a measure of the extent of deviation from Arrhenius dependence and T_0 is the Vogel temperature. In contrast, TEHP contained in nanopores exhibits a confinement effect, i.e., when approaching the glass transition temperature the relaxation rate becomes faster than in the bulk. This increase is—counterintuitively—highest for the smallest pores (Fig. 7). Donth [65] explained this scenario in terms of hindered dynamic glass transition, which is implicitly based on Adams and Gibbs free volume theory. This model postulates that the glassy dynamics in pores becomes faster if its growing characteristic length-scale exceeds the pore sizes. However, this explanation is premised on the assumption that the temperature dependencies of ω_α in bulk and confined molecules are identical. The possible surface effects are also not considered. A refined possibility to analyze the observed change in the type of thermal activation of ω_α, shown in Fig. 7, is to determine the difference quotient of the *measured* relaxation rate with respect to 1/T and to compare it with the *calculated* derivative Eq. (12)

$$\mathrm{d}\left(\frac{\log \omega_\alpha}{T^{-1}}\right) = -DT_0\left(1 - \frac{T_0}{T}\right)^2 \log e \tag{12}$$

For the VFT dependence, a plot of $(-\mathrm{d}\log \omega_\alpha/\mathrm{d}T^{-1})^{-1/2}$ versus $(1/T)$ results in a straight line [37]. This method is sensitive to the functional form of ω_α(T) irrespective of the pre-factor. Pronounced deviations are observed, which are attributed to the

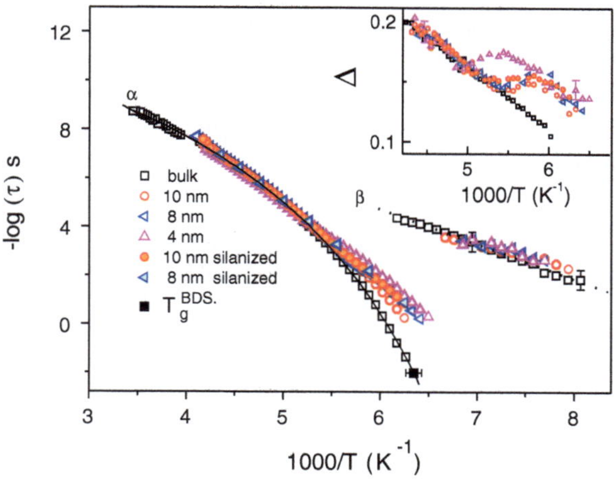

Fig. 7 Mean relaxation rates versus inverse temperature for bulk and confined TEHP showing α- and β-processes. *Open square symbols* represent bulk data; *open circles* and *triangles* denote relaxation of TEHP confined in silanized silica membranes. The *solid line* is a fit by the Vogel–Fulcher–Tammann (VFT) equation to the bulk data (α-process). The fit parameters are: $\omega_\infty = 2.53 \times 10^{12}\ \mathrm{s}^{-1}$, $D = 13.31$, and $T_0 = 112\,\mathrm{K}$. The *dotted line* is a fit by the Arrhenius equation (AE) to the bulk data. The fit parameter ω_∞, and activation energy E_a obtained from AE are: $\omega_\infty = 1.85 \times 10^{13}\ \mathrm{s}^{-1}$ and 27.5 kJ/mol respectively. *Inset* shows derivative plots for the α-relaxation rates versus inverse temperature where $\Delta = (-\mathrm{d}\log\omega_\alpha/\mathrm{d}T^{-1})^{-1/2}$. The *error bars* are indicated if not smaller than the *symbols*. Adopted from [33]

interplay between intermolecular and host/ guest interactions. The derivative plots (inset in Fig. 7) show a peculiar temperature dependence. The confinement effect sets in for 4 nm at the relatively highest temperature. This is attributed to changes in packing density. Slow molecular relaxations are sensitive to small changes in density [66]. From computer simulations, density of fluids decreases by 2–5 % when confined in nanopores [67, 68]. Kawasaki [69] and Spiess [70] separately showed that when mass density/or packing density changes by merely 2 %, the relaxation time spans a range of 8 orders of magnitude. On the other hand the observed temperature dependencies for 4, 8, and 10.4 nm show that the changes in molecular dynamics are more intricate. At higher temperatures, the dynamic glass transition of bulk and confined TEHP has identical thermal activation. Conversely, at lower temperatures, a shift to second VFT dependence occurs (for confined molecules) at characteristic temperatures related to the pore diameters. Certainly, the intramolecular mobility of the guest molecules besides the counterbalance between surface and confinement effects has to be considered. In order to investigate the effects of the guest/host interactions, similar measurements were done using silica pores that were coated with a hydrophobic silane layer. We found that closer to T_g, the rotational diffusion of TEHP confined in silanized pores is slightly faster compared to uncoated pores. A realistic analysis would require atomistic simulations to unravel the details of guest/host interactions, which is at the moment not possible.

The rates of the secondary β-relaxation of confined molecules coincide with that of the bulk within the limits of experimental accuracy (see Fig. 7). This is because the β-process is assigned to librational fluctuations of the C–O moiety close to the molecular core of TEHP. The relaxation rate ω_β exhibits an Arrhenius-type of thermal activation described by

$$\log \omega_\beta(T) = \log \omega_\infty - \left(\frac{E_\mathrm{a}}{k_\mathrm{B}T}\right) \log e \tag{13}$$

where E_a is the activation energy, ω_∞ is the relaxation rate in the high temperature limit. We obtain an activation energy $E_\mathrm{a} = 27.5$ KJ/mol which is similar to the values obtained for other low molecular weight glass forming liquids [37].

To analyze the effective dielectric strength ($\Delta\varepsilon_\alpha$) for confined molecules, a model of two impedances in parallel is employed: one due to the guest molecules and the other due to the confining matrix (silica) for the case where there is no interfacial relaxation. Equations (14) and (15) give the measured real and imaginary dielectric permittivities:

$$\varepsilon'_\mathrm{m} = \phi\varepsilon'_\mathrm{c} + (1-\phi)\varepsilon'_{\mathrm{SiO_2}} \tag{14}$$

$$\varepsilon''_\mathrm{m} = \phi\varepsilon''_\mathrm{c} + (1-\phi)\varepsilon''_{\mathrm{SiO_2}} \tag{15}$$

where ε'_c, $\varepsilon'_{\mathrm{SiO_2}}$, ε''_c, $\varepsilon''_{\mathrm{SiO_2}}$ are real and imaginary permittivities of confined molecules and silica matrix, respectively, and ϕ is the porosity. Considering Eq. (15) the dielectric loss for confined molecules becomes

$$\varepsilon''_\mathrm{c} = \frac{\varepsilon''_\mathrm{m}}{\phi} - \left(\frac{1-\phi}{\phi}\right)\varepsilon''_{\mathrm{SiO_2}} \tag{16}$$

From Kramers/Kronig relations, dielectric strength $\Delta\varepsilon_\mathrm{c}$ is obtained by integrating Eq. (16). In Fig. 8, the measured dielectric relaxation strength scaled with porosity using Eq. (16) and the calculated dielectric loss assuming for the confined liquid *bulk* properties are compared to the measured values in different pore sizes. Within experimental uncertainty *no* differences are observed proving that the confined liquid behaves bulk-like as measured by $\Delta\varepsilon$, being proportional to the number density of the interacting dipoles. $\Delta\varepsilon$ is measured only with an accuracy of $\pm 5\,\%$, hence the subtle assumed decrease in the density on $\sim 0.1\,\%$ scale under conditions of geometrical confinement cannot be resolved by this quantity. As discussed above, the structural relaxation time, which is more sensitive to changes in density, shows deviations from bulk closer to T_g.

3.2.2 Rotational Diffusion of 7BT

Liquid crystals exhibit various mesophases with phase transitions that are sensitive to perturbations. Consequently, they are ideal candidates to study rotational diffusion

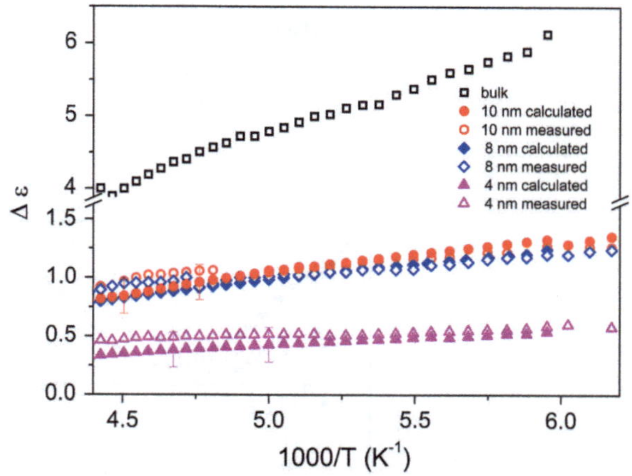

Fig. 8 Dielectric relaxation strength for bulk and confined TEHP. *Open symbols* represent measured data while *filled symbols* are calculated $\Delta\varepsilon$ for confined TEHP assuming bulk-like properties of $\varepsilon''(\omega)$. The porosities for membranes with pore diameters of 4, 8 and 10 nm are 0.09, 0.2, and 0.23 respectively. Adopted from [33]

when confined in porous media and to quantify the surface and confinement effects. The structure of LC, 7BT is shown in Fig. 9a. The dielectric loss (ε'') spectra of 7BT in bulk and in nanopores of mean diameter 7.5 nm at selected temperatures are depicted in Fig. 9b–d.

The dielectric properties of bulk 7BT in the smectic-E (SmE) and isotropic phases have been studied before [71] and they are dominated by contributions from one polar group, i.e. isothiocyanate (NCS) with dipole moment of $\mu = 3.5$ D oriented roughly along the para-axis of the benzene ring [56]. In the SmE phase of bulk 7BT, a single Debye-type relaxation process around the short axis (δ-relaxation) [72] is detected in the low frequency regime. In nanopores, two relaxation processes are observed at high temperatures (see Fig. 9). One process has the same origin as that of the bulk response and corresponds to the flip-flop motions of molecules around their short axis (Fig. 9). The second process is attributed to librational motions (β_{LC}-relaxation) of molecules [27, 73]. In the case of polar molecules an interfacial re-order is induced by the pore walls [27]. The existence of polar layers has been confirmed by Nuclear Magnetic Resonance (NMR) [32, 74, 75] and neutron experiments [44, 76]. Due to the symmetry of 7BT, the molecular jumps around the short axis are more probable if the applied electric field is parallel to the molecular axis. The β_{LC}-process is detected under geometrical confinement if the external electric field is parallel to the pore walls and molecular dipole moments are inclined to the surface. With decreasing temperature both processes merge and the resultant relaxation rate exhibits an Arrhenius-type of thermal activation. On subsequent heating, only the β_{LC}-relaxation process is detected. The influence of the pore size on the dielectric response is shown in Fig. 10. It is evident that as the pore size decreases, thermal hysteresis becomes less pronounced. In the pores of mean diameter 4 nm, the δ-relaxation does not show up and hysteresis is not observed.

The activation plots of the various relaxation processes observed for 7BT in bulk and nanopores of different diameters are shown in Fig. 11. The observed enhancement of molecular mobility around the short axis (7BT) compared to the bulk is typical for

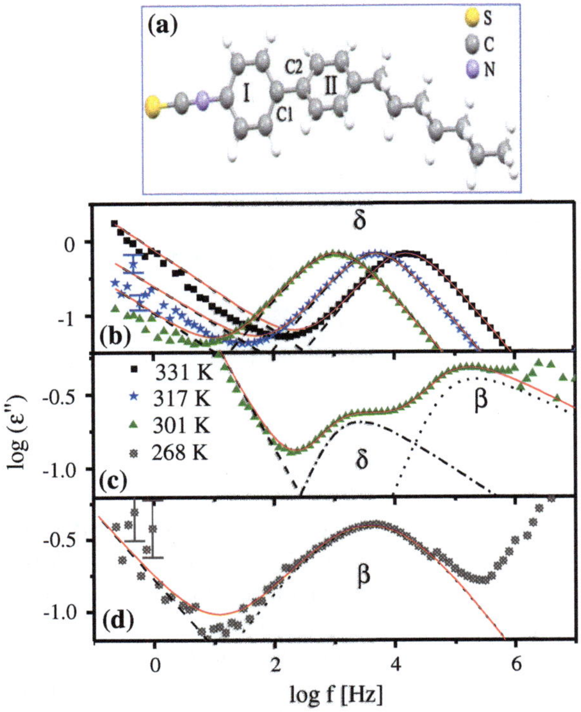

Fig. 9 **a** Scheme of the structure of 4-heptan-4′-isothiocyanatobiphenyl (7BT). Geometry of the lowest energy conformation of the molecule optimized by a quantum mechanical method (DFT at the B3LYP/6-31 level). Dielectric loss of 7BT at different temperatures as indicated in the SmE phase for bulk (**b**) and confined in the pores of diameter 7.5 nm (**c**) and (**d**). The *solid lines* are a fit of a sum of two Havriliak–Negami functions to the experimental data. The *dashed-dotted* and *dotted lines* indicate δ and β_{LC} relaxations, respectively. The *dashed line* corresponds to conductivity contribution. Adopted from [35]

confined LCs [27–31, 37, 73, 77, 78]. It indicates that the molecules in nanopores have more free volume. The temperature dependence of the δ-relaxation rates is Arrhenius-like for bulk as well as confined molecules. The values of activation energy of this process are slightly higher for the sample confined in pores compared to the bulk of the SmE phase, which is similar to previous observations [27]. The second process is attributed to a librational motion around the long axis of the LC molecules and appears at relaxation rates faster by two decades than the δ-relaxation.

Based on this information, we sketched the morphology of the molecules in pores (see Fig. 12). At the center of pores, the preferred direction of molecular orientation is parallel to the pore walls, and the δ-relaxation is observed. On decreasing temperature, a layer of partially immobilized molecules develops on pore walls. The molecules inclined to the pore walls liberate resulting in the observed β_{LC}-relaxation process. The interfacial layer increases progressively and becomes comparable to the pore radius at lower temperatures [37]. As the pore size decreases the surface effect

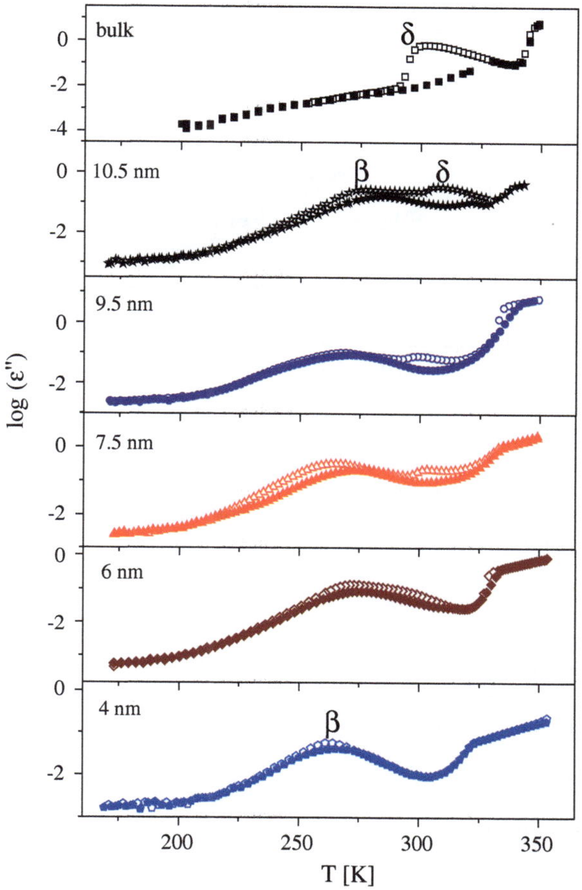

Fig. 10 Dielectric loss measured at 1.218 kHz as a function of temperature on cooling (*empty symbols*) and subsequent heating (*solid symbols*) of the sample in the bulk and in pores of mean diameters of 4, 6, 7.5, 9.5, and 10.5 nm. Adapted from [35]

becomes more pronounced, consequently the δ-relaxation is suppressed in smaller pores. Quantitative information regarding the density distribution of the molecules in nanopores is not directly accessible from the experiments described in the current work.

4 Conclusions

In this study, we used porous silica having uni-directional nanopores in the range 4–10.5 nm to investigate rotational diffusion of a glass forming liquid (TEHP) and a liquid crystal (7BT). By use of these nanopores, quantitative information regarding

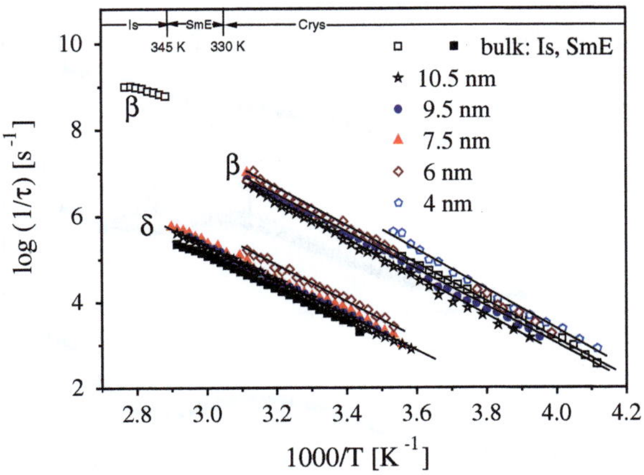

Fig. 11 Activation plot of 7BT upon cooling in bulk and confined in pores of mean diameters 4–10.5 nm as indicated. The activation energies for the δ process is 73 kJ mol^{-1} in the bulk and 76 kJ mol^{-1} in pores. For the β_{LC} process, it is 30 kJ mol^{-1} in bulk and 80 kJ mol^{-1} in pores. The phase transition temperatures are indicated. Adapted from [35]

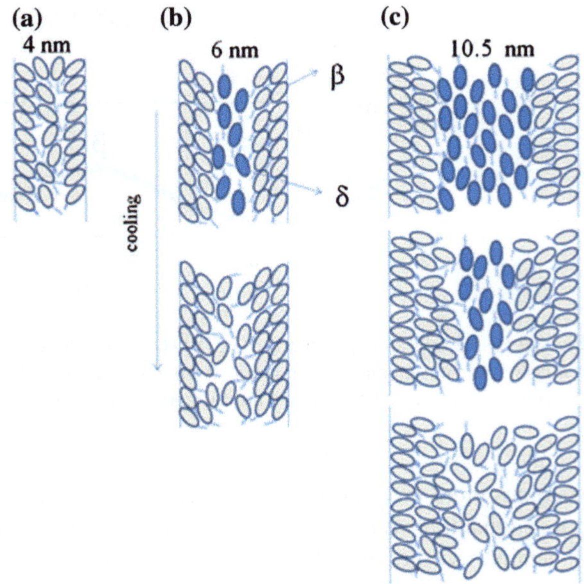

Fig. 12 Scheme of the molecular orientational order distribution of 7BT in nanopores of varying diameter. **a** The molecules are partially immobilized and only librational fluctuations take place. **b** and **c** Upon cooling, a layer of ordered LC molecules is formed at the pore walls, the number of which increases with decreasing temperature. Adapted from [35]

the surface and confinement effects on the rotational diffusion of guest molecules is reliably obtained. From the ω_α(T), $\Delta\varepsilon$(T) and RTD, of TEHP confined silica nanopores, we conclude that these molecules do not strongly interact with the pore walls maybe because of steric hindrance by nonpolar terminal groups. The mean α-relaxation rates of TEHP increases with decreasing pore sizes as the glass transition temperature is approached. This is attributed to a slight decrease in density caused by confinement. The study of liquid crystalline 7BT under similar confinement shows remarkable difference in rotation diffusion compared to the bulk molecules. The smetic E phase observed in bulk is replaced by a molecular order imposed by the interaction of the molecules with the pore walls. In nanopores of mean diameter 10–6 nm, two relaxation processes are observed in low frequency regime, which are assigned to the rotation of molecules around their short axis (δ-relaxation) and the libration of molecules close to the pore walls (β_{LC}-relaxation). Decreasing the pore diameters further to 4 nm leads to the suppression of the δ-relaxation. This shows the intricate counterbalance between surface and confinement effects for different guest molecules under conditions of geometrical confinement.

Acknowledgments Financial support by the DFG (Germany), within IRTG "Diffusion in Porous Materials," SFB/TRR 102 within the project "Polymers Under Multiple Constrains," Alexander von Humboldt Foundation and Leipzig School of Natural Sciences, "Building with Molecules and Nano-Objects" (BuildMoNa) is gratefully acknowledged.

References

1. Crupi V, Magazù S, Majolino D, Maisano G, Migliardo P (1999) Dynamical response and H-bond effects in confined liquid water. J Mol Liq 80(2–3):133–147
2. Patkowski A, Ruths T, Fischer EW (2003) Dynamics of supercooled liquids confined to the pores of sol–gel glass: a dynamic light scattering study. Phys Rev E 67(2):021501
3. Morishige K, Nobuoka K (1997) X-ray diffraction studies of freezing and melting of water confined in a mesoporous adsorbent (MCM-41). J Chem Phys 107(17):6965–6969
4. Crupi V, Majolino D, Migliardo P, Venuti V (2002) Neutron scattering study and dynamic properties of hydrogen-bonded liquids in mesoscopic confinement. 1. The water case. J Phys Chem B 106(42):10884–10894. doi:10.1021/jp020503m
5. Luo R-S, Jonas J (2001) Raman scattering study of liquid ethylene glycol confined to nanoporous silica glasses. J Raman Spectrosc 32(11):975–978. doi:10.1002/jrs.786
6. Stapf S, Kimmich R, Seitter RO (1995) Proton and deuteron field-cycling NMR relaxometry of liquids in porous glasses: evidence for Lévy–Walk statistics. Phys Rev Lett 75(15):2855–2858
7. Brandani S, Ruthven DM, Kärger J (1995) Concentration dependence of self-diffusivity of methanol in NaX zeolite crystals. Zeolites 15(6):494–495
8. Buntkowsky G, Breitzke H, Adamczyk A, Roelofs F, Emmler T, Gedat E, Grunberg B, Xu Y, Limbach H-H, Shenderovich I, Vyalikh A, Findenegg G (2007) Structural and dynamical properties of guest molecules confined in mesoporous silica materials revealed by NMR. Phys Chem Chem Phys 9(35):4843–4853. doi:10.1039/b707322d
9. Li Y, Ishida H (2002) A differential scanning calorimetry study of the assembly of hexadecylamine molecules in the nanoscale confined space of silicate galleries. Chem Mater 14(3):1398–1404. doi:10.1021/cm0103747
10. Elamin K, Jansson H, Kittaka S, Swenson J (2013) Different behavior of water in confined solutions of high and low solute concentrations. Phys Chem Chem Phys 15(42):18437–18444. doi:10.1039/c3cp51786a
11. Farrer RA, Fourkas JT (2003) Orientational dynamics of liquids confined in nanoporous Sol-gel glasses studied by optical kerr effect spectroscopy. Acc Chem Res 36(8):605–612. doi:10.1021/ar0200302
12. Loughnane BJ, Scodinu A, Fourkas JT (1999) Extremely slow dynamics of a weakly wetting liquid at a solid/liquid interface: CS2 confined in nanoporous glasses. J Phys Chem B 103(29):6061–6068. doi:10.1021/jp991176u

13. Arndt M, Stannarius R, Gorbatschow W, Kremer F (1996) Dielectric investigations of the dynamic glass transition in nanopores. Phys Rev E: Stat Phys Plasmas Fluids Relat Interdisc Top 54(5):5377–5390
14. Arndt M, Stannarius R, Groothues H, Hempel E, Kremer F (1997) Length scale of cooperativity in the dynamic glass transition. Phys Rev Lett 79(11):2077–2080
15. Brás AR, Dionísio M, Schönhals, A (2008) Confinement and surface effects on the molecular dynamics of a nematic mixture investigated by dielectric relaxation spectroscopy. J Phys Chem B 112(28):8227–8235. doi:10.1021/jp802133e
16. Brás AR, Frunza S, Guerreiro L, Fonseca IM, Corma A, Frunza L, Dionísio M, Schönhals A (2010) Molecular mobility of nematic E7 confined to molecular sieves with a low filling degree. J Chem Phys 132(22):224508
17. Schönhals A, Goering H, Schick C, Frick B, Zorn R (2003) Glassy dynamics of polymers confined to nanoporous glasses revealed by relaxational and scattering experiments. Eur Phys J E 12(1):173–178. doi:10.1140/epje/i2003-10036-4
18. Schönhals A, Goering H, Schick C, Frick B, Zorn R (2004) Glass transition of polymers confined to nanoporous glasses. Colloid Polym Sci 282(8):882–891. doi:10.1007/s00396-004-1106-3
19. Schönhals A, Goering H, Schick C, Frick B, Zorn R (2005) Polymers in nanoconfinement: What can be learned from relaxation and scattering experiments? J Non-Cryst Solids 351(33,36):2668–2677
20. Kremer F (2002) Dielectric spectroscopy, yesterday, today and tomorrow. J Non-Cryst Solids 305(1,3):1–9
21. Frunza L, Kosslick H, Pitsch I, Frunza S, Schönhals A (2005) Rotational fluctuations of water inside the nanopores of SBA-type molecular sieves. J Phys Chem B 109(18):9154–9159. doi:10.1021/jp044503t
22. Sinha GP, Aliev FM (1998) Dielectric spectroscopy of liquid crystals in smectic, nematic, and isotropic phases confined in random porous media. Phys Rev E 58(2):2001–2010
23. Streck C, Mel'nichenko YB, Richert R (1996) Dynamics of solvation in supercooled liquids confined to the pores of sol–gel glasses. Phys Rev B 53(9):5341
24. Ryabov Ya, Gutina A, Arkhipov V, Feldman Y (2001) Dielectric relaxation of water absorbed in porous glass. J Phys Chem B 105(9):1845–1850
25. Prisk TR, Tyagi M, Sokol PE (2011) Dynamics of small-molecule glass formers confined in nanopores. J Chem Phys 134:114506(114501–114509)
26. Cerveny S, Mattsson J, Swenson J, Bergman R (2004) Relaxations of hydrogen-bonded liquids confined in two-dimensional vermiculite clay. J Phys Chem B 108(31):11596–11603. doi:10.1021/jp037346r
27. Aliev FM, Nazario Z, Sinha GP (2002) Broadband dielectric spectroscopy of confined liquid crystals. J Non-Cryst Solids 305(1,3):218–225
28. Schönhals A, Frunza S, Frunza L, Unruh T, Frick B, Zorn R (2010) Vibrational and molecular dynamics of a nanoconfined liquid crystal. Eur Phys J Spec Top 189(1):251–255. doi:10.1140/epjst/e2010-01329-5
29. Frunza L, Frunza S, Kosslick H, Schönhals A (2008) Phase behavior and molecular mobility of n-octylcyanobiphenyl confined to molecular sieves: dependence on the pore size. Phys Rev E 78(5):051701
30. Cramer C, Cramer T, Arndt M, Kremer F, Naji L, Stannarius R (1997) NMR and dielectric studies of nano-confined nematic liquid crystals. Molecular crystals and liquid crystals science and technology. Mol Cryst Liq Cryst Sect A 303(1):209–217. doi:10.1080/10587259708039426
31. Massalska-Arodz M, Gorbachev VY, Krawczyk J, Hartmann L, Kremer F (2002) Molecular dynamics of the liquid crystal 6O2OCB in nanopores. J Phys: Condens Matter 14(36):8435
32. Crawford GP, Ondris-Crawford R, Žumer S, Doane JW (1993) Anchoring and orientational wetting transitions of confined liquid crystals. Phys Rev Lett 70(12):1838–1841
33. Kipnusu WK, Kossack W, Iacob C, Jasiurkowska M, Sangoro JR, Kremer F (2012) Molecular order and dynamics of tris(2-ethylhexyl)phosphate confined in uni-directional nanopores. In: Zeitschrift für Physikalische Chemie international journal of research in physical chemistry and chemical physics, vol 226(7–8), p 797

34. Iacob C, Sangoro JR, Kipnusu WK, Valiullin R, Karger J, Kremer F (2012) Enhanced charge transport in nano-confined ionic liquids. Soft Matter 8(2):289–293. doi:10.1039/c1sm06581e
35. Jasiurkowska M, Kossack W, Ene R, Iacob C, Kipnusu WK, Papadopoulos P, Sangoro JR, Massalska-Arodz M, Kremer F (2012) Molecular dynamics and morphology of confined 4-heptyl-4′-isothiocyanatobiphenyl liquid crystals. Soft Matter 8(19):5194–5200. doi:10.1039/c2sm07258k
36. Richert R (2010) Dielectric spectroscopy and dynamics in confinement. Eur Phys J Spec Top 189(1):37–46. doi:10.1140/epjst/e2010-01308-x
37. Kremer F, Schönhals A (2003) Broadband dielectric spectroscopy. Springer, Berlin
38. Richert R (2011) Dynamics of nanoconfined supercooled liquids. Ann Rev Phys Chem 62:65–84
39. Cordoyiannis G, Zidanšek A, Lahajnar G, Kutnjak Z, Amenitsch H, Nounesis G, Kralj S (2009) Influence of confinement in controlled-pore glass on the layer spacing of smectic-A liquid crystals. Phys Rev E 79(5):051703
40. Kutnjak Z, Kralj S, Lahajnar G, Žumer S (2004) Influence of finite size and wetting on nematic and smectic phase behavior of liquid crystal confined to controlled-pore matrices. Phys Rev E 70(5):051703
41. Bellini T, Radzihovsky L, Toner J, Clark NA (2001) Universality and scaling in the disordering of a smectic liquid crystal. Science 294(5544):1074–1079. doi:10.1126/science.1057480
42. Guégan R, Morineau D, Loverdo C, Béziel W, Guendouz M (2006) Evidence of anisotropic quenched disorder effects on a smectic liquid crystal confined in porous silicon. Phys Rev E 73(1):011707
43. Kityk AV, Wolff M, Knorr K, Morineau D, Lefort R, Huber P (2008) Continuous paranematic-to-nematic ordering transitions of liquid crystals in tubular silica nanochannels. Phys Rev Lett 101(18):187801
44. Chahine G, Kityk AV, Knorr K, Lefort R, Guendouz M, Morineau D, Huber P (2010) Criticality of an isotropic-to-smectic transition induced by anisotropic quenched disorder. Phys Rev E 81(3):031703
45. Clark NA, Bellini T, Malzbender RM, Thomas BN, Rappaport AG, Muzny CD, Schaefer DW, Hrubesh L (1993) X-ray scattering study of smectic ordering in a silica aerogel. Phys Rev Lett 71(21):3505–3508
46. Nambiar DC, Gaudh JS, Shinde VM (1994) Tris(2-ethylhexyl)phosphate as an extractant for trivalent gallium, indium and thallium. Talanta 41(11):1951–1955
47. Rizos AK, Petihakis L, Ngai KL, Wu J, Yee AF (1999) A dielectric relaxation study of the γ-relaxation in tetramethylbisphenol A polycarbonate plasticized by tris(2-ethylhexyl) phosphate. Macromolecules 32(23):7921–7924. doi:10.1021/ma980204o
48. Katsu T, Ido K, Kataoka K (2002) Poly(vinyl chloride) membrane electrode for a stimulant, phentermine, using tris(2-ethylhexyl) phosphate as a solvent mediator. Sens Actuators B: Chem 81(2–3):267–272
49. Ueda K, Rei Y, Komagoe K, Masuda K, Hanioka N, Narimatsu S, Katsu T (2006) Tris(2-ethylhexyl)phosphine oxide as an effective solvent mediator for constructing a serotonin-selective membrane electrode. Anal Chim Acta 565(1):36–41
50. Feng JK, Sun XJ, Ai XP, Cao YL, Yang HX (2008) Dimethyl methyl phosphate: a new non-flammable electrolyte solvent for lithium-ion batteries. J Power Sources 184(2):570–573
51. Lalia BS, Fujita T, Yoshimoto N, Egashira M, Morita M (2009) Electrochemical performance of nonflammable polymeric gel electrolyte containing triethylphosphate. J Power Sources 186(1):211–215
52. Czupryński K, Januszko A (1992) Preparation of nematic mixtures from smectic compounds. Molecular crystals and liquid crystals science and technology. Mol Cryst Liq Cryst Sect A 215(1):199–204. doi:10.1080/10587259208038525
53. Czupryński K (1990) Phase diagrams of mixtures consisting of polar compounds SE and SA d. Mol Cryst Liq Cryst Incorporating Nonlinear Opt 192(1):47–52. doi:10.1080/00268949008035604

54. Jasiurkowska M, Ściesiński J, Czub J, Massalska-Arodź M, Pełka R, Juszyńska E, Yamamura Y, Saito K (2009) Infrared spectroscopic and X-ray studies of the 4-Propyl-4-n-alkyl-4′ isothiocyanatobiphenyl (nTCB). J Phys Chem B 113(21):7435–7442. doi:10.1021/jp901339c
55. Jasiurkowska M, Budziak A, Czub J, Massalska-Arodź M, Urban S (2008) X-ray studies on the crystalline E phase of the 4-n-alkyl-4′-isothiocyanatobiphenyl homologous series (nBT, n = 2–10). Liq Cryst 35(4):513–518. doi:10.1080/02678290801989975
56. Jasiurkowska M, Zieliński PM, Massalska-Arodź M, Yamamura Y, Saito K (2011) Study of polymorphism of 4-Hexyl-4′-isothiocyanatobiphenyl by complementary methods. J Phys Chem B 115(43):12327–12335. doi:10.1021/jp201936x
57. Beale MIJ, Benjamin JD, Uren MJ, Chew NG, Cullis AG (1985) An experimental and theoretical study of the formation and microstructure of porous silicon. J Cryst Growth 73(3):622–636
58. Smith RL, Collins SD (1992) Porous silicon formation mechanisms. J Appl Phys 71(8):R1–R22. doi:10.1063/1.350839
59. Lehmann V, Gösele U (1991) Porous silicon formation: a quantum wire effect. Appl Phys Lett 58(8):856–858
60. Zhang G (2006) Porous silicon: morphology and formation mechanisms. In: Vayenas CG, White R, Gamboa-Adelco M (eds) Modern aspects of electrochemistry, vol 39. Springer, US, pp 65–133
61. Zhang XG, Collins SD, Smith RL (1989) Porous silicon formation and electropolishing of silicon by anodic polarization in HF solution. J Electrochem Soc 136(5):1561–1565. doi:10.1149/1.2096961
62. Valiullin R, Khokhlov A (2006) Orientational ordering of linear n -alkanes in silicon nanotubes. Phys Rev E 73(5):051605
63. Johari GP, Andersson O (2006) On the nonlinear variation of dc conductivity with dielectric relaxation time. J Chem Phys 125(12):124501
64. Demus D, Goodby J, Gray GW, Spiess HW (1998) Handbook of liquid crystals: fundamentals, vol 1. Wiley-VCH, New York
65. Donth E-J (2001) The glass transition: relaxation dynamics in liquids and disordered materials. In: Hull R, Jagadish C, Osgood RM, Parisi J, Wang ZM, Uchida S-I (eds), vol 48. Springer series in materials science. Springer, Heidelberg
66. Floudas G, Mpoukouvalas K, Papadopoulos P (2006) The role of temperature and density on the glass-transition dynamics of glass formers. J Chem Phys 124(7):74905
67. Thomas JA, McGaughey AJH (2008) Density, distribution, and orientation of water molecules inside and outside carbon nanotubes. J Chem Phys 128(8):084715
68. Shi W, Sorescu DC (2010) Molecular simulations of CO2 and H2 sorption into ionic liquid 1-n-Hexyl-3-methylimidazolium Bis(trifluoromethylsulfonyl)amide (hmim[Tf2N]) confined in carbon nanotubes. J Phys Chem B 114(46):15029–15041. doi:10.1021/jp106500p
69. Kawasaki T, Takeaki A, Tanaka H (2007) Correlation between dynamic heterogeneity and medium-range order in two-dimensional glass-forming liquids. Phys Rev Lett 99(21):215701
70. Spiess HW (2010) Interplay of structure and dynamics in macromolecular and supramolecular systems. Macromolecules 43(13):5479–5491. doi:10.1021/ma1005952
71. Stenzel O (2005) The physics of thin film optical spectra. Springer, Berlin
72. Urban S, Czuprynski K, Dąbrowski R, Gestblom B, Janik J, Kresse H, Schmalfuss H (2001) Dielectric studies of the 4-n-alkyl-4′-thiocyanatobiphenyl (nBT) homologous series (n=2-10) in the isotropic and E phases. Liq Cryst 28(5):691–696. doi:10.1080/02678290010023343
73. Róźańskia SA, Kremer F, Groothues H, Stannarius R (1997) The dielectric properties of nematic liquid crystal, 5CB confined to treated and untreated anopore membranes. Molecular crystals and liquid crystals science and technology. Mol Cryst Liq Cryst Sect A 303(1):319–324. doi:10.1080/10587259708039441
74. Iannacchione GS, Crawford GP, Qian S, Doane JW, Finotello D, Zumer S (1996) Nematic ordering in highly restrictive Vycor glass. Phys Rev E 53(3):2402–2411
75. Ziherl P, Vilfan M, Vrbancic-Kopac N, Žumer S, Ondris-Crawford RJ, Crawford GP (2000) Substrate-induced order in the isotropic phase of a smectogenic liquid crystal: a deuteron NMR study. Phys Rev E 61(3):2792–2798

76. Lefort R, Morineau D, Guégan R, Guendouz M, Zanotti J-M, Frick B (2008) Relation between static short-range order and dynamic heterogeneities in a nanoconfined liquid crystal. Phys Rev E 78(4):040701
77. Sinha G, Leys J, Glorieux C, Thoen J (2005) Dielectric spectroscopy of aerosil-dispersed liquid crystal embedded in Anopore membranes. Phys Rev E 72(5):051710
78. Rozanski SA, Stannarius R, Groothues H, Kremer F (1996) Dielectric properties of the nematic liquid crystal 4-n-pentyl-4′-cyanobiphenyl in porous membranes. Liq Cryst 20(1):59–66. doi:10.1080/02678299608032027

Rotational and Translational Diffusion of Ionic Liquids in Silica Nanopores

Ciprian Iacob, Joshua Sangoro, Wycliffe Kipnusu and Friedrich Kremer

Abstract Diffusion in ionic liquids (ILs) contained in silica nanopores is investigated in a wide frequency and temperature range by a combination of Broadband Dielectric Spectroscopy (BDS) and Pulsed Field Gradient Nuclear Magnetic Resonance (PFG NMR). By applying the Einstein-Smoluchowski relations to the dielectric spectra, diffusion coefficients are obtained in quantitative agreement with independent PFG NMR. More than tenfold systematic decrease in the effective diffusion coefficient (for [HMIM] [PF_6]) from the bulk value is observed in the silica nanopores. A model assuming a reduced mobility at the IL/porous matrix is proposed and shown to provide quantitative explanation for the remarkable decrease of effective transport quantities (such as diffusion coefficient, DC conductivity and consequently, the dielectric loss) of the IL in bare porous silica membranes. This approach is supported by the observation that silanization of silica nanopores results in significant increase of the effective diffusion coefficient, which approaches the value for the bulk liquid. For a different IL ([BMIM] [BF_4]), it is observed that ionic mobility at lower temperatures is enhanced by more than two decades under nanoconfinement in comparison to the bulk value. This increase in the diffusivity is attributed to reduced packing density of the ions in the nanopores. In summary, the resultant macroscopic transport properties of glass-forming ILs in confining space are determined by a subtle interplay between surface- and confinement-effects.

C. Iacob
Department of Materials Science and Engineering, Penn State University, University Park, State College, PA 16802, USA

J. Sangoro (✉)
Department of Chemical and Biomolecular Engineering, University of Tennessee, Knoxville, TN 37996-1600, USA
e-mail: jsangoro@utk.edu

W. Kipnusu · F. Kremer
Institute of Experimental Physics I, University of Leipzig, Linné str. 5, 04103 Leipzig, Germany

F. Kremer (ed.), *Dynamics in Geometrical Confinement*, Advances in Dielectrics,
DOI: 10.1007/978-3-319-06100-9_6,

Keywords Ionic liquids · Diffusion · Confinement and surface effects · Broadband dielectric spectroscopy · Silica nanopores · Ionic mobility

Abbreviations

BDS	Broadband Dielectric Spectroscopy
BMIM BF_4	1-Butyl-3-methylimidazolium tetrafluoroborate
FTIR	Fourier Transform Infrared
HMDS	Hexamethyldisilazane
HMIM PF_6	1-hexyl-3-methylimidazolium hexafluorophosphate
NMR	Nuclear Magnetic Resonance
PFG NMR	Pulsed Field Gradient Nuclear Magnetic Resonance
SEM/TEM	Scanning and Tunneling Electron Microscopy
VFT	Vogel-Fulcher-Tammann

1 Introduction

Ionic liquids (ILs) are promising for manifold technological as well as fundamental applications because they exhibit unique features such as low melting points, low vapor pressures, wide liquids ranges, high thermal stability, high conductivity, and wide electrochemical windows. They are under intense investigation for use as reaction media, in batteries and supercapacitors, in solar and fuel cells, for electrochemical deposition of metals and semiconductors, for protein extraction and crystallization, in nanotechnology applications, in physical chemistry, and many others. Thus, detailed knowledge of diffusion in ILs is imperative for their optimal utilization. Furthermore, since many ILs can easily be supercooled, they offer a rare opportunity to address basic questions regarding the correlation between ion conduction (translational diffusion) and the dynamic glass transition (rotational diffusion) in the broadest length—and time scales as well as localized molecular fluctuations (secondary relaxations). Due to its ability to measure the complex dielectric function (and consequently, the complex conductivity) over many orders of magnitude in frequency and in a wide temperature interval, Broadband Dielectric Spectroscopy (BDS) turns out to be an ideal experimental tool for this pursuit [1].

Recent experimental studies of some imidazolium-based ILs under nanoconfinement have demonstrated reduction of molecular mobility and conductance. For instance, using specialized nuclear magnetic resonance (NMR) techniques, Le Bideau et al. found a decrease in the diffusivity of ILs in monolithic silica matrices [2]. Davenport et al. observed about threefold reduction in the conductance of ILs in small single-pores (<10 nm) and explained their results in terms of the "hindered transport theory" [3]. Recently, the current authors found that the decrease in diffusivity of a selected IL in directed silica mesopores (7.5 nm pore diameter) could be reversed by silanization [4]. These reports emphasize the significance of the pore sizes, pore wall-IL interaction, as well as the physico-chemical properties of the IL investigated. Whereas experimental results have shown slower dynamics in nanopores, theoretical methods have predicted both slower and faster dynamics of ILs in nanopores depending on the molecular details of the confining space as well as the IL. For example, Wang et al. performed classical molecular dynamics simulations of confined ILs in contact with graphite surface and noted remarkable adsorption and formation of dense double-layers leading to reduction of the ionic mobility. Conversely, while investigating sorption of CO_2 and H_2 into ILs confined in carbon nanotubes by a combination of molecular dynamics and Monte-Carlo simulations, Shi and Sorescu found an increase of self-diffusion coefficients by 1–2 orders of magnitude in comparison to corresponding bulk IL values [5, 6]. In a recent study, the current authors obtained results confirming this prediction for a tetrafluoroborate-based imidazolium IL in nanopores [7]. A quantitative microscopic explanation of the origin of the experimentally observed enhancement of ionic mobility in some confined ILs and decrease in other ILs are not yet understood. This chapter presents detailed experimental studies of two representative ILs, one exhibiting faster and the other slower diffusivity in nanopores. The impact of the ILs/pore-walls interactions on ionic mobility of the confined ILs is also explored.

Rotational and translational diffusion in confining space are determined by a subtle interplay between surface- and confinement-effects. Strong interactions between the host system and the guest molecules lead to slower dynamics while spatial restriction of certain molecules in nanometric length scales may have the reverse effect of enhancing the molecular dynamics. In nanopores, the resultant transport properties depend critically on the microscopic properties of the confined molecule, extent of confinement, nature of the inner surfaces, as well as the architecture of the molecules with respect to the pore walls. The impact of each of these factors on diffusion in confined ILs forms the main subject of this chapter.

2 Experimental Details

The ILs (1-butyl-3-methylimidazolium tetrafluoroborate [BMIM BF_4] and 1-hexyl-3-methylimidazolium hexafluorophosphate [HMIM PF_6], purity $>99\,\%$) purchased from Iolitec GmbH were systematically studied in bulk as well as in silica nanopores with mean pore diameters between 7.5 and 10.4 nm. The membranes of porous silicon with nonintersecting nanopores were prepared by electrochemical etching of highly doped p-type $\langle 100 \rangle$ oriented silicon wafers with resistivity less than $5\,\mathrm{m}\,\Omega\,\mathrm{cm}$. This process was followed by thermal oxidization at 1153 K for 3 h resulting in unidirectional silica nanopores ($PSiO_2$). The insulating oxidized membranes (glassy SiO_2 membranes) were annealed in an evacuated chamber (10^{-5}) mbar at 470 K for 24 h prior to filling with the IL. For the case of silanized SiO_2 membranes,

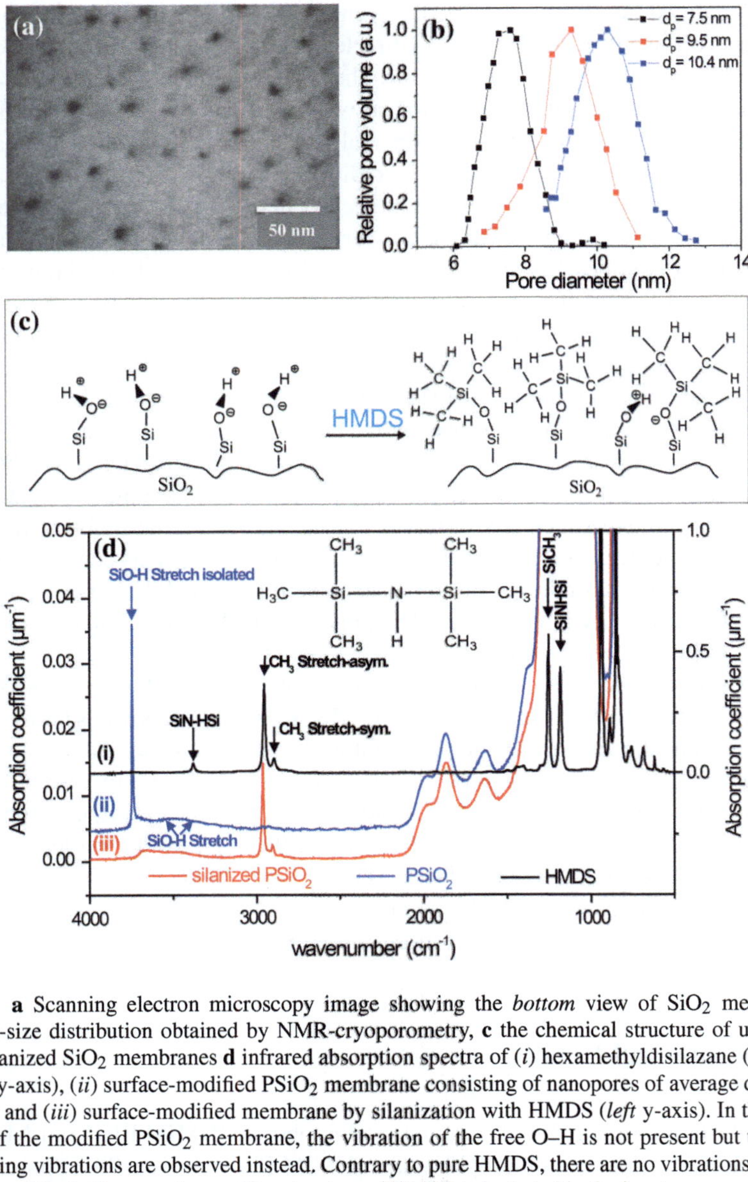

Fig. 1 **a** Scanning electron microscopy image showing the *bottom* view of SiO_2 membrane, **b** pore-size distribution obtained by NMR-cryoporometry, **c** the chemical structure of untreated and silanized SiO_2 membranes **d** infrared absorption spectra of (*i*) hexamethyldisilazane (HMDS) (*right* y-axis), (*ii*) surface-modified $PSiO_2$ membrane consisting of nanopores of average diameter 7.5 nm and (*iii*) surface-modified membrane by silanization with HMDS (*left* y-axis). In the spectrum of the modified $PSiO_2$ membrane, the vibration of the free O–H is not present but the CH_3 stretching vibrations are observed instead. Contrary to pure HMDS, there are no vibrations of N–H in the modified silica membrane. The structure of HMDS is included in the inset

hexamethyldisilazane (HMDS) (purity >99.0 %, purchased from Sigma-Aldrich) was injected into the vacuum preparation chamber containing the SiO_2 membrane at 350 K. After 30 min, the chamber with silica membrane was again evacuated for 3 h at 350 K. By treating the membrane with HMDS, the silanol groups were replaced with trimethylsilyl groups. The inner surface consequently became hydrophobic. The ILs were injected into the membranes in a vacuum chamber and the pores were filled by capillary wetting for approximately 48 h at room temperature.

In order to characterize the silica nanopores, scanning and tunneling electron microscopy (SEM/TEM) as well as NMR cryoporometry were employed (see Fig. 1a, b). To suppress the formation of an interfacial layer due to the presence of dangling hydrogen bonds, the inner surfaces of the $PSiO_2$ membranes were modified by silanization using HMDS (Fig. 1c). Trimethylsilyl groups replaced the silanol groups and the inner surfaces became hydrophobic as confirmed by the resulting Infrared (IR) spectra (Fig. 1d). A Varian FTIR-Spectrometer equipped with a UMA 500 microscope working in transmission mode was employed to obtain the infrared spectra. To quantify the effect of surface interaction between guest molecules and walls of host membranes, ILs in both surface-modified and bare nanoporous silica membranes were studied.

The untreated silica membrane exhibits a characteristic band at $\sim$3740 cm^{-1} corresponding to nonhydrogen-bonded OH groups. The intensity of this band dramatically decreases after the membrane is treated with HDMS indicating successful exchange of silanol with trimethylsilyl groups. This is further evident from the C–H stretching bands attributed to methyl groups in the region 2,800–3,000 cm^{-1}.

The concentration of CH_3 groups in the silanized membranes can be calculated from the IR absorption spectra using the Lambert-Beer law. As a reference system, we chose HMDS, where the concentration of methyl groups is known and equal to 6[HMDS] [4]. The concentration can be estimated by $[CH_3] = 6\frac{\int A_{\text{membrane}}}{\int B_{\text{HMDS}}} = 6\,[\text{HMDS}]\frac{\int A_{\text{membrane}}}{\int A_{\text{HMDS}}}$ where: A and B denote the absorption and the molar absorption coefficients, respectively. The integrals indicate that the whole area under the C–H band must be considered. Substitution of the experimental values gives $[CH_3] = 6\frac{0.76\,\text{gcm}^{-3}}{161.4\,\text{gmol}^{-1}} \cdot 0.033 = 0.93\,\text{M}$. Since the pore surface area is much larger than that of the outer surface, the distance between the silane groups can be estimated by $[CH_3] \sim \frac{3}{N_A}\frac{4p}{d}\frac{1}{d_{\text{silane}}^2}$, where p is the porosity and d is the pore diameter, since. Using $p = 0.07$ and $d = 7.5$ nm, one obtains $d_{\text{silane}} = 0.45$ nm. For comparison, the Si–Si distance in the HMDS molecule is 0.304 nm whereas the distance between methyl groups bonded to different Si atoms ranges from 0.38 to 0.53 nm. Therefore, we prove that the pore surface is fully silanized and covered by a hydrophobic layer with thickness $\sim$0.5 nm (the geometrical parameters used in these calculations are determined using the ArgusLab software).

Broadband dielectric measurements were carried out using a Novocontrol high-resolution alpha dielectric spectrometer (0.1 Hz–10 MHz) in the temperature range between 110 and 380 K with accuracy better than $\pm$0.1 K achieved by QUATRO temperature controllers. All the measurements were performed under dry nitrogen atmosphere. In addition, a 125 MHz pulsed field gradient NMR spectrometer with home-built gradient unit having gradients $q \leq 35\,\text{Tm}^{-1}$ was used to perform diffusion coefficient measurements. Both proton and fluoride nuclei were probed. The diffusion coefficients presented in this chapter are the sum of the diffusivities of both the cations and anions.

3 Results and Discussion

Due to its unique ability to probe molecular fluctuations and charge transport over a broad frequency and temperature range, broadband dielectric spectroscopy (BDS) has proven indispensable in the quest to understand the underlying mechanisms of charge transport and dynamic glass transition in ion conducting glass-forming systems [1]. BDS measures the complex dielectric function, $\varepsilon^* = (\varepsilon' - i\varepsilon'')$, as well as the complex conductivity function, $\sigma^* = (\sigma' + i\sigma'')$. At low applied electric fields (within the linear response regime), the two functions are given by $\sigma^*(\omega, T) = i\varepsilon_0\omega\varepsilon^*(\omega, T)$, implying that $\sigma' = \varepsilon_0\omega\varepsilon''$ and $\sigma'' = \varepsilon_0\omega\varepsilon'$ where ε_0, and ω denote the permittivity of vacuum, and radial frequency, respectively. The real part of the complex conductivity function σ' in bulk ILs is characterized on the intermediate frequency regime by a plateau the value of which directly yields the DC conductivity, σ_0, as well as the characteristic frequency, f_c, at which dispersion sets in and turns into a power law at higher frequencies. The typical conductivity spectra of the IL 1-butyl-3-methylimidazolium tetrafluoroborate—([BMIM] [BF_4]) in bare SiO_2 nanopores of mean pore diameter of 7.5 nm are shown in Fig. 2. In the current case, we assign σ_0 to the value of σ' at f_c. The spectra in the entire accessible temperature range investigated coincide upon scaling with respect to σ_0 and f_c as demonstrated in the insets of Fig. 2. This indicates identical thermal activation of the underlying mechanisms of charge transport and electrode polarization even in nanopores.

Einstein and Smoluchowski [8, 9] proposed a microscopic description of diffusion enabling the determination of diffusion coefficients from dielectric spectra. Within this framework, the diffusion coefficient, D, is given by $D = \left(\lambda^2\omega_c/2\right)$, where λ denotes the mean ion jump length in the timescale of $\omega_c = (2\pi/f_c)$.

Figure 3 shows a comparison of diffusion coefficients from BDS and PFG NMR spanning more than 12 orders of magnitude. A value of $\lambda = 0.16$ nm was used to determine D from the dielectric spectra according to the procedure described in our previous articles [10]. In order to gain deeper insight into the mechanisms of charge transport in ILs in nanopores, PFG NMR, a well-established technique for probing bulk diffusion, is used to probe the ILs both in bulk and nanopores. At higher temperatures, the measured diffusion coefficients from BDS and PFG NMR coincide for both bulk and confined ILs. Remarkable increases are observed in the low temperature regime for the confined ILs. For nanopores with mean diameter of 7.5 nm, the diffusion coefficients are more than two orders of magnitude higher than their corresponding bulk values (see Fig. 3). It should be noted that pore-surface modification by silanization (Fig. 3) does not lead to any observable changes in the diffusion coefficients for the current [BMIM BF_4], in contrast to the hexafluorophosphate-based IL discussed in the next paragraphs. This indicates that the pore wall/IL interactions associated with adsorption and desorption processes only play a minimal role and cannot explain the present results.

The temperature dependence of bulk transport properties of ILs is described by the empirical Vogel-Fulcher-Tammann (VFT) equation expressed as: $D(\mathrm{T}) =$

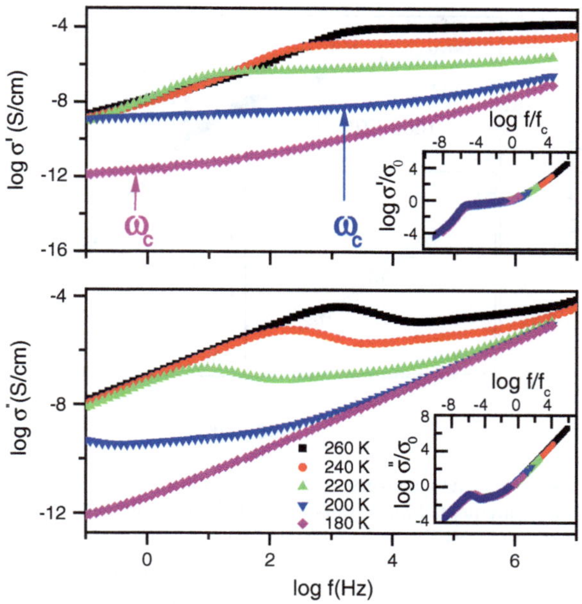

Fig. 2 Complex conductivity function ($\sigma'(\omega) = \sigma'(\omega) + i\sigma''(\omega)$) of the IL BMIM BF_4 versus frequency at different temperatures in bare 7.5 nm silica nanopores as indicated. *Inset* scaling with respect to the characteristic frequency f_c and DC-conductivity σ_0. The error bars are comparable to the sizes of the symbols, if not specified otherwise

$D_\infty \exp[-B/(T - T_0)]$, where D_∞, B, and T_0 denote the diffusion coefficient in the high temperature limit, a constant, and the Vogel temperature, respectively. It is instructive to check how well the diffusion coefficient follows the VFT equation under confinement. The derivative technique often provides a means of verifying the subtle features of the transport quantities spanning several orders of magnitude [1]. Within this description, the VFT equation is given by $\frac{d \log \sigma_0}{d(1/T)} = -B\left(1 - \frac{T_0}{T}\right)^{-2} \log e$. When the resultant derivative of the diffusion coefficient is plotted against inverse temperature, the VFT equation is transformed into an Arrhenius-type relation (expressed as: $\log D(T) = \log D_\infty - \frac{E_A}{k_B T} \log e$, where E_A denotes the activation energy) with a single slope as demonstrated in the inset of Fig. 3. In the nanopores, a change from a VFT-like into an Arrhenius-like thermal activation is observed, resulting in an enhancement of diffusivities by more than two orders of magnitude [11]. The precise origin of this remarkable behavior of some ILs in nanopores remains unclear.

Our conjecture is that the increase of ionic mobility in nanopores stems from subtle changes in ion packing due to the reduction of the density of the ILs in small pores. A simple physical explanation can be obtained by considering the problem of packing density of cylindrical balls in cylindrical containers. Both experimental and theoretical studies of spherical balls in cylindrical containers indicate that the mass density decreases by up to 7 % when the radii of the balls become comparable to that

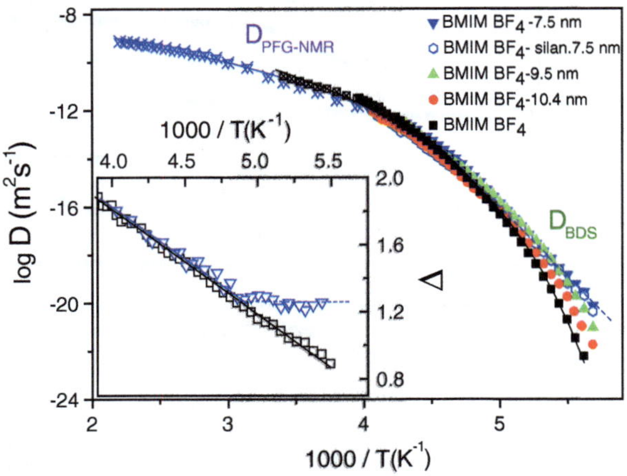

Fig. 3 Diffusion coefficients versus inverse temperature as obtained by PFG NMR (*crossed open symbols*) and BDS measurements (*filled symbols*) for bulk and confined ([BMIM] [BF_4] as indicated. The *lines* denote fits by VFT (*continuous*) and Arrhenius (*dotted*) equations. *Inset* difference quotients $\Delta = \left[\frac{\log D(1/T_i) - \log D(1/T_{i+1})}{1/T_i - 1/T_{i+1}}\right]^{-1/2}$ versus $1/T$ for the bulk BMIM BF4 (*square symbols*) and confined in porous silica with mean diameter of 7.5 nm (denoted by *triangles*) as determined experimentally and $\Delta = \left[d \log(D)/d(1/T)L\right]^{-1/2}$) as obtained from the above VFT (*continuous lines*)—and Arrhenius (represented by *dotted lines*)—fit functions

of the confining cylinder. Neglecting surface effects, the mass density ρ_c under confinement can be expressed by the empirical relation: $\rho_c = \rho_b - E/R - K/H$, where ρ_b, R, and H denote the bulk mass density, radius of the cylinder, and height of the cylinder, respectively. E and K are geometrical constants. More rigorous approaches involving density functional theory and Voronoi-Delaunay methods deliver comparable results concerning the decrease of mass density under confinement [12–16]. This approach has been extended to explain molecular properties with considerable success. Recently, Shi and Sorescu employed atomistic simulations to study diffusion in the IL [HMIM NTf_2] confined in carbon nanotubes of diameter 4.5 nm and observed significant decrease in the mass density compared to the bulk value [5, 6]. As a result, the diffusion coefficients increased by about two orders of magnitude. Our present results confirm this picture at lower temperatures for BF_4-based IL (see Fig. 3). In a recent review article, it was pointed out that 2.1 % change in the mass density of bulk poly (vinyl) acetate corresponds to variation of the structural relaxation rate over eight orders of magnitude [17]. This underscores the role of density in determining the dynamic properties of glass-forming materials. Regarding the coinciding plots of diffusion coefficients at higher temperatures, we argue that due to the higher thermal energy of the molecules, the effects of density on dynamics are minimal in the high temperature regime. We expect that these effects should, however, show up in static quantities like the dielectric strength. Experiments in this direction are in progress for low molecular-weight glass-forming liquids.

In contrast to [BMIM BF_4], the diffusivity of the [HMIM PF_6] in the unmodified silica nanoporous membranes decreases by about one decade compared to the bulk IL. Notably, this decrease holds on the time scales from nanoseconds to tens of seconds, probed by both BDS and PFG NMR (Fig. 4). However, the IL in the silanized membrane shows only small deviation from bulk values. In general, two effects may contribute to the slowing down of the diffusivity of confined molecules, namely the interaction with the pore walls and the so-called tortuosity effect [18]. In what follows, we analyze their possible impact in our experiments.

There is sufficient evidence in the literature that even very carefully grown linear pores in porous silicon possess some degree of mesoscale disorder [19, 20]. This inevitably results in a distribution of pore sizes. Presumably, this distribution reflects the variation of the pore diameter along the pore axis. In turn, molecular propagation in such a disordered structure can be slowed down and computer simulations can easily quantify the reduction of molecular diffusivities. To estimate the disorder effect upon molecular diffusivities, Monte Carlo simulations of a point-like particle in linear pores composed of small, statistically distributed cylindrical sections with different pore diameters—are employed. Distribution $P(d)$ of pore diameters is taken to be Gaussian $P(d) \propto \exp\left\{-\frac{(d-d_0)}{2\Sigma^2}\right\}$, where d_0 is the mean pore diameter and Σ the standard deviation. In the simulations, the mean square displacements $\langle(z-z_0)^2\rangle$ along the pore axis direction are determined as function of Monte Carlo steps number n. At sufficiently large n, the function $\langle(z-z_0)^2\rangle$ is found to grow proportionally to n, as expected for random processes. Thereafter, using the Einstein relation $\langle(z-z_0)^2\rangle = 2Dn$, the diffusivity D can easily be calculated. The normalized diffusivities D/D_0, where D_0 is the bulk diffusivity, for different structures with different pore size distributions (i.e., different values of Σ) are evaluated. The results obtained reveal that with the experimentally estimated value of $\Sigma = 0.26$, the diffusivity is expected to decrease only slightly compared to D_0 (by ~5 %), suggesting that geometrical disorder can by no means explain the observed results in Fig. 4.

The ability of hydrogen-bonded liquids to attach to untreated SiO_2 surface (due to strong hydrophilicity of the walls of SiO_2 membranes) is well established from previous studies of the role of hydrogen-bond (H-bond) formation on molecular dynamics. In such cases, a layer of reduced mobility is expected and the remaining volume is filled with bulk-like molecules—an effect not yet investigated for ILs. Based on this consideration, the measured dielectric response can be modelled in terms of three contributions; "bulk-like" molecules in the pores, adsorbed molecules at the pore-matrix interface, and the insulating porous silica matrix. The simple model consists of the three impedances in parallel (Z_b^*—bulk impedance, Z_a^*—impedance of the adsorption layer and $Z_{SiO_2}^*$—impedance of silicon dioxide matrix) corresponding to the three main components. The total measured impedance Z_m^* is given by $Z_m^* = \frac{1}{i\omega\varepsilon_m^* C_m}$ where ε_m^*—represents the measured complex dielectric function and C_m denotes the vacuum capacitance of the measurement cell. Thus, the real and imaginary parts of the measured dielectric function can be expressed as $\varepsilon_m' = f_b\varepsilon_b' + f_a\varepsilon_a' + f_{SiO_2}\varepsilon_{SiO_2}'$ and $\varepsilon_m'' = f_b\varepsilon_b'' + f_a\varepsilon_a'' + f_{SiO_2}\varepsilon_{SiO_2}''$, where ε_b^*, ε_a^*, $\varepsilon_{SiO_2}^*$

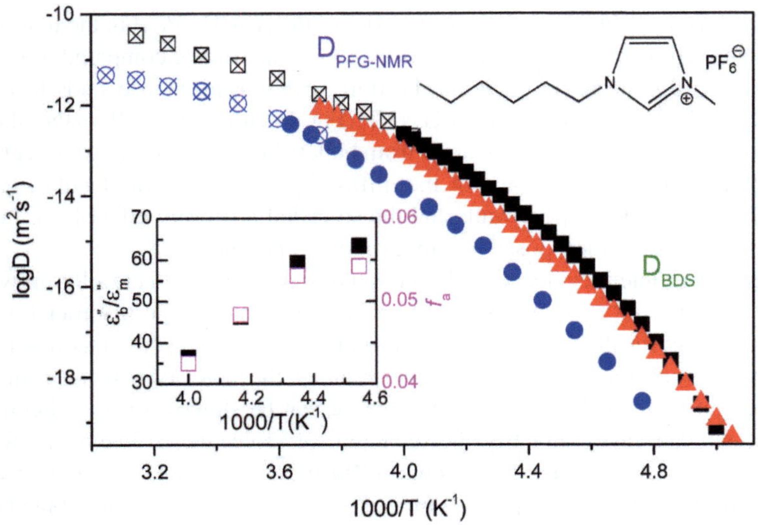

Fig. 4 Diffusion coefficient determined by applying the Einstein-Smoluchowski equation to the dielectric spectra of [HMIM PF_6] in bulk (*solid squares*) untreated nanopores (*solid circles*) and silanized nanopores (*solid triangles*). The respective values obtained by PFG NMR are shown by crossed symbols. *Inset* Temperature dependence of the ratio of the dielectric loss in bulk to the total dielectric loss in nanopores (*solid symbols*). The surface area fraction corresponding to the adsorption layer (f_a) as a function of temperature is also given (*open symbols*). The structure of [HMIM PF_6] is included in the inset

represent the complex dielectric functions of bulk, adsorption layer and silica membrane, respectively. f_b, f_a and f_{SiO_2} represent the volume fractions of the bulk liquid, adsorption layer and SiO_2 matrix. With porosity of 7 %, one obtains $f_b + f_a = 0.07$. The oxidized SiO_2 membrane was observed to exhibit very low conductivity with $\varepsilon''_{SiO_2} \leq 10^{-3}$. It is therefore evident that $\varepsilon''_{SiO_2} < 10^{-3} \Rightarrow f_{SiO_2}\varepsilon''_{SiO_2} < 0.01$ and thus, $\frac{\varepsilon''_b}{\varepsilon''_m} \cong \frac{1}{f_b}$. From this simple analysis, the following experimental features are obtained: (i) the total measured dielectric loss (and consequently, characteristic charge transport rate and diffusion coefficient) in the unmodified nanopores is about 1.4 decades less than the bulk value, (ii) ε'_m is dominated by the contribution from the SiO_2 membrane, thereby leading to the absence of the step at frequency corresponding to ω_c as compared to the bulk spectra, (iii) the deviation of transport quantities (diffusion coefficient, charge transport rate ω_c, DC conductivity, dielectric loss) from the bulk value increases with decreasing temperature since the adsorption contribution becomes more pronounced at lower temperatures. Thus, $f_a(T_1) < f_a(T_2)$ if $T_1 > T_2$, in quantitative agreement with the values of the size of the adsorption layer at different temperatures as shown in the inset Fig. 4. With decreasing temperature, the bulk-like contribution from the pores dramatically decreases [21]. The lack of H-bonding and the decreasing dipole–dipole interactions might explain the increased diffusivity in the silanized nanopores. Recent studies on a series of alkylcitrates [22] illustrate the role of dipole–dipole interactions on diffusivity. The conductivity in the treated pores is much higher, in accordance with the fact that the diffusion coefficient increases. The dielectric loss ratio $\varepsilon''_{\text{bulk}}/\varepsilon''_{\text{silanized}} \sim 8$ in the entire bulk temperature range. This would lead to a negative value for f_a, implying that there is no adsorbed layer of ILs in the surface-modified membranes, which also means that the effective conductivity is higher than in the bulk.

Understanding the mechanism of the formation and the extent of the adsorption layer as well as its temperature dependence enables one to devise means of eliminating (or countering) it. Thus, we proceed to modify the surface of the SiO_2 membrane by silanization in order to probe the nature of the adsorbed layer. It turns out that silanization of SiO_2 pore walls significantly reduces this effect (see Fig. 4). This finding is easily understood by considering our model presented above. Due to the absence of hydrogen bonding between the IL and the membrane, we expect that the adsorption layer is both thinner and more mobile, compared to the untreated membrane case.

4 Conclusions

Rotational and translational diffusion in glass-forming ILs is investigated by a combination of Broadband Dielectric Spectroscopy (BDS) and Pulsed Field Gradient Nuclear Magnetic Resonance (PFG-NMR). Detailed analysis of the dielectric spectra delivers the mean ion jump rate and enables one to deduce—using the Einstein-Smoluchowski equation—the translational diffusion coefficient of the charge carriers in quantitative agreement with PFG-NMR measurements. To understand the impact of two-dimensional nanoconfinement of charge transport and glassy dynamics in ILs, unidirectional nanoporous silica with mean pore diameters between 7.5 and 10.4 nm are prepared in a home-built anodization set-up by electrochemical etching of ⟨100⟩ p-type silicon followed by well-controlled thermal oxidation. Diffusion in tetrafluoroborate- and hexafluorophosphate-based ILs contained in the nanoporous membranes is investigated using BDS and PFG-NMR. This enables one to determine the diffusion coefficient and the diffusion rate over more than 13 decades and to trace its temperature dependence. In nanopores, a change from a Vogel-Fulcher-Tammann into an Arrhenius-like thermal activation is observed, accompanied by 100-fold enhancement of diffusivities at lower temperatures. The effect becomes more pronounced with decreasing pore diameter. It is attributed to changes in molecular packing and hence in density leading to higher mobility and electrical conductivity, a view supported by recent atomistic simulations.

For 1-hexyl-3-methylimidazolium-based ILs with the hexafluorophosphate anion, more than tenfold systematic decrease in the effective diffusion coefficient from the bulk value is observed in the silica native unsilanized nanopores. The differences of diffusivity in the nanopores compared to the bulk are explained within a theoretical framework that considers, in addition to the bulk response, the contribution from the layer of reduced mobility at the pore—porous matrix interface. In agreement with this simple model, a remarkable increase of the diffusion coefficient of the IL in nanopores is observed upon silanization of the silica membranes. Thus, the resultant macroscopic transport properties of ILs in confining space are determined by a subtle interplay between surface- and confinement-effects.

Strong interactions between the host system and the guest molecules lead to lower diffusion coefficients while spatial restriction of certain molecules in nanometric length-scales may have the reverse effect of enhancing the diffusivities. It remarkable that the dielectric properties of the ILs based on tetrafluoroborate and hexafluorophosphate anions which, appear similar at first glance, are dominated by different mechanisms, namely, confinement and surface effects, respectively. More extensive work involving combination of experimental and computational studies for systematic series of ILs is still required to unravel the origin of differences in diffusivity of ILs in nanopores.

Acknowledgments We thank Dr. Sergej Naumov for help in conducting the PFG-NMR measurements and Dr. Periklis Papadopoulos for FT-IR measurements. Financial support from DFG (Germany), NOW (The Netherlands) within IRTG "Diffusion in Porous Materials" and DFG Priority Program SPP 1191 on Ionic Liquids is gratefully acknowledged. J. R. S. thanks the University of Tennessee-Knoxville for financial support through tenure-track faculty research start-up funds. Ciprian Iacob thanks the Penn State University and St. Jude Medical for financial support.

References

1. Kremer F, Schönhals A (2003) Broadband dielectric spectroscopy. Springer, Berlin
2. Le Bideau J et al (2007) Effect of confinement on ionic liquids dynamics in monolithic silica ionogels: 1H NMR study. Phys Chem Chem Phys 9(40):5419–5422
3. Davenport M et al (2009) Squeezing Ionic liquids through nanopores. Nano Lett 9(5):2125–2128
4. Iacob C et al (2010) Charge transport and diffusion of ionic liquids in nanoporous silica membranes. Phys Chem Chem Phys 12(41):13798–13803
5. Shi W, Sorescu DC (2010) Molecular simulations of CO_2 and H_2 sorp tion into ionic liquid 1-n-Hexyl-3-methylimidazolium Bis(trifluoromethylsulfonyl)amide ([hmim] [Tf2N]) confined in carbon nanotubes. J Phys Chem B 114(46):15029–15041
6. Wang S, et al (2009) Molecular dynamic simulations of ionic liquids at graphite surface. J Phys Chem C 114(2):990–995
7. Sangoro JR, Kremer F (2012) Charge transport and glassy dynamics in ionic liquids. Acc Chem Res 45(4):525–532
8. Einstein A (1905) On the motion of small particles suspended in liquids at rest, required by the molecular-kinetic theory of heat. Ann Phys 17:549–560
9. Smoluchowscki M (1906) Essay on the theory of Brownian motion and disor dered media. Ann Phys 21:755–780
10. Sangoro JR et al (2011) Diffusion in ionic liquids: the interplay between mo lecular structure and dynamics. Soft Matter 7(5):1678–1681
11. Iacob C, Sangoro JR, Kipnusu WK, Valiullin R, Kärger J, Kremer F (2011) Enhanced charge transport in nanoconfined ionic liquids. Soft Matter 8:289–293
12. Scott GD, Kilgour DM (1969) The density of random close packing of spheres. J Phys D: Appl Phys 2(6):863–866
13. Roth R (2010) Fundamental measure theory for hard-sphere mixtures: a re view. J Phys Condens Matter 22(6):063102

14. Luchnikov VA, Medvedev NN, Gavrilova ML (2001) The Voronoi- Delaunay approach for modeling the packing of balls in a cylindrical container. In: Proceedings of the International Conference on Computational Sciences-Part I2001. Springer-Verlag, New York, pp 748–752
15. Finney JL (1970) Random packings and the structure of simple liquids. I. The geometry of random close packing. Proc R Soc Lond A 319(1539):479–493
16. Luchnikov VA et al (1999) Voronoi-Delaunay analysis of voids in systems of nonspherical particles. Phys Rev E 59(6):7205
17. Spiess HW (2010) Interplay of structure and dynamics in macromolecular and supramolecular systems. Macromolecules 43(13):5479–5491
18. Sen PN (2004) Time-dependent diffusion coefficient as a probe of geometry. Concepts Magn Reson Part A 23:1
19. Naumov S, Khokhlov A, Valiullin R, Karger J, Monson PA (2008) Understanding capillary condensation and hysteresis in porous silicon: network effects within independent pores. Phys Rev E: Stat Nonlinear Soft Matter Phys 78:060601
20. Wallacher D, Künzner N, Kovalev D, Knorr N, Knorr K (2004) Capillary condensation in linear mesopores of different shape. Phys Rev Lett 92:19
21. Gratz M, Wehring M, Galvosas P, Stallmach F (2009) Multidimen sional NMR diffusion studies in microporous materials. Microporous Mesoporous Mater 125:30–34
22. Kipnusu WK, Kossack W, Iacob C, Zeigermann P, Jasiurkowska M, Sangoro JR, Valiullin R, Kremer F (2013) The interplay between inter- and intramolecular dynamics in a series of alkylcitrates. Soft Matter 9:4681–4686

Polymer Nanofluidics by Broadband Dielectric Spectroscopy

Anatoli Serghei

Abstract A multitude of experimental techniques can be employed to investigate the dynamics of polymers under condition of geometric confinement. Among them, Broadband Dielectric Spectroscopy has proven its strength in this research field, being used in numerous investigations on polymer dynamics in samples having one, two, or three dimensions on the nanometer length scale. However, in most cases, the investigated polymers were subjected to constraints arising only from a "static" confinement, where the dimensionality of the confining geometry and the interfacial interactions played the most important role. In addition to these static factors, kinetic constraints and frustrations arise when polymers are flowing in nanometric confinement. Measuring polymer dynamics during flow in nanoconfinement is a challenging task, which requires the development of new experimental procedures and methods. Here, we show that different aspects of the flow process of polymers in cylindrical nanopores can be investigated by means of Broadband Dielectric Spectroscopy, using a recently developed experimental approach, which employs highly ordered nanoporous media as nanofluidics cell. No significant shifts in the glassy dynamics of poly-2-vinylpyridine (P2VP) investigated during capillary flow into nanopores are found down to a pore diameter of 18 nm. In the case of phase-separated polystyrene-block-poly-4-vinylpyridine (PS-b-P4VP) flowing into nanopores, no shifts in the dynamic glass transition of the P4VP blocks are observed down to a confinement size allowing the flow of only one (single) copolymer domain.

Keywords Nanofluidics · Capillary flow · Polymers · Flow kinetics · Segmental dynamics · AAO porous membranes · Broadband dielectric spectroscopy · Nanopores

---

A. Serghei (✉)
Ingénierie des Matériaux Polymères, Université Lyon 1, CNRS, UMR 5223,
69622 Villeurbanne, France
e-mail: anatoli.serghei@univ-lyon1.fr

F. Kremer (ed.), *Dynamics in Geometrical Confinement*, Advances in Dielectrics,
DOI: 10.1007/978-3-319-06100-9_7,

Abbreviations

AAO	Anodized aluminum oxide
°C	Degree(s) Celsius
g	Gram(s)
h	Hour(s)
Hz	Hertz
K	Degree(s) kelvin
kHz	Kilohertz
μm	Micrometer(s)
min	Minute(s)
Mn	Number-averaged molecular weight
mol	Mole(s)
Mw	Weight-averaged molecular weight
NaOH	Sodium hydroxide
nm	Nanometer(s)
P2VP	Poly-2-vinylpyridine
P4VP	Poly-4-vinylpyridine
PCDF-TrFE	Polyvinylidene fluoride trifluoroethylene
PDI	Polydispersity index
PS	Polystyrene
PS-b-P4VP	Polystyrene-block-poly-4-vinylpyridine
PVDF	Polyvinylidene fluoride
SAXS	Small angle X-ray scattering
TEM	Transmission electron microscopy
Tg	Glass transition temperature
TrFE	Trifluoroethylene
V	Volt(s)

1 Introduction

Polymer dynamics, measured at different time and length scales under condition of geometric confinement, remains a topic of large scientific interest in the recent years [1–11]. In addition to "static" factors related to the dimensionality and morphology of the confining geometry as well as to the influence of the interfacial interactions, kinetic aspects come into play when the flow of polymers is investigated in nanometric confinement. The flow process can give rise to additional constraints and frustrations, especially in the case when complex fluids are investigated.

While numerous scientific studies have been reported on polymer dynamics under "static" confinement [1–8], carrying out investigations during the flow process is a challenging task, due to the fact that only a few experimental methods can be adapted to measure—in situ—the kinetics of the flow process. Among them, Broadband

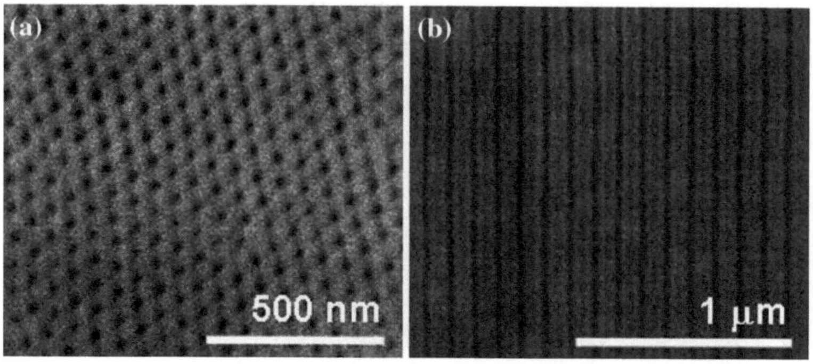

Fig. 1 Images by scanning electron microscopy of highly ordered nanoporous membranes of aluminum oxide prepared by electrochemical anodization. **a** *top view*; **b** *cross section view*

Dielectric Spectroscopy has proven, in a recent development [10, 11], its ability to perform this kind of investigation. In order to enable electrical measurements during a flow process, a crucial requirement must be fulfilled: as opposed to the conventional approach where the sample is placed between two metal electrodes, in situ investigations under flow imply electrical measurements without employing a direct contact between the sample and at least one electrode. This can be done due to a fundamental principle underlying electrical properties: the equivalence between the direct and the displacement current, as stipulated to the 4th Maxwell equation. Based on this equivalence, it has been proven in several recent developments that electrical properties can be measured without placing the electrode in direct contact with the sample under investigation. This has allowed, for instance, avoiding metallization in dielectric studies on nanometric thin polymer films [12–14], which opened the perspective for measuring the molecular dynamics of single (isolated) polymer chains [3]. The same principle has been used, as well, to develop a novel experimental approach in the field of nanofluidics [10, 11], which enables investigations on flow processes by means of dielectric measurements.

2 Experimental

As experimental cells for nanofluidics, nanoporous membranes of aluminum oxide (AAO membranes—anodized aluminum oxide) have been used [15, 16]. These membranes have been prepared in a two-step electrochemical anodization process of pure aluminum. The AAO membranes exhibit a highly ordered array of parallel (hexagonally packed) nanopores, with a narrow distribution of pore diameters and interpore distances (Fig. 1).

The pore diameters and the porosity of the AAO membranes can be controlled by adjusting the preparation conditions: the concentration of the anodization solution, the type of acid, the temperature, and the applied voltage. A schematical

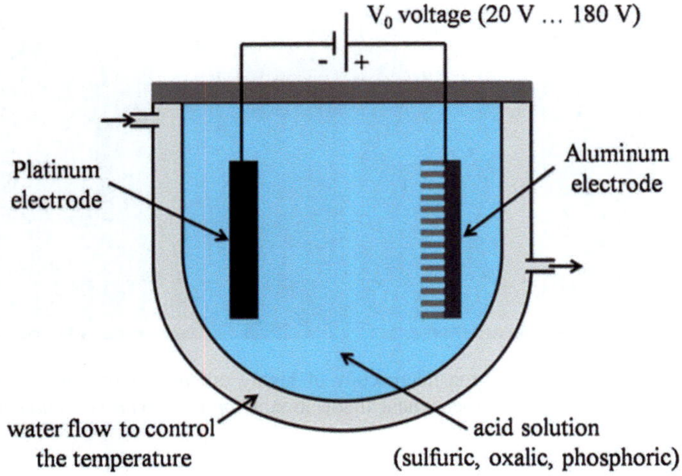

Fig. 2 Schematic representation of the anodization cell used to prepare nanoporous membranes of aluminum oxide. The pores diameter and the distance between the pores can be controlled by adjusting the anodization conditions: type of electrolyte, concentration of the electrolyte solution, temperature, and the applied voltage

representation of the anodization cell is shown in Fig. 2. The electrochemical anodization process is carried out in a two-step procedure, in order to ensure a highly ordered morphology of nanopores. After the preparation, the oxidized nanoporous layer, still in contact with the underlying aluminum substrate, can be detached by applying a voltage in a solution of 2,3-butanedione [17]. This procedure leads to high quality AAO membranes exhibiting ordered nanopores having both their ends open.

3 Nanofluidics by Optical Microscopy: The Scaling Law of the Capillary Flow

The kinetics of the capillary flow of polymers into nanoporous AAO membranes can be easily investigated by optical microscopy. A thick polymer layer is prepared first by hot pressing and annealed well above its melting point (or its glass transition temperature) to release the residual stresses and equilibrate the sample. After that, the polymer layer is placed on top of the AAO membrane and the cell is heated up to initiate the flow process. After keeping the cell at a constant temperature for a certain time t, the flow process is stopped by quenching the sample to room temperature. The layer of bulk polymer present on top of the AAO membrane can be removed by immersing the sample in liquid nitrogen. At the end, the cell is broken to take optical images of the cross section. The kinetics of the flow process is clearly revealed in these images, due to the optical contrast between the empty nanopores and the nanopores filled with the polymer under investigation. A typical example is given

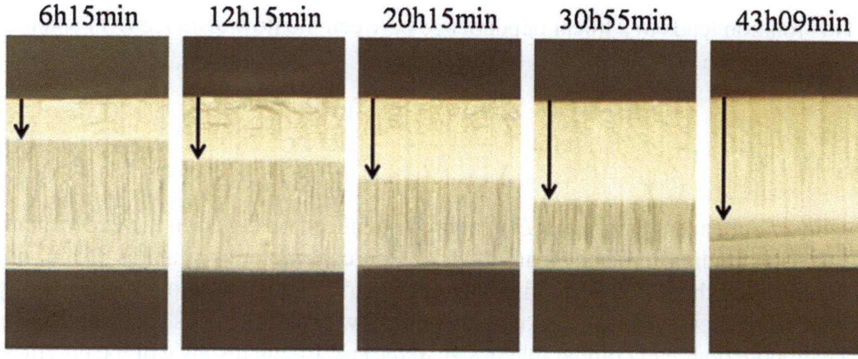

Fig. 3 Images by optical microscopy (in cross section) showing the time evolution of the capillary flow of polystyrene (Mw = 32,000 g/mol) into nanopores of 15 nm diameter. The flow process was carried out for different time periods, as indicated, at a temperature of 156 °C. After stopping the flow by quenching the sample to room temperature, the bulk polystyrene was removed from the upper interface by immersing the samples in liquid nitrogen. The porous membranes have a total thickness of 112 μm

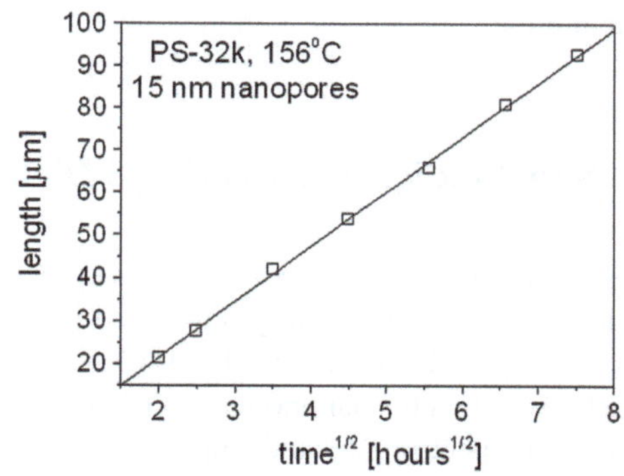

Fig. 4 Length of the capillary filling of the nanopores (15 nm in diameter) with polystyrene (Mw = 32,000 g/mol) as a function of square root of time. A linear dependence is observed

in Fig. 3, showing the capillary flow of polystyrene (Mw = 35,000 g/mol, purchased from Polymer Source) in nanopores of 15 nm diameter.

The experiment was carried out at a temperature of 156 °C. After a flow time of 4 h, the polystyrene has infiltrated a distance of 21.2 μm into the nanopores, 27.6 μm after 6 h 15 min, 41.8 μm after 12 h 15 min, 53.7 μm after 20 h 15 min, 65.8 μm after 30 h 55 min, 80.8 μm after 43 h 09 min, 92.9 μm after 56 h 27 min. The dependence between the flow distance and flow time is shown in Fig. 4. According to the theory of the dynamics of capillary flow developed by Washburn [18], the velocity of the capillary flow, defined as a first derivative of the flow distance $L(t)$ in respect of the flow time t, is directly proportional to the pore radius R, the interfacial tension γ and the wetting angle $\cos(\theta)$, and inversely proportional to the polymer viscosity η and the flow distance:

$$\frac{\mathrm{d}L(t)}{\mathrm{d}t} = \frac{R}{\eta}\frac{\gamma}{4L(t)}\cos(\theta) \quad (1)$$

This relation leads to a linear dependence between the flow distance $L(t)$ and the square root of the flow time:

$$L(t) = \left(\frac{\gamma R \cos(\theta)}{2\eta}\right)^{1/2} \sqrt{t} \tag{2}$$

A linear dependence between $L(t)$ and $\sqrt{t}$ is indeed observed in the experiment (Fig. 4). This finding indicates that, down to a pore diameter as small as 15 nm, no deviations from the expected scaling law are observed in the case of capillary flow in nanoconfinement. One can also note that this flow experiment has been conducted for a nanopore diameter comparable in size to the end-to-end distance of the polymer used in this study. According to Eq. 2, the slope of the linear dependence shown in Fig. 4 is inversely proportional to the square root of the polymer viscosity $\sqrt{\eta}$. This opens the perspective of using the simple optical detection routine described here as a method to determine the polymer viscosity in nanoconfinement. It offers furthermore a simple approach to determine the characteristic timescale of the kinetics of the flow process, which provides an accurate estimation of the filling factor of the porous membrane in dependence on the flow time.

4 Nanofluidics by Broadband Dielectric Spectroscopy

To monitor the flow process by means of dielectric measurements, a special nanofluidics cell has been recently developed [10, 11] (Fig. 5). The approach is based on metalizing both surfaces of the AAO porous membrane with thin layers of gold in such a way that the nanopores remain still open. The gold layers are connected to the dielectric spectrometer and employed as electrodes to allow dielectric measurements in real-time during the flow process. Since the electrodes are deposited on the empty nanofluidics cell (before the flow process is initiated), no contribution from the bulk polymer layer—acting as a flow reservoir—is measured by this approach. The only signals contributing to the global response arise from the polymer flowing into the nanopores as well as from the empty nanopores and the aluminum oxide of which the AAO membrane is made. Taking into account the hexagonal arrangement of nanopores, the global complex permittivity of the nanofluidics cell can be expressed as

$$\varepsilon^*_{\text{measured}} = (1 - p)\,\varepsilon^*_{\text{alumina}} + p\,\varepsilon^*_{\text{nanopores}} \tag{3}$$

with

$$\frac{L}{\varepsilon^*_{\text{nanopores}}} = \frac{l}{\varepsilon^*_{\text{polymer}}} + L - l \tag{4}$$

and

$$p = \frac{\pi\sqrt{3}}{9}\left(\frac{d_P}{d_{cc}}\right)^2 \tag{5}$$

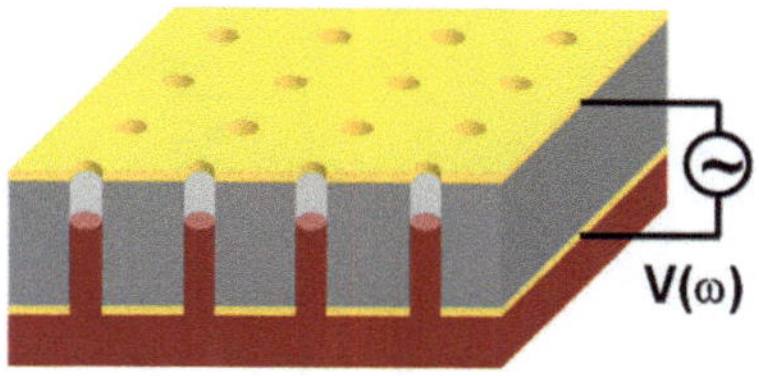

Fig. 5 Nanofluidics cell for dielectric measurements. Both sides of an empty AAO membrane are metalized by thin layers of gold without closing the nanopores. The gold layers, employed as electrodes, are connected to a dielectric spectrometer and used to monitor the flow process by means of a permittivity measurement

where p represents the porosity of the AAO membrane, d_{cc} the center-to-center distance between the nanopores, d_p the nanopores diameter, L the total thickness of the AAO membrane, and l the length of the capillary flow.

A typical example of a nanofluidics experiment monitored by means of dielectric measurements is given in Fig. 6, showing the capillary flow of a ferroelectric polymer (polyvinylidene fluoride co trifluoroethylene, PVDF-TrFE, 75 % PVdF and 25 % TrFE, Mn = 79,000 g/mol, PDI = 1.37) into nanopores of 40 nm diameter. The polymer PVDF-TrFE, in a solid state at room temperature, was placed on top of the empty nanofluidics cell and heated up above its melting point to initiate the flow process. The global permittivity of the cell was continuously measured during heating as well as isothermally at 440 K to monitor the flow process. At the beginning of the heating program, as long as the temperature is still smaller than the melting point of the polymer, the measurement reflects only the response of the empty sample cell, which does not show a significant time or temperature dependence. At the moment when the temperature is reaching the melting point of PVDF-TrFE (410 K), the polymer begins to flow into the nanopores leading to a pronounced increase in the permittivity of the sample cell. The kinetics of the flow process can be readily investigated by monitoring the time evolution of the global permittivity of the sample cell. According to Eqs. 3–5, the increase in the global permittivity measured during the flow process is related to the increasing length of the flow distance l.

Frequency-dependent measurements of the complex dielectric permittivity, and in particular of the dielectric loss, open the possibility to investigate molecular relaxations during the flow process. One essential condition for this type of investigations is that the flow experiment is carried out at a temperature at which a molecular relaxation is detectable in the frequency window of the measurement. The dielectric response of the nanofluidics cell is measured, as a function of frequency and time, to monitor the changes in the dielectric spectra taking place during the flow process. As the flow proceeds in time, an increase in the dielectric loss is observed, which emerges into a relaxation peak of increasing intensity. Two fundamentally different aspects of the flow process are captured in a single experiment: (a) the kinetics of the capillary flow, reflected by the rate at which the dielectric signals are changing in time and (b) molecular relaxations investigated during the flow process.

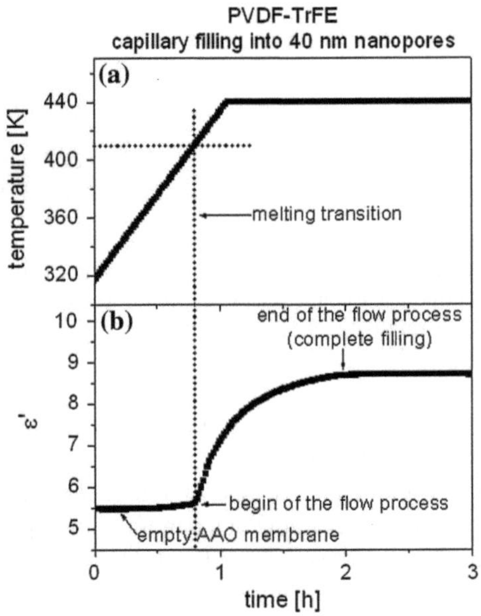

Fig. 6 **a** The temperature program of a nanofluidics experiment. The polymeric material (PVDF-TrFE), in a solid state at room temperature, is placed on top of the nanofluidics cell and the sample is heated above its melting point to initiate the flow process; **b** the kinetics of the capillary flow is monitored in real-time by measuring the global permittivity of the nanofluidics cell

An example of molecular relaxation measured during the capillary flow is given in Fig. 7, showing the alpha relaxation process of poly-2-vinylpyridine (P2VP, Mw = 35,000 g/mol, PDI = 1.04, purchased from Polymer Source Inc.) flowing into nanopores of 200, 40, and 18 nm diameter. The weak relaxation peaks observed at the beginning of the nanofluidics experiment (at time $t = 0$, defined as the time when the sample reached the setpoint temperature) is due to the fact that the P2VP begins to flow into the nanopores before the setpoint temperature is attained. The slight increase in the dielectric loss observed at high frequencies (i.e., above 100 kHz) is due to the contact resistance of the electrodes employed in this study. On the low frequency side, the linear increase in the dielectric loss with decreasing frequency is due to the conductivity contribution of the polymer sample. The relaxation peak of increasing intensity corresponds to the segmental dynamics of P2VP measured during the flow process. Down to a pore diameter as small as 18 nm, no considerable shifts in the position of the alpha relaxation are observed, as compared to the segmental dynamics of P2VP in the bulk. This indicates that the length scale of the interfacial interactions of P2VP with the walls of the nanopores is much shorter than the pore radius of 9 nm. It indicates furthermore that possible changes in the chain conformation induced by the influence of the capillary force which promotes the flow process do not significantly affect the segmental dynamics down the length scales of geometrical confinement comparable in size to the end-to-end distance of the polymer chains.

In addition to investigations on molecular dynamics during flow, the data presented in Fig. 7 can be also used to characterize another important aspect related to the flow process: the kinetics of the capillary filling. This is exemplified in Fig. 8, which shows

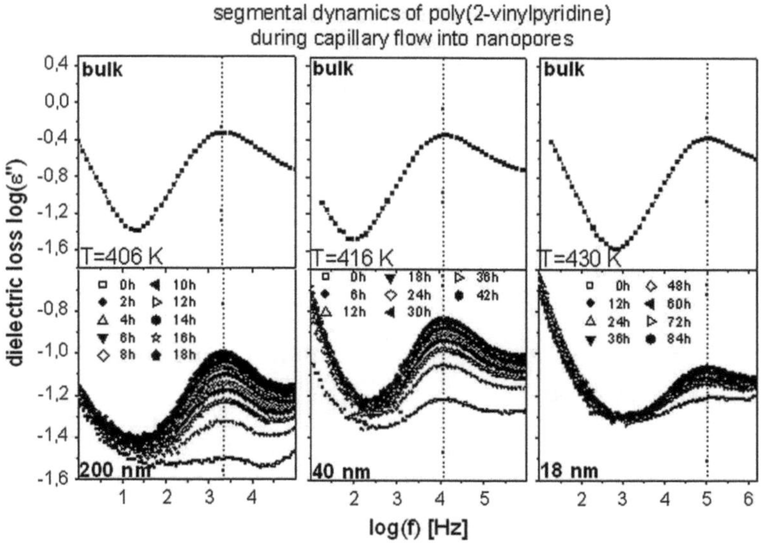

Fig. 7 Segmental dynamics of P2VP during capillary flow into nanopores of different diameters, as indicated. The results are compared to the segmental relaxation of P2VP in the bulk (*upper row*). The increasing intensity of the relaxation peaks is related to the kinetics of the capillary flow while the peak positions give the mean relaxation rate of the dynamic glass transition measured during the flow process

the time evolution of the intensity of the relaxation peaks, taken at the frequency position of the peak maxima. Although the flow process has been carried out at different temperatures for different pore diameters, it is clear that, in accordance to Eq. 1, the kinetics of the capillary filling becomes much slower with decreasing the pore diameter.

A similar nanofluidics approach can be employed, as well, to investigate the flow process of complex fluids. The phase separation phenomenon of block copolymers, for instance, leads to the formation of specific morphologies on the nanometric length scale. Depending on the chemical composition of the block copolymer, different types of nanodomains can be formed: lamellar, cylindrical, spherical. Above the glass transition temperatures of the two blocks and below the order–disorder transition, the block copolymer is in a viscoelastic state and therefore it can flow, while it still retains its nanostructured morphology induced by the phase separation. As compared to the case of homogeneous fluids, phase-separated block copolymers are exposed to multiple static and kinetic constraints during the flow process: the break in symmetry between the copolymer morphology and the cylindrical nanopores, the incommensurability between the pore diameter and the natural period of the phase-separated domains, selective interfacial interactions, the flow frustration caused by the large difference in the viscosity of the copolymer blocks, which, being covalently linked, are constrained to flow together, although they would have, ordinarily, largely different flow velocities.

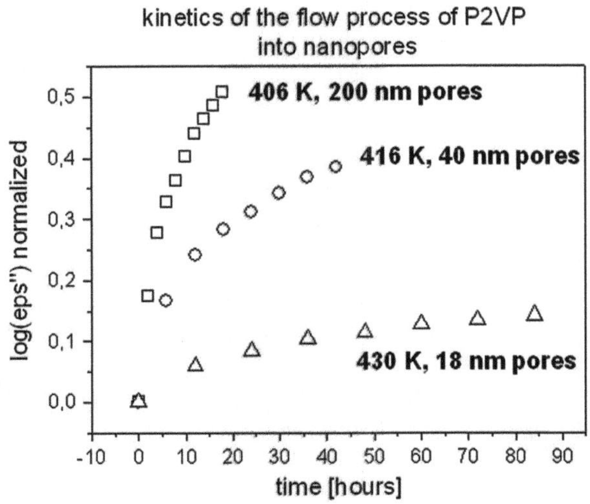

Fig. 8 Kinetics of the capillary flow of P2VP into nanopores of different diameters, as reflected in the increasing intensity of the dielectric loss measured at the frequency position of the peak maxima (see Fig. 7)

An example of nanofluidics experiment with complex (nanostructured) fluids is given in Fig. 9, for polystyrene-block-poly-4-vinylpyridine (PS-b-P4VP, Mw = 9.8 kg/mol for PS, Mw = 10 kg/mol for P4VP, polydispersity index PDI = 1.06, purchased from Polymer Source Inc.). Due to its symmetric chemical composition, this copolymer exhibits a lamellar morphology in the bulk. The formation of this morphology with a domain separation distance of 25 nm has been proved by means of SAXS measurements (data not shown).

During flow of PS-b-P4VP, a first constraint is arising from the geometrical mismatch between the symmetry of the block copolymer (lamellar) and the symmetry of the nanoporous media (cylindrical). Second, between the two blocks of the copolymer there is a considerable difference in their glass transition temperature (Tg(P4VP) − Tg(PS) ≅ 40 °C). This corresponds to a difference by several orders of magnitude in the viscosity of the two blocks that would ordinarily imply—according to Eq. 1—largely different flow velocities. But, due to the chemical bonds that link the PS and the P4VP domains, the two polymer blocks are forced to flow together, at the same flow velocity. One block is therefore flowing much faster while the other one much slower than their "natural" flow velocities. This leads to an additional kinetic "frustration" appearing during the flow process. Additionally, due to the fact that P4VP is much more polar than PS, an interfacial segregation of the copolymer is expected at the walls of the nanopores as a result of the interfacial interactions. Under the influence of these multiple constraints, the segmental dynamics of the P4VP block of the phase-separated PS-b-P4VP was investigated during capillary flow into nanopores of 100 and 50 nm diameter. The contribution of the segmental dynamics of the PS block is not detectable because it is much weaker than that of P4VP. For both pore

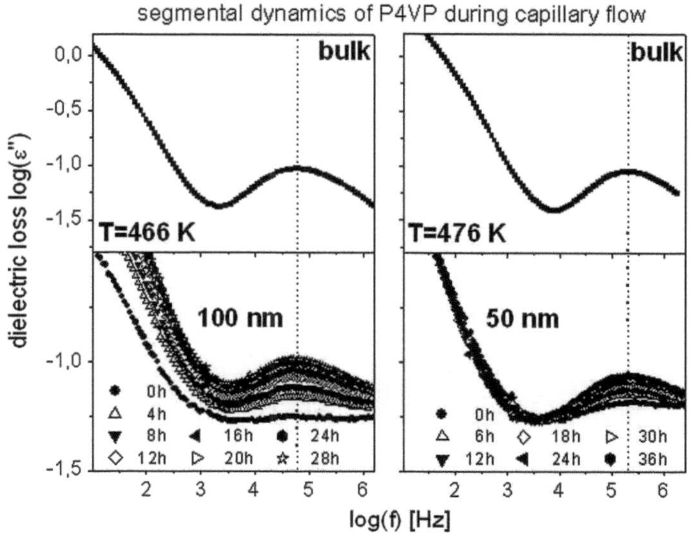

Fig. 9 Segmental dynamics of the P4VP block of phase-separated PS-b-P4VP during capillary flow into nanopores of different diameters, as indicated. The results are compared to the molecular dynamics of the P4VP block in the bulk (*upper row*)

diameters used in this study, no significant shifts in the glassy dynamics of P4VP are observed during the flow process, as compared to the glassy dynamics in the bulk.

To visualize the morphology of the polymer nanowires formed by the flow process, the AAO membranes were dissolved in an aqueous solution of NaOH (5 %) and the solution was filtered to deposit the resulting polymer nanowires on a TEM grid. After that, the samples were stained by iodine vapor to enhance the electronic contrast, followed by investigations by Transmission Electron Microscopy (Fig. 10).

These investigations reveal a change in the polymer morphology taking place under condition of geometrical confinement: the lamellar structure characteristic to the morphology of PS-b-P4VP in the bulk was converted into a morphology consisting of concentric cylinders. This structural transition is induced by the break in symmetry and the curvature imposed on the block copolymer by the walls of the nanopores. In the case of nanopores of 50 nm diameter, it is additionally observed that only one single domain (one cylinder consisting of one P4VP shell with a PS core) is flowing into the nanopores.

5 Conclusions

In addition to "static" factors affecting the polymer dynamics in confinement, kinetic aspects are coming into play when polymer samples are investigated under flow. Additionally, in the case of structured fluids, such as for instance, phase-separated

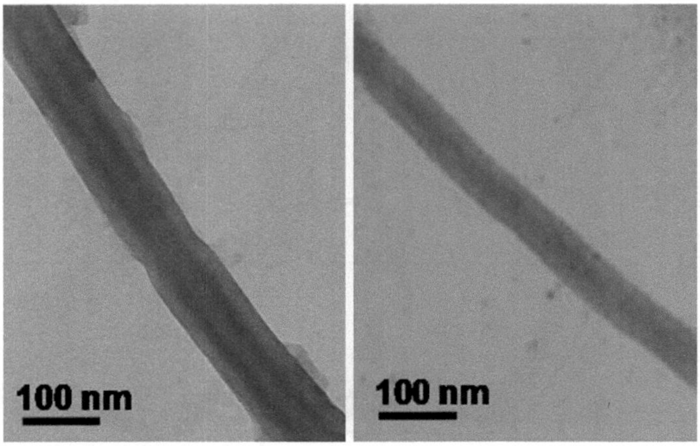

Fig. 10 Images by transmission electron microscopy of the polymer nanowires fabricated by the capillary flow experiment presented in Fig. 9

block copolymers, multiple constraints arise during the flow process in geometrical nanoconfinement. Here, we have shown that different aspects of polymer nanofluidics can be investigated by means of dielectric measurements, using a recently developed experimental approach. The method employs highly ordered nanoporous media as experimental cells and can be used to investigate, in a single experiment, the kinetics of the flow process as well as the molecular dynamics during the capillary flow. It is found that, down to a pore diameter as small as 18 nm, no significant shifts in the molecular dynamics of P2VP are observed, as compared to the molecular dynamics in the bulk. Similar conclusions are drawn for the flow process of phase-separated PS-b-P4VP, down to a geometrical constraint allowing the flow of only one (single) domain of the block copolymer.

References

1. Lagrené K, Zanotti J-M, Daoud M, Farago B, Judeinstein P (2010) Eur Phys J ST 189:231
2. Ok S, Steinhart M, Serbescu A, Franz C, Vaca Chávez F, Saalwächter K (2010) Macromolecules 43:4429
3. Tress M, Mapesa EU, Kossack W, Kipnusu WK, Reiche M, Kremer F (2013) Science 341:1371
4. Mapesa EU, Tress M, Schulz G, Huth H, Schick C, Reiche M, Kremer F (2013) Soft Matter 9:10592–10598
5. Suzuki Y, Duran H, Steinhart M, Butt H-J, Floudas G (2013) Soft Matter 9:2621
6. Krutyeva M, Wischnewski A, Monkenbusch M, Willner L, Maiz J, Mijangos C, Arbe A, Colmenero J, Radulescu A, Holderer O, Ohl M, Richter D (2013) Phys Rev Lett 110:11
7. Suzuki Y, Duran H, Akram W, Steinhart M, Floudas G, Butt H-J (2013) Soft Matter 9:9189
8. Dimitrov DI, Milchev A, Binder K (2007) Phys Rev Lett 99:054501
9. Schönhals A, Goering H, Schick C, Frick B, Zorn R (2003) Eur Phys J E Soft Matter 12:173
10. Serghei A, Chen D, Lee DH, Russell TP (2010) Soft Matter 6:1111

11. Serghei A, Zhao W, Wei X, Chen D, Russell TP (2010) Eur Phys J Spec Top 189:95
12. Serghei A, Huth H, Schick C, Kremer F (2008) Macromolecules 41:3636
13. Serghei A, Kremer F (2008) Rev Sci Instrum 79:026101
14. Serghei A, Kremer F (2006) Rev Sci Instrum 77:116108
15. Masuda H, Fukuda K (1995) Science 268:1466
16. Masuda H, Hasegwa F, Ono S (1997) J Electrochem Soc 144:127
17. Zhao S, Roberge H, Yelon A, Veres T (2006) J Am Chem Soc 128:12352
18. Washburn EW (1921) Phys Rev 17:273

Heterogeneous Dynamics of Multilayered Thin Polymer Films

Koji Fukao, Hirokazu Takaki and Tatsuhiko Hayashi

Abstract The glass transition and related dynamics of two types of multilayered thin films were investigated using differential scanning calorimetry and dielectric relaxation spectroscopy, to clarify the nature of heterogeneous dynamics in thin polymer films. First, the α-process of multilayered thin films of poly(2-chlorostyrene) (P2CS) and polystyrene (PS) with various geometries was investigated during annealing process. The relaxation rate of the P2CS layer increases near the upper electrode with annealing, while that near the bottom electrode remains almost constant or slightly decreases with annealing. The relaxation strength for the α-process of the P2CS layer increases with annealing near the upper electrode, while it decreases near the bottom electrode. A distinct positional dependence of the α-process could be observed in the multilayered films. Second, the glass transition temperature, T_g, and the dynamics of the α- and β-processes for stacked thin films of poly(methyl methacrylate) (PMMA) were investigated during the annealing process. The T_g and the dynamics of the α-process of as-stacked PMMA thin films exhibit thin-film-like behavior. Annealing at high temperature causes the T_g to increase from the reduced value, and causes the dynamics of the α-process to become slower approaching those of the bulk. Contrary to the α-process, the relaxation time of the β-process is almost equal to that of the bulk PMMA, and is unaffected by the annealing process. The fragility index increases with annealing, which suggests that the glassy state of the stacked thin films changes from strong to fragile.

Keywords Glass transition · Fragility · Heterogeneous dynamics · Interfacial interaction · Multilayered thin films · Stacked thin polymer films

---

K. Fukao (✉) · H. Takaki · T. Hayashi
Department of Physics, Ritsumeikan University, Noji-Higashi 1-1-1, Kusatsu 525-8577, Japan
e-mail: fukao.koji@gmail.com

F. Kremer (ed.), *Dynamics in Geometrical Confinement*, Advances in Dielectrics,
DOI: 10.1007/978-3-319-06100-9_8,

Abbreviations

Al	Aluminum
α_{ps}	Shape parameter of dielectric spectrum of PS
α_{p2cs}	Shape parameter of dielectric spectrum of P2CS
β_{ps}	Shape parameter of dielectric spectrum of PS
β_{p2cs}	Shape parameter of dielectric spectrum of P2CS
β_K	Stretching parameter of KWW function
C^*	Complex electric capacitance
C'	Real part of C^*
C''	Imaginary part of C^*
C^*_{ps}	Complex electric capacitances of PS layer
C^*_{p2cs}	Complex electric capacitances of P2CS layer
DRS	Dielectric relaxation spectroscopy
DSC	Differential scanning calorimetry
d	Thickness
ΔC_{ps}	$\Delta\varepsilon_{ps}$ multiplied by geometrical capacitance of PS
ΔC_{p2cs}	$\Delta\varepsilon_{p2cs}$ multiplied by geometrical capacitance of P2CS
$\Delta\varepsilon_\alpha$	Dielectric relaxation strength of α-process
$\Delta\varepsilon_\beta$	Dielectric relaxation strength of β-process
$\Delta\varepsilon_{ps}$	Dielectric relaxation strength of α-process of PS
$\Delta\varepsilon_{p2cs}$	Dielectric relaxation strength of α-process of P2CS
ΔT_α	Width of dielectric loss peak of α-process
ΔT_β	Width of dielectric loss peak of β-process
ΔT^l_β	Value of low temperature side of ΔT_β
ΔT^r_β	Value of high temperature side of ΔT_β
ε^*	Complex dielectric constant
ε'	Real part of ε^*
ε''	Imaginary part of ε^*
$\varepsilon''_{\max}$	Peak height of dielectric loss of α-process
ε_0	Permittivity of a vacuum
ε^*_{ps}	Complex dielectric constant of PS layer
ε^*_{p2cs}	Complex dielectric constant of P2CS layer
$\varepsilon_{ps,\infty}$	Dielectric permittivity at very high frequency of PS
$\varepsilon_{p2cs,\infty}$	Dielectric permittivity at very high frequency of P2CS
f	Frequency of applied electric field
f_α	Relaxation rate of α-process
f_β	Relaxation rate of β-process
f_{g}	Relaxation rate of α-process at T_{g}
$f_{\max}$	Peak frequency at which dielectric loss shows a maximum
HN eq.	Havriliak-Negami equation

KWW	Kohlrauch-Williams-Watts relaxation function
k_B	Boltzmann constant
LCST	Lower critical solution temperature
ℓ	Total thickness
ℓ_{ps}	Thickness of PS layer
ℓ_{p2cs}	Thickness of P2CS layer
M_{w}	Weight-averaged molecular weight
M_{n}	Number-averaged molecular weight
m	Fragility index
μ	Effective dipole moment
N	Number density of relaxing dipoles
P2CS	Poly(2-chlorostyrene)
PS	Polystyrene
PMMA	Poly(methyl methacrylate)
R	Equivalent electrical resistance in sample
S	Area of electrode
σ_{dc}	Electric conductivity
t_a	Effective annealing time
τ	Characteristic time of temporal change in T_α
τ_α	Relaxation time of α-process
τ_{g}	Relaxation time of α-process at T_{g}
τ_{ps}	Relaxation time of α-process of PS
τ_{p2cs}	Relaxation time of β-process of P2CS
τ_K	Relaxation time of KWW function
T_a	Annealing temperature
T_α	Temperature at which the dielectric loss shows a peak due to α-process
T_β	Temperature at which the dielectric loss shows a peak due to β-process
T_{g}	Glass transition temperature
T_α^0	Initial value of T_α
T_α^∞	Final value of T_α
T_V	Vogel temperature
U	Apparent activation energy
VFT law	Vogel-Fulcher-Tammann law
$\phi(t)$	Relaxation function
ω	Angular frequency

1 Introduction

Amorphous materials exhibit glass transition if the material is cooled from a high temperature to a lower temperature under an appropriate cooling condition. The physical mechanism of glass transition has long been investigated by many researchers [1]. However, the mechanism is still in debate. One of the most important key words for understanding the mechanism of glass transition is dynamical heterogeneity

[2, 3]. Recent experiments and simulations [4–6] clearly show the existence of the dynamical heterogeneity, where the molecular motion related to glass transition shows cooperative behavior with respect to spacial correlation in the dynamics. The existence of dynamical heterogeneity can be directly related to the characteristic length scale, which governs the mechanism of glass transition [7].

For the purpose of approaching the characteristic length scale experimentally, the glass transition and related dynamics of thin polymer films have been intensively investigated by various experimental techniques. It has been reported that the glass transition behavior of thin polymer films shows a large deviation from that of the bulk [8–10], with few exceptions [11, 12]. The physical origins for the depression of the glass transition temperature, T_g, are suggested to include surface or interfacial effects, and confinement effects in thin film geometry [13]. With regard to interfacial effects, there have been many reports of surface or interfacial mobile regions in thin films [14–19]. From these results, it can be elucidated that there is mobile surface and/or interfacial regions which play a crucial role especially in determining the glass transition dynamics in thin film geometry.

Several measurements have also revealed a distribution of T_g in polymer films, where T_g is varied with the distance from the surface or the substrate. Labeling method is a powerful measurement technique for the investigation of T_g distributions. This involves the insertion of a thin polymer layer labeled with a dye or deuterated atom into unlabeled layers to form multilayered thin films so that only the signal from the labeled layer can be extracted selectively from the overall observed signal [10, 20–26]. Ellison and Torkelson measured T_g at a given distance from the substrate or from the free surface using fluorescence measurements for a labeled thin polymer layer inserted into two unlabeled polymer layers [20]. Inoue and coworkers observed a strong positional dependence of T_g in layered thin films of poly(methyl methacrylate) (PMMA) and polystyrene (PS) using neutron reflectivity measurements [23–25]. Napolitano and coworkers showed that the deviation of T_g and the relaxation strength from the bulk values were dependent on the position of the labeled thin layer in the multilayered thin films using dielectric relaxation spectroscopy (DRS) measurements [10, 26].

In stacked films of many thin polymeric layers, each layer with a thickness of several tens of nanometers, it is expected that the interfacial interaction between the thin polymer layers is much more enhanced when compared to the bulk system. Hence, the stacked thin polymer film can be regarded as an ideal system, in which the effect of interfacial interaction on the dynamics of molecular motion of the polymer films can be investigated. Recent investigations revealed that stacked thin polymer films exhibit glass transition behavior typical of ultra thin polymer films and also that of the bulk systems, but this is dependent on the annealing conditions [27, 28]. This could provide a possibility for the control of T_g for thin polymer films by a change in the interfacial interaction.

In our previous studies, we have investigated the effect of annealing on T_g and the dynamics of the α-process for stacked thin films consisting of PS or poly(2-chlorostyrene) (P2CS) have been investigated using differential scanning calorimetry (DSC) and DRS during the isothermal annealing process above T_g [29–31]. We have successfully observed the depression of T_g and the relaxation time of the α-process, τ_α, for as-stacked thin films of PS or P2CS, similar to the single thin polymer films [8, 32, 33]. Furthermore, an increase of T_g and τ_α up to the values of the bulk system was observed with increasing annealing time, which suggests that the interfacial interactions between thin polymer layers play an important role in the depression of T_g for thin polymer films. Although the change in the interfacial interactions of stacked thin polymer films by annealing should depend on the interface position in the stack, the observed results are averaged along the direction normal to the surface of each single layer. Hence, the positional dependence of the annealing effects on the interfacial interactions in stacked thin polymer films is required to discuss the mechanism of the change in T_g for stacked thin polymer films with respect to the interfacial interactions.

In complex polymeric systems, there are several dynamical modes in addition to the α-process, which is related to the segmental motion of polymer chains. In PMMA, the β-process is observed in a frequency-temperature region separated from that of the α-process. The dynamics of PMMA has been investigated using various experimental techniques, such as nuclear magnetic resonance [34], dynamics light scattering [35], and DRS [36–40]. On the basis of the experimental evidence, the molecular origin of the β-process is thought to be due to the hindered rotation of the –$COOCH_3$ group about the C–C bond attached to the main chain, but including an intermolecular cooperativity. Here, we question whether different dynamical modes are affected by the interfacial interaction in similar or different ways. It has been reported from the measurements using a photobleaching technique that PMMA has a mobile surface layer of 4 nm thickness at the bulk T_g [41, 42]. Hence, it is expected that the as-stacked PMMA thin films will keep a mobile interfacial region and have a thin-film-like glass transition behavior, as observed in stacked PS or P2CS thin films [30].

In our present study, thermal and dielectric measurements have been made on two different types of multilayered thin polymer films during annealing process using DSC and DRS. The results and discussions obtained from such measurements will be described in this review. First, the glass transition behavior of multilayered films of P2CS and PS was investigated using DRS to determine positional dependence of the annealing effects on interfaces between two different polymers [43]. If the dynamics of the α-process have positional dependence along the direction normal to the surface of the thin polymer films, then it can be expected that the effect of annealing on the α-process of stacked thin polymer films should exhibit a similar positional dependence. In this study, we have attempted to elucidate a positional dependence on the annealing effects related to the α-process. Such information will be required to discuss the origin of T_g depression in thin polymer films. Second, the T_g and the dynamics of stacked PMMA thin films during the annealing process have been investigated using DSC and DRS, to elucidate how the interfacial interaction affects the T_g and the dynamics of segmental and local motions [44]. Better understanding of the mechanisms of the glass transition in thin polymer films will be promoted through the two measurements.

In this review, the experimental details are described in Sect. 2. Results and discussions on the topics of "the α-process of multilayered films of P2CS and PS" will be given in Sect. 3. Then, those on the topics of "Segmental and local dynamics of stacked thin films of PMMA" will be given in Sect. 4. Finally, conclusions and remarks will be done in order to compare the results of the present works with those of the previous ones in Sect. 5.

2 Experiments

2.1 Sample Preparation for Multilayered Thin Films of P2CS and PS

The P2CS and PS specimens used in this study were purchased from Polymer Source, Co. Ltd. The molecular weights and molecular weight distributions are as follows: $M_w = 3.62 \times 10^5$ and $M_w/M_n = 2.11$ for P2CS, and $M_w = 2.24 \times 10^6$ and $M_w/M_n = 1.18$ for PS. Five different P2CS/PS multilayered film geometries were prepared, as shown in Fig. 1. For the upper type, a 195 nm thick PS layer is prepared on an Al-deposited glass substrate by spin coating from toluene solution. A 19 nm thick P2CS layer is also prepared on a glass substrate in the same manner and the film is then floated on the water surface and transferred onto the pre-mounted PS layer on the Al-deposited glass substrate. Al is then vacuum deposited to serve as an upper electrode. For the upper− type, a second 7 nm thick PS layer is an addition to the structure of the upper type geometry and is placed between a 15 nm thick P2CS layer and the upper Al electrode. The thickness of the PS layer beneath the P2CS layer is 147 nm. For the middle type, a 16 nm thick P2CS layer is placed between two 78 nm thick PS layers, and Al electrodes are located on both sides of the PS/P2CS/PS tri-layered film. For the bottom type, a 13 nm thick P2CS layer is prepared on an Al-deposited glass substrate by spin coating, upon which a 127 nm thick PS layer is then mounted before Al is deposited to serve as an upper electrode. For the bottom+ type, a 7 nm PS layer is added to the bottom structure between a 14 nm thick P2CS layer and the lower Al electrode. The thickness of the PS layer above the P2CS layer is 137 nm. In this study, the thickness is controlled by changing the concentration of the solution and the rotational speed of the spin coater. The total thickness of the multilayered thin film is evaluated from the electrical capacitance at 100 kHz and 273 K. The thicknesses of some of the thin films were directly measured using atomic force microscopy. In these five samples, the P2CS layer has larger polarity than the PS layer does, so that the P2CS layer can be regarded as a labeled layer in the multilayered thin polymer films. The positional dependence of the glass transition dynamics can be evaluated by extracting the dielectric signal of the P2CS component from the overall signal.

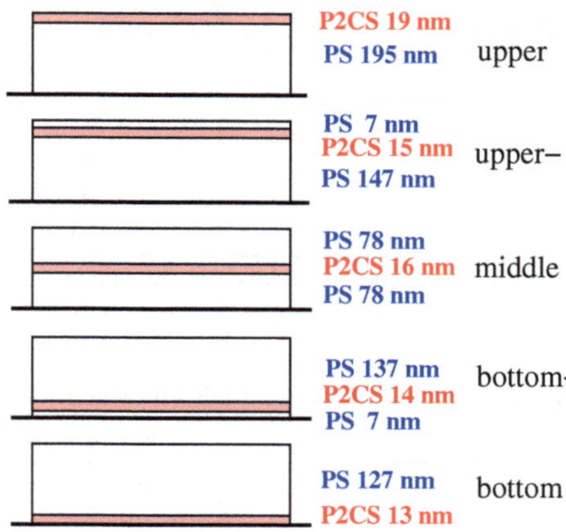

Fig. 1 Five different P2CS/PS multilayered film geometries employed in the present study. (1) *upper*: substrate+PS (195 nm)+P2CS (19 nm), (2) *upper−*: substrate+PS (147 nm)+P2CS (15 nm)+PS (7 nm), (3) *middle*: substrate+PS (78 nm)+P2CS (16 nm)+PS (78 nm), (4) *bottom+*: substrate+PS (7 nm)+P2CS (14 nm)+PS (137 nm), (5) *bottom*: substrate+P2CS (13 nm)+PS (127 nm)

2.2 Sample Preparation for Stacked PMMA Thin Films

The PMMA used in this study was purchased from General Science, Co. Ltd. The weight-averaged molecular weight is $M_{\mathrm{w}} = 5.4 \times 10^5$. Stacked thin films are prepared as follows [30]: single ultra thin films of various thicknesses are prepared on a glass substrate by spin coating from a toluene solution. The thickness of a single thin layer, d, is controlled by changing the concentration of the solution and evaluated from direct measurements using atomic force microscope [32, 45]. The film is floated onto the surface of water and transferred on to the top of the substrate or previously stacked thin films of polymers on the substrate. This procedure using ultra thin films of the same thickness is repeated until the total number of the stacked thin layers reaches tens to a several hundred, as listed in Tables 1 and 2. For the dielectric measurements, the first layer of PMMA is prepared directly onto the gold-deposited glass substrate, and the succeeding thin layers are prepared according to the above mentioned procedure. Gold is then vacuum-evaporated again to serve as the upper electrode. For the DSC measurements, a Teflon plate is used as the substrate. The stacked thin films consisting of several hundred thin layers are detached from the Teflon plate using a razor.

2.3 Dielectric Spectroscopy Measurements

LCR meters (Agilent Technology, 4284A and/or E4980A) are used to conduct dielectric measurements, and obtain the complex dielectric constant ε^* and the complex electric capacitance C^*. Here, $\varepsilon^* = \varepsilon' - i\varepsilon''$ and $C^* = C' - iC''$. One measurement to cover the frequency, f, ranging from 20 Hz to 1 MHz, required approximately 50 s.

Table 1 Samples prepared for DSC measurements

d (nm)	No. of layers	T_g(K)
bulk	1	389.6
80	60	386.7
48	100	384.0
15	300	380.8

The thickness of the single layer, d, the number of stacked layers, and the glass transition temperature, T_g, determined from the DSC measurements of stacked PMMA thin films

Table 2 Samples prepared for DRS measurements

d (nm)	No. of layers	T_α (K) (100 Hz)
65	10	402.8
41	30	401.5
15	50	399.5

The thickness of the single layer, d, the number of the stacked layers, the temperature where the dielectric loss shows a peak due to the α-process, T_α, at 100 Hz

Multilayered thin films of P2CS and PS:

Dielectric measurements are performed repeatedly for the following temperature cycles. One temperature cycle consists of heating from 273 to 423 K at a rate of 1 K/min and then cooling from 423 to 273 K at the same rate. It can be expected that effective annealing proceeds linearly with the number of thermal cycles. Hence, the number of thermal cycles can be regarded as the effective annealing time. The observed dielectric data are corrected for further analysis, according to the procedures detailed in our previous study [32].

Stacked PMMA thin films:

The stacked PMMA thin films for DRS measurements are prepared, as shown in Table 2. The temperatures of the samples are then changed according to the program in Fig. 2. Before taking any dielectric measurements, the temperature of the as-stacked PMMA thin films is changed between 273 and 350 K (or 380 K) at the rate of 1 K/min in order to stabilize the samples for the dielectric measurements. The following temperature cycle is then repeated many times. One temperature cycle consists of two successive parts:
(1) Cooling and heating processes between 409 and 273 K at the rate of 1 K/min, carried out twice.
(2) After the ramping process, the temperature is kept at 409 K for 1 h (isothermal annealing at T_a=409 K).
Throughout all the temperature cycles, dielectric measurements for the frequency range from 1 MHz to 20 Hz are repeated. As the thermal change in stacked PMMA thin films during the ramping process between 273 and 409 K can be much smaller than that during the isothermal annealing process (except for the first ramping process), the total time for which the sample is kept at 409 K can be regarded as the effective annealing time t_a in these measurements.

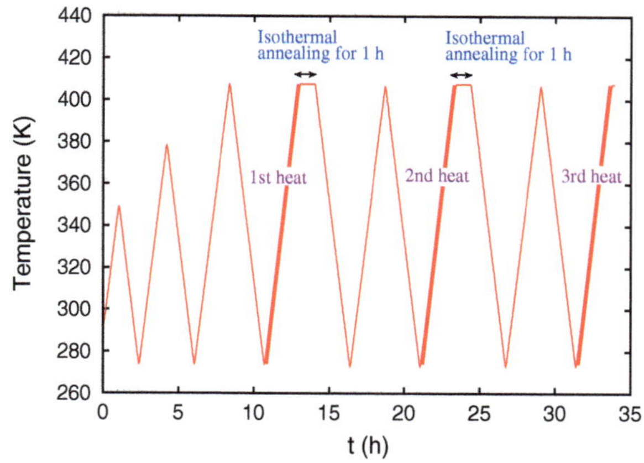

Fig. 2 Temperature cycle assigned to stacked PMMA thin films for DRS measurements. The dielectric data measured during the heating process labeled as 1st heat, 2nd heat, and so on are used for further analysis in this chapter

3 The α-Process of Multilayered Films of P2CS and PS

Experimental results of the α-process of multilayered films of P2CS and PS observed by DRS during annealing process will be discussed in this section.

3.1 Dielectric Loss Spectra

Figure 3 shows the frequency dependence of the dielectric loss observed for a 225 nm thick single PS thin film at various temperatures above T_g. Both the horizontal and vertical axes are normalized with respect to the maximum position ($\log_{10} f_{\max}, \varepsilon''_{\max}$). The loss peak observed in Fig. 3 is due to the segmental motion of PS chains, i.e., the α-process. The solid curve is drawn for the model function of the Havriliak-Negami equation (Eq. (5)) accompanied by the contribution of the DC conductivity with the best fitting parameters [46], $\alpha_{ps} = 0.66 \pm 0.01$ and $\beta_{ps} = 0.35 \pm 0.01$. Hereafter, these values are used as the two shape parameters α_{ps} and β_{ps} for the dielectric loss spectra of the components only from the PS layer included in the multilayered films.

Figure 4a shows the frequency dependence of the imaginary part of the complex dielectric constant ε'', observed during the first cooling process for the upper− type multilayered thin film. A distinct dielectric loss peak is evident, which is associated with the α-process due to the PS layer. In addition to this peak, an additional contribution can be detected in the lower frequency side of the peak. Normalization of the observed data with respect to the peak position and height of the α-process of the PS layer gives the curves shown in Fig. 4b. The observed dielectric loss and the best-fit curves for the neat 225 nm thick PS film is also shown (gray symbol and curve) for comparison. Figure 4b shows that normalization of the observed data works well in the higher frequency range, and the slope at this side agrees well with that of the 225 nm thick PS thin film. However, there is an appreciable deviation from the gray

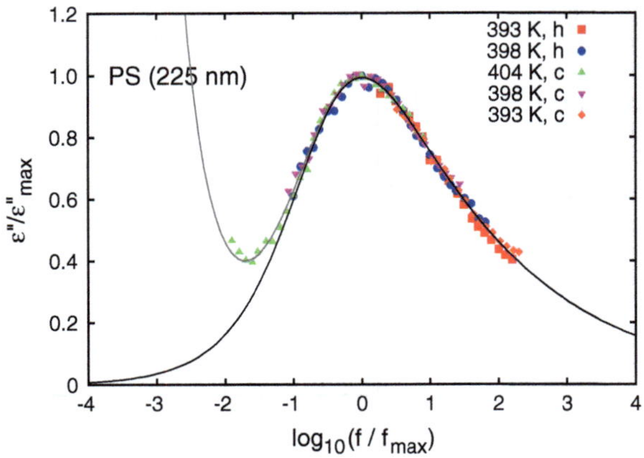

Fig. 3 Frequency dependence of the dielectric loss at various temperatures above T_g for a 225 nm thick single PS thin film. Both the frequency and dielectric loss are normalized with respect to the maximum position due to the α-process. The *solid curves* are obtained using the Havriliak-Negami equation with the DC conductivity component. The best-fitting parameters were: $\alpha_{ps} = 0.66 \pm 0.01$, $\beta_{ps} = 0.35 \pm 0.01$, $\tau_{ps} f_{\max} = 0.68 \pm 0.01$, $\sigma_{dc} = (2.1 \pm 0.7) \times 10^{-15}$ (S/m) and $\Delta\varepsilon_{ps}/\varepsilon''_{\max} = 5.55 \pm 0.01$

curve in the lower frequency range, the extent of which is strongly dependent on the temperature. This suggests that there is a contribution to the dielectric loss from the α-process of the P2CS layer in the multilayered films. A similar frequency dependence of the dielectric loss is also observed for the other upper, middle, bottom+, and bottom types of multilayered films.

Figure 5 shows the frequency dependence of the imaginary part of the complex dielectric constant for the *bottom* and *upper* type multilayered thin films at 413 K for the 1st to the 30th (35th) annealing cycles. Similar results were observed for the other types of PS/P2CS multilayered films. The frequency dependence of the dielectric loss clearly changes with the progress of effective annealing. To extract the change in the contribution to the dielectric loss spectrum from only the P2CS layer in the multilayered films, it is assumed that the overall frequency dependence of the observed electric capacitance can be described by an equivalent serial circuit of two condensers, each of which corresponds to the contribution from PS and that from P2CS, accompanied by the contribution from the DC conductivity. With this assumption, the observed complex electric capacitance, $C^*(\omega)$, can be expressed as:

$$C^*(\omega) = \frac{C^*_{ps}(\omega) C^*_{p2cs}(\omega)}{C^*_{ps}(\omega) + C^*_{p2cs}(\omega)} + \frac{1}{i\omega R}, \tag{1}$$

where C^*_{ps} and C^*_{p2cs} are the complex electric capacitances of the PS layer(s) and P2CS layer, respectively, R is the equivalent electrical resistance in the sample condenser, and $\omega = 2\pi f$. These quantities are given as:

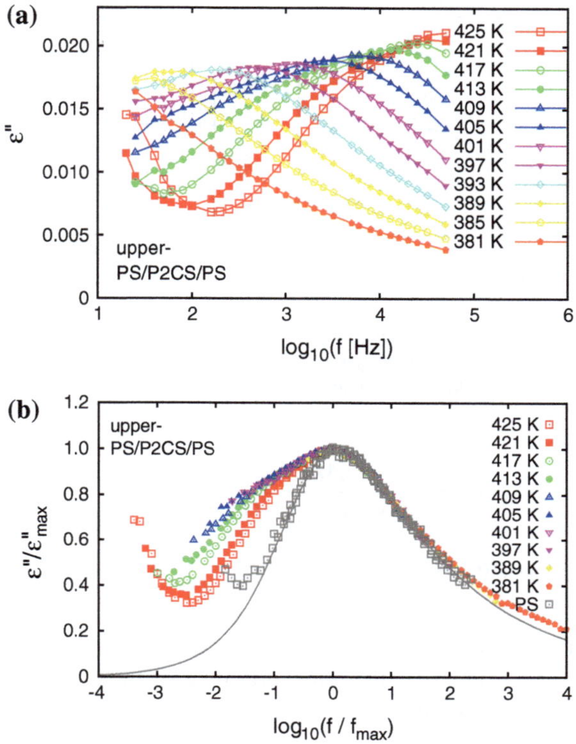

Fig. 4 **a** Frequency dependence of the imaginary part of the complex dielectric constant at various temperatures during the first cooling process for the upper− type multilayered thin film of P2CS and PS. **b** Normalized dielectric loss as a function of normalized frequency for the upper− type multilayered thin film of P2CS and PS. The data for the 225 nm thick single PS thin film are also shown as a *gray curve*

$$C^*(\omega) = \varepsilon^*(\omega)\varepsilon_0 \frac{S}{\ell} \tag{2}$$

$$C^*_{ps}(\omega) = \varepsilon^*_{ps}(\omega)\varepsilon_0 \frac{S}{\ell_{ps}}, \quad C^*_{p2cs}(\omega) = \varepsilon^*_{p2cs}(\omega)\varepsilon_0 \frac{S}{\ell_{p2cs}} \tag{3}$$

$$R = \frac{1}{\sigma_{dc}} \frac{\ell}{S}, \quad \ell = \ell_{ps} + \ell_{p2cs}, \tag{4}$$

where $\varepsilon^*(\omega)$ is the total complex dielectric constant, ε_0 is the permittivity of a vacuum, S is the area of the electrode, ℓ is the total thickness, ℓ_{ps} and ℓ_{p2cs} are the thicknesses of the PS and P2CS layers, respectively, and σ_{dc} is the electric conductivity. $\varepsilon^*_{ps}(\omega)$ and $\varepsilon^*_{p2cs}(\omega)$ are the complex dielectric constants of the PS and P2CS layers at angular frequency ω, respectively, and are given by the Havriliak-Negami equations (HN eq.) [46]:

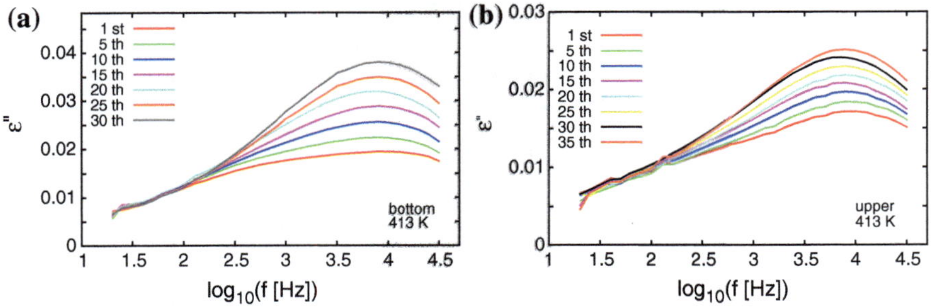

Fig. 5 Frequency dependence of the imaginary part of the complex dielectric function ε'' at 413 K during annealing of the bottom and upper type multilayered thin films. The curves show the evolution of $\varepsilon''(\omega)$ from the 1st to the 30th (35th) annealing cycle

$$\varepsilon^*_{ps}(\omega) = \frac{\Delta\varepsilon_{ps}}{(1+(i\omega\tau_{ps})^{\alpha_{ps}})^{\beta_{ps}}} + \varepsilon_{ps,\infty}, \tag{5}$$

$$\varepsilon^*_{p2cs}(\omega) = \frac{\Delta\varepsilon_{p2cs}}{(1+(i\omega\tau_{p2cs})^{\alpha_{p2cs}})^{\beta_{p2cs}}} + \varepsilon_{p2cs,\infty}, \tag{6}$$

where $\Delta\varepsilon_i$ is the dielectric relaxation strength, α_i and β_i are the shape parameters of the dielectric spectrum, τ_i is the relaxation time of the corresponding dynamical process, and $\varepsilon_{i,\infty}$ is the dielectric permittivity at very high frequency. Here, i represents *ps* or *p2cs*.

Using Eqs. (1)–(6), we can reproduce the imaginary part of the observed dielectric constant at 413 K for the multilayered thin films of the bottom type during the annealing process, as shown in Fig. 6. The blue and green curves represent the contributions from the PS and P2CS layers, respectively. It should be noted that the dielectric loss of the P2CS component is reduced by a factor of 1/50. The purple curve indicates the contribution from the DC conductivity. The red circles are the observed values of the imaginary part of the complex dielectric constant and the red curve is that calculated using Eq. (1). Figure 6 shows that the equivalent serial circuit well reproduces the observed dielectric constant as a function of frequency.

Annealing effects on dielectric loss spectra of the α-process in the upper type multilayered thin films of P2CS and PS are also shown in Fig. 7, in order to investigate dependence on the position of the α-process of the P2CS layer in the stacked films. The observed dielectric loss spectra over all the stages of the annealing process can be well reproduced using the model functions given in this section. The annealing effect on the dielectric loss spectra for the P2CS layer is clearly different between the dielectric loss spectra of the bottom and upper type multilayered thin films of P2CS and PS. Although data for the other types of multilayered thin films are not shown here, similar good reproduction was also obtained for the upper−, middle, and bottom+ type multilayered thin films. Thus, the dielectric loss spectrum of the P2CS layer and also that of the PS layer change as the effective aging time elapses for all multilayered thin film types. To evaluate the evolution of the dielectric loss spectrum

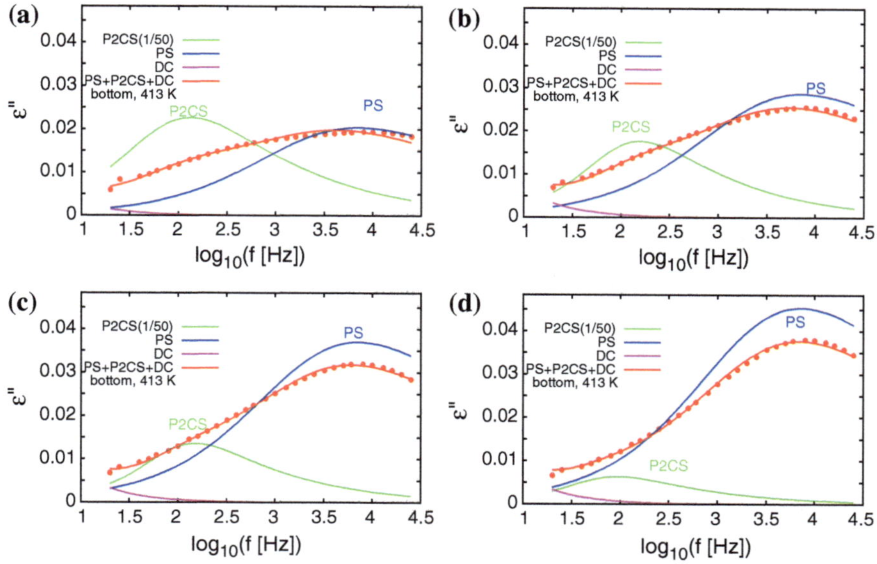

Fig. 6 Frequency dependence of the dielectric loss at 413 K for the *bottom* type multilayered films of P2CS and PS at various stages of the annealing process: **a** 1st, **b** 10th, **c** 20th, and **d** 30th cycles. The *green curve* shows the contribution from the P2CS layer where the vertical axis is $\varepsilon''_{p2cs}\frac{\ell}{50\ell_{p2cs}}$, while the *blue curve* is the contribution from the PS layer where the vertical axis is $\varepsilon''_{ps}\frac{\ell}{\ell_{ps}}$. Here, $\ell_{p2cs} = 13$ nm, $\ell_{ps} = 127$ nm, and $\ell = 140$ nm. The red circles indicate the observed total values of ε'', and the *red curve* shows the calculated ε'' evaluated by data fitting using Eqs. (1)–(6). The parameters $\alpha_{ps} = 0.66$, $\beta_{ps} = 0.35$, $\tau_{ps} = 9 \times 10^{-5}$ s, and $\beta_{p2cs} = 0.50$ were used for all fitting procedures

for the P2CS layer at various positions in the multilayered films, the dependence of several parameters of the dielectric loss spectra for the contributions from the P2CS layer on the number of annealing cycles for the five multilayered thin film types are shown in Sect. 3.2.

3.2 Annealing Effects on Relaxation Rate of the α-Process for the P2CS Layer and Its Distribution

In this section, parameters such as the shape parameter α_{p2cs}, the relaxation rate of the α-process $f_{\max}$, and the relaxation strength $\Delta\varepsilon_{p2cs}$ that describe the dielectric loss spectra of the P2CS layer in the multilayered thin films are evaluated to extract the molecular motion of P2CS chains for various positions in the films.

Figure 8 shows the time evolution of the HN-shape parameter α_{p2cs}, which corresponds to the distribution of the relaxation time of the α-process for the P2CS layer. The upper and middle types of multilayered thin films have smaller

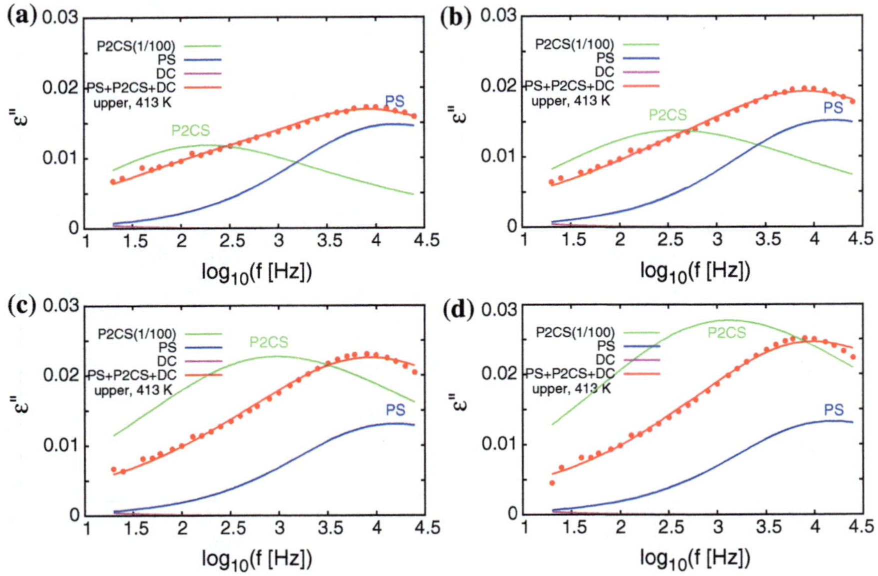

Fig. 7 Frequency dependence of the dielectric loss at 413 K for the *upper* type multilayered films of P2CS and PS at various stages of the annealing process: **a** 1st, **b** 10th, **c** 25th, and **d** 35th cycles. The caption of Fig. 6 is also valid for Fig. 6

α_{p2cs} than the other types, and hence a broader distribution of the relaxation times for the α-process. It should be noted that the other shape parameter, β_{p2cs}, is fixed to be 0.5, so that there is a distribution of the relaxation times, even for $\alpha_{p2cs} = 1.0$. The relation $\beta_K = (\alpha_{p2cs}\beta_{p2cs})^{1/1.23}$ [47] is used to evaluate the exponent of the Kohlrauch-Williams-Watts relaxation function $\phi(t) = \exp(-(t/\tau_K)^{\beta_K})$ as follows: $\beta_K = 0.57$ and 0.32 for the first cycle with the bottom and upper types of multilayered thin films, respectively. The smaller value of the exponent β_K indicates a broader distribution of relaxation times for the α-process.

Next, to evaluate the positional dependence of the α-relaxation rate for the segmental motion in the P2CS layer of the multilayered thin films, the peak frequency $f_{\max}$, at which only the dielectric loss due to the α-process of the P2CS layer shows a maximum, was evaluated using the relation [48]:

$$f_{\max} = \frac{1}{2\pi\tau_{p2cs}} \left(\frac{\sin\frac{\pi}{2}\frac{\alpha_{p2cs}}{\beta_{p2cs}+1}}{\sin\frac{\pi}{2}\frac{\alpha_{p2cs}\beta_{p2cs}}{\beta_{p2cs}+1}} \right)^{\frac{1}{\alpha_{p2cs}}}. \tag{7}$$

Figure 9 shows $f_{\max}$ for various positions in the multilayered thin films as a function of the annealing cycle number. The relaxation rate of the α-process for the 225 nm thick single PS thin film and that for a 20 nm thick single P2CS thin film (from Ref. [30]) are also shown for comparison with that of the P2CS layer at various

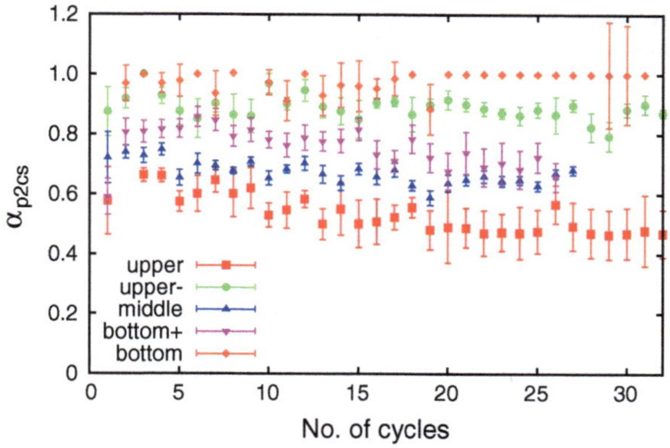

Fig. 8 Dependence of the shape parameter α_{p2cs}, on the number of annealing cycles (at 413 K) for various types of P2CS and PS multilayered thin films

positions in the various multilayered film types. The following results are obtained from Fig. 9:

(1) The relaxation rates of the α-process for P2CS layers with thicknesses of 13–19 nm in all types of the multilayered thin films are almost equal to or slightly larger than that for the 20 nm thick single P2CS film, and smaller than that of the 225 nm thick single PS films.
(2) For the annealing process up to the 10th cycle, the relaxation rate remains almost constant or slightly decreases with the cycle number for all five types of multilayered films. The absolute value of the relaxation rate for the bottom type multilayered thin film is smaller than that for the other types.
(3) At approximately the 10th cycle, the relaxation rate for the upper type multilayered thin film starts to increase monotonically and approach 2×10^3 Hz as the annealing proceeds. For the other four types of multilayered thin films, the relaxation rate remains almost constant or slightly increases. The absolute value of the relaxation rate for the bottom type multilayered thin film is smaller than that for the other types, even after the 10th cycle.

According to the result item (1), the relaxation rate of the α-process for the P2CS layer of an as-stacked multilayered film is larger than that for bulk P2CS, and even larger than that of the 20 nm thick P2CS thin films. However, the relaxation rate for the P2CS layer is smaller than that for bulk PS, which suggests that the thin-film-like dynamics of the α-process and glass transition behavior are maintained under as-stacked conditions or in the earlier stage of annealing, independent of the position of the P2CS layer in the multilayered films.

Furthermore, it was reported in the literature that PS/P2CS is a lower critical solution temperature (LCST) type of a miscible polymer blend with the critical temperature $T_C \approx 410$ K [49], and that the dielectric loss profile of the α-process for PS/P2CS blend at 423 K is broader than that of neat P2CS, while that at 443 K becomes narrower and is almost equal to that of neat P2CS [50]. Here, the narrowing

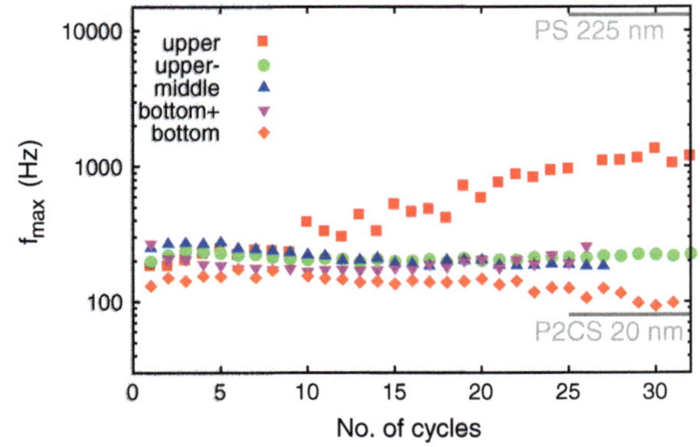

Fig. 9 Evolution of the α-process relaxation rate for the P2CS layer in the PS/P2CS multilayered thin films observed at 413 K

of the dielectric loss spectra can be regarded as a signal that indicates immiscibility. Hence, the annealing between 273 and 423 K is determined to be mainly in the miscible region, or that the effect of immiscibility on the α-process of the P2CS layer in the immiscible region can be negligible, because no narrowing was observed in the present measurements. Therefore, it is expected that the annealing process can promote interdiffusion at the interface between the two layers of different polymers, PS and P2CS. The T_g of PS is lower than that of P2CS; therefore, interdiffusion at the interface can decrease the T_g of the P2CS layer. This could be the reason for the increase in f_{max} during the late stage of annealing observed for all types of multilayered thin films except the bottom type.

For the upper type multilayered thin film, f_{max} begins to increase at approximately the 10th annealing cycle. Hence, the molecular motion at the interface for the upper type is more pronounced than that for the other types. In contrast, for the bottom type, the interaction due to the substrate may compete with the effect of interdiffusion at the interface between P2CS and PS, so that the increase in f_{max} is more suppressed than for the other types. This can also be interpreted as a reduction in the molecular mobility near the substrate. The observed relaxation rate of the α-process for the P2CS layer in the multilayered thin films can be explained in this way.

Here, we will compare the results from this study with previously published results. Ellison and Torkelson measured the fluorescence signal from only the dye-labeled-PS thin layer located between two PS layers with various thickness to selectively obtain the glass transition behavior at various positions in the films [20]. Their results indicate that the T_g of the upper layer mounted on the bulk neat PS layer is lower than the T_g of the bulk, while the T_g of the middle and bottom layers is almost equal to that of the bulk. Inoue et al. showed that the T_g of a deuterated PS layer near the substrate is higher than that of the bulk, while the T_g near the free surface is lower than that of the bulk [23]. In their measurements, the sample preparation conditions for the multilayered thin films were controlled to suppress mutual diffusion as much as possible and confine the labeled polymer chains in the ultra thin layer. Therefore, we can compare these results only with our present results in the early stage of annealing. Although faster dynamics near the upper electrode region are

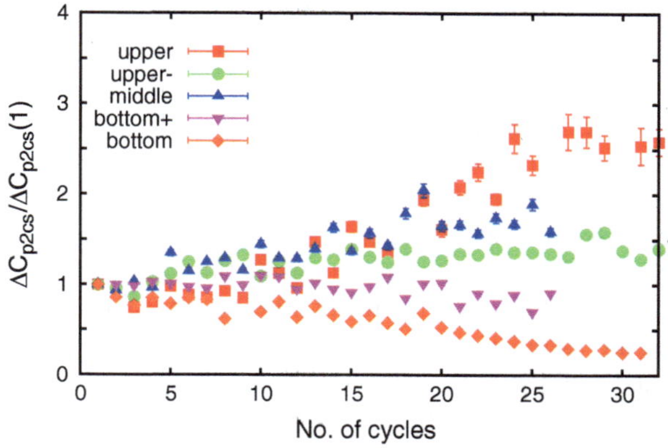

Fig. 10 Dependence of the dielectric strength of the P2CS layer in various types of P2CS/PS multilayered thin films at 413 K, multiplied by the geometrical capacitance at each annealing cycle number. The vertical axis is normalized according to the value at the first cycle

commonly observed in these three studies, the previous results seem to be different from our present results in that the dynamics of the α-process for the middle and bottom type multilayered thin films in this study are faster than those of the bulk, even in the early stage of annealing. The comparison suggests that the as-stacked thin films of our present study can maintain thin-film-like glass transition dynamics for longer annealing times than those in the other studies. Furthermore, studies on stacked PS thin films conducted by Koh and Simon [28] shows that the stacked PS thin films can maintain thin-film-like T_g for a relatively longer time, similar to this study. Hence, the sample preparation conditions may affect the lifetime of the thin-film-like glass transition dynamics; therefore, more detailed studies will be required to confirm this.

3.3 Annealing Effects on Relaxation Strength of the α-Process for the P2CS Layer

Figure 10 shows the evolution of the relaxation strength of the α-process for the P2CS layer in the various types of multilayered thin films, $\Delta C_{p2cs} \equiv \Delta\varepsilon_{p2cs}\varepsilon_0 \frac{S}{\ell_{p2cs}}$. ΔC_{p2cs} increases with increasing number of annealing cycles for the upper and middle types, while it decreases for the bottom type. It should be noted that ΔC_{p2cs} is controlled by two factors; the variation in the dielectric strength $\Delta\varepsilon_{p2cs}$, and the variation in the effective thickness of the P2CS layer ℓ_{p2cs} in the multilayered thin film. During the annealing process, interdiffusion between the P2CS and PS layers is promoted, which increases the effective thickness of the P2CS layer. Hence, the second factor should decrease the value of ΔC_{p2cs} with annealing. On the other hand, the value of $\Delta\varepsilon_{p2cs}$ can be controlled by the relation:

$$\Delta\varepsilon_{p2cs} = \frac{N\mu^2}{3k_B T}, \tag{8}$$

where N is the number density of relaxing dipoles, μ is the effective dipole moment, and k_B is the Boltzmann constant. Therefore, if the change in $\Delta\varepsilon_{p2cs}$ competes with the change in ℓ_{p2cs}, then the value of ΔC_{p2cs} increases or remains constant; otherwise, it decreases with annealing.

For the upper and middle types of multilayered thin films, ΔC_{p2cs} increases with annealing, as shown in Fig. 10. This result suggests that the density of the dipole moments, which are associated with the orientational polarization, becomes larger with annealing and overcomes the effect due to the change in the effective thickness of the P2CS layer. However, the reason for the increase in the density of relaxing dipole moments for the upper type is different from that for the middle type. For the upper type, *higher mobility* induces relaxation of a larger amount of the dipole moments under an applied electric field, because the relaxation rate of the α-process for the upper type is larger than that of the other types, especially during the late stages of annealing.

In contrast, the relaxation rate of the α-process for the middle type is smaller than that for the upper type. However, the P2CS layer is sandwiched by two 78 nm thick PS layers. Therefore, the dipole moments located at the interfaces of both sides can participate in the relaxation phenomena of the α-process for the P2CS layer. As a result, the density of the relaxing dipole moments becomes sufficiently large to increase ΔC_{p2cs} in the same way as for the upper type.

For the bottom type, ΔC_{p2cs} decreases with annealing. In this case, the change in the density of the relaxing dipole moments cannot compete with the change in the effective thickness of the P2CS layer. Therefore, the density of the relaxing dipole moments at the interface should decrease or remain constant. As Napolitano et al. proposed [10], the adsorption of polymer chains onto the Al-deposited glass substrate can reduce the dielectric relaxation strength and the relaxation rate of the α-process. This scenario may also be valid for the present bottom type of multilayered thin films.

4 Segmental and Local Dynamics of Stacked PMMA Thin Films

Experimental results of T_g and segmental and local dynamics of stacked PMMA thin films observed by DSC and DRS during annealing process will be discussed in this section.

4.1 Thermal Properties Determined by DSC

A commercial instrument, Q200 (TA Instruments) is used for DSC measurements. First, the DSC measurements on as-stacked PMMA thin films are done over a heating process from 303 to 420 K at the rate of 10 K/min and then for the subsequent cooling process down to 303 K at the same rate. These measurements are repeated three times in order to obtain the reproducibility of the measurements. Following that, the stacked samples used for the measurements are annealed at 453 K for 12 h in vacuo. The same DSC measurements are then carried out for the heating and cooling processes between 303 and 420 K. A nitrogen gas flow of 30 ml/min is maintained during the DSC measurements.

Figure 11 shows the temperature dependence of the total heat flow observed during the heating process at the rate of 10 K/min for the stacked PMMA films consisting of thin layers of various thicknesses, as shown in Table 1. In Fig. 11a, there is a temperature region where the total heat flow shows anomalous temperature dependence. The averaged location of the anomalous signal can be regarded as T_g, and is determined by DSC. The value of T_g obtained in this manner is 381 K for the as-stacked PMMA thin films consisting of 15 nm thick layers. Because the T_g of the bulk sample is 390 K, the T_g of the as-stacked PMMA thin films of 15 nm layer-thickness is depressed by about 9 K from the bulk value. Table 1 shows that the values of T_g for the as-stacked PMMA thin films of $d = 48$ and 80 nm are 384 and 387 K, respectively. Hence, the T_g of the as-stacked PMMA thin films decreases from the bulk T_g with decreasing thickness of the single layer in the stacked PMMA thin films. Keddie et al. showed that the T_g of PMMA thin films supported on a gold-deposited substrate decreases with decreasing film thickness [51]. On the basis of these experimental results, it can be concluded that the glass transition of the as-stacked PMMA thin films has a thin-film-like character.

Figure 11b shows the temperature dependence of the total heat flow observed for stacked PMMA thin films after annealing at 453 K for 12 h. In Fig. 11b, the T_g for the stacked PMMA thin films is clearly shifted to the higher temperature side after annealing, and converges with the value of the bulk system, although there is a scatter in the measured T_g values. It is expected that annealing at a high temperature above T_g could lead to a reduction and an eventual disappearance in the density gap at the interfaces between the thin layers stacked onto each other. Therefore, the decrease in T_g of the as-stacked PMMA thin films is strongly associated with the existence of the interfaces between the stacked thin layers. It should be noted that the observed results for the stacked PMMA thin films in this study are consistent with our previous experimental results observed for the stacked PS or P2CS thin films [30].

4.2 Dynamics of Stacked PMMA Thin Films Observed by DRS

As shown in Sect. 4.1, the as-stacked PMMA thin films show a thin-film-like glass transition behavior, which changes to the bulk-like glass transition behavior by annealing above T_g. In Sect. 4.2, experimental results observed by DRS for the stacked PMMA thin films during the annealing process are shown, to elucidate the temporal changes in the molecular dynamics of PMMA chains in stacked thin films.

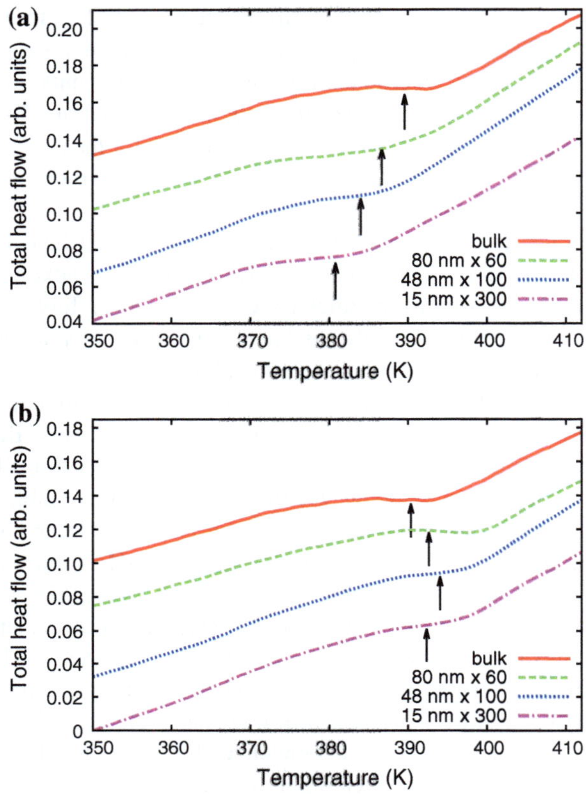

Fig. 11 **a** Temperature dependence of the total heat flow observed by DSC measurements during the heating process for as-stacked thin films of PMMA layers of 80, 48, and 15 nm thickness and bulk PMMA films. The arrows show the glass transition temperatures, which are listed in Table 1. **b** Temperature dependence of the total heat flow observed in the same manner as in (*a*) for the same stacked PMMA thin films and bulk PMMA films after annealing at 453 K for 12 h [44]

4.2.1 Temperature Dispersion of Dielectric Loss Spectra

Figure 12 shows the temperature dependence of the dielectric loss at 100 Hz for the stacked PMMA thin films consisting of 41 nm thick layers during the annealing process at 409 K. In Fig. 12, there are two dielectric loss processes; one is the α-process located around 400 K, and the other is the β-process located at 333 K. The microscopic origin of the α-process is segmental motion, and that of the β-process is local motion, or the motion of the side branch attached to the PMMA main chain, as mentioned in Sect. 1. In Fig. 12, the arrows show that the dielectric behavior of both processes changes with increasing effective annealing time t_a during the annealing process.

In order to evaluate the t_a dependence of the dielectric behavior of the α- and β-processes in more detail, we evaluate several physical quantities such as peak

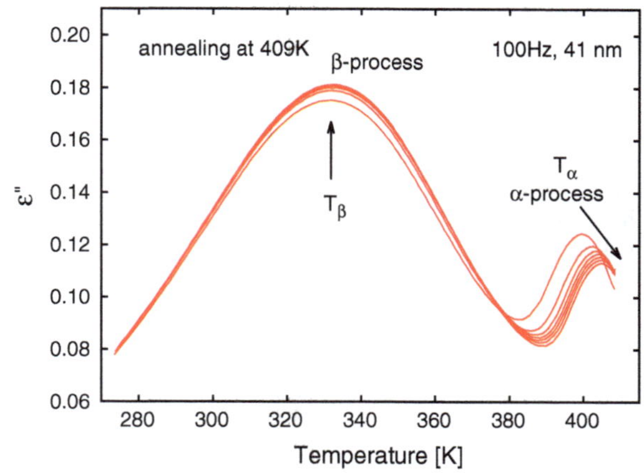

Fig. 12 Dielectric loss spectra in the temperature domain at 100 Hz for stacked thin films of 30 PMMA layers of 41 nm thickness, observed during the annealing process at 409 K for t_a = 0, 1, 4, 8, 12, 16, and 24 h

temperatures, peak widths, and relaxation strengths of the two processes from the observed dielectric loss in the temperature domain. For this purpose, it is assumed that the temperature dependence of the dielectric loss spectra in the temperature domain at a given frequency can be described by the following formula:

$$\varepsilon''(\omega, T) = \sum_{i=\alpha,\beta} \frac{\Delta\varepsilon_i}{1 + ((T - T_i)/\Delta T_i)^2}, \tag{9}$$

where i is α or β, $\Delta\varepsilon_i$ is the relaxation strength of the i-process, T_i is the temperature at which the dielectric loss shows a peak due to the i-process, and ΔT_i is the width of the i-process, which is a measure of the distribution of the relaxation times. Here, it should be noted that the shape of the β-process shows a small deviation from a symmetrical shape in the temperature domain, as shown in Fig. 12, although it is difficult to judge the symmetry of the α-process. In order to take into account the asymmetry of the β-process, we assume that ΔT_β is given by

$$\Delta T_\beta = \begin{cases} \Delta T_\beta^l : T < T_\beta \\ \Delta T_\beta^r : T > T_\beta, \end{cases} \tag{10}$$

while ΔT_α has the same value both above and below T_α.

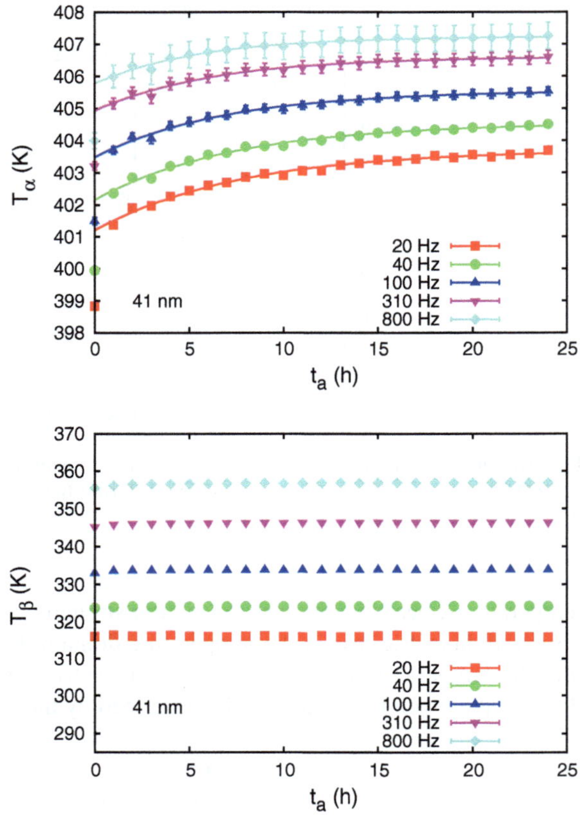

Fig. 13 The dependence on annealing time t_a of the peak temperatures T_α and T_β at which the dielectric loss shows a peak due to the α- and β-processes, respectively, at various frequencies from 20 to 800 Hz for the stacked PMMA films consisting of 41 nm thick layers. The curves given for T_α are evaluated using Eq. (11)

4.2.2 T_α and T_β as a Function of t_a

Figure 13 shows the dependence on annealing time of the peak temperatures, (a) T_α and (b) T_β, at various frequencies from 20 to 800 Hz during the annealing process at 409 K for the stacked PMMA thin films consisting of 41 nm thick layers. In Fig. 13a, T_α increases monotonically from 400.6 to 404.1 K with increasing t_a from 0 to 24 h at 20 Hz. As the frequency increases from 20 to 800 Hz, the absolute value of T_α increases at a given value of t_a, but the T_α has a similar t_a dependence. On the other hand, in Fig. 13b, the peak temperature T_β due to the β-process shows almost no change during the annealing process and remains constant at a fixed frequency. Hence, we can judge that the t_a dependence of the β-process is distinctly different from that of the α-process. As the frequency increases, the absolute value of T_β also increases at a given value of t_a. The temperature and frequency dependence of T_α and T_β of the two processes will be discussed in Sect. 4.2.4.

In order to investigate the universal behavior of the change observed during the annealing process, it is assumed that the t_a dependence of T_α is given by the following simple exponential law:

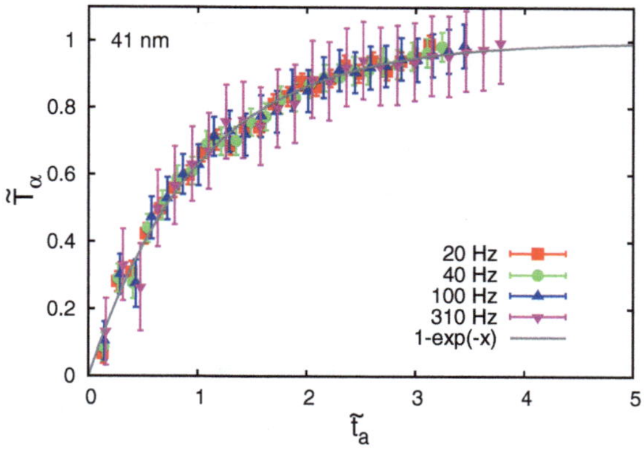

Fig. 14 The dependence of the reduced α-peak temperature $\tilde{T_\alpha}$ on the reduced annealing time $\tilde{t_a}$ at various frequencies for the stacked PMMA thin films consisting of 41 nm thick layers during the annealing process at 409 K. The curve is evaluated from Eq. (12)

$$T_\alpha(t_a) = T_\alpha^0 + (T_\alpha^\infty - T_\alpha^0)\left(1 - \exp\left(-\frac{t_a}{\tau}\right)\right), \tag{11}$$

where $T_\alpha^0 \equiv T_\alpha(0)$, $T_\alpha^\infty \equiv T_\alpha(\infty)$, and τ is the characteristic time of the change in T_α during the annealing process at 409 K. From Eq. (11), the following master equation can be obtained:

$$\tilde{T_\alpha}(t_a) = 1 - \exp(-\tilde{t_a}), \tag{12}$$

where $\tilde{T_\alpha}$ and $\tilde{t_a}$ are defined by

$$\tilde{t_a} \equiv \frac{t_a}{\tau}, \quad \tilde{T_\alpha}(t_a) \equiv \frac{T_\alpha(t_a) - T_\alpha^0}{T_\alpha^\infty - T_\alpha^0}. \tag{13}$$

The curves in Fig. 13 are drawn using Eq. (11), and are very close to the observed values of T_α for a wide range of t_a values, except for the initial value at $t_a = 0$. After rescaling the observed values of T_α using the normalized $\tilde{t_a}$ and $\tilde{T_\alpha}$ defined by Eq. (13), the data for all frequencies between 20 and 310 Hz can well be reduced to a master curve given by Eq. (12), as shown in Fig. 14. This scaling behavior can be observed not only for stacked PMMA thin films consisting of 41 nm thick layers, but also for those of 65 nm and 15 nm thick layers. From fitting the observed values of T_α to Eq. (11), the characteristic time τ can be evaluated as a function of frequency. In Fig. 15, the value of τ for a change in T_α during the annealing process is a decreasing function of f.

These results suggest that there is a universal mechanism for the annealing effects on the α-process over a large frequency range. Only the characteristic time τ changes with the frequency of the applied electric field and the annealing time. In this case, it can be concluded that the change in the dynamics of the α-process occurs from the high frequency side to the low frequency one, i.e., small-scale motion changes earlier than large-scale motion does.

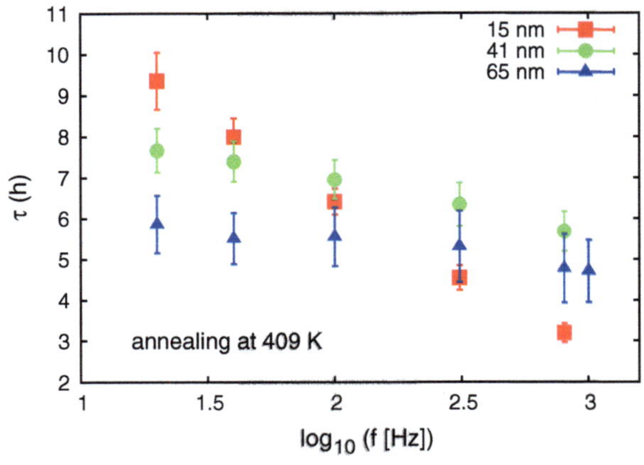

Fig. 15 The frequency dependence of the characteristic time τ of the change in T_α during the annealing process for the stacked thin films consisting of PMMA layers of 15, 41, and 65 nm thickness [44]

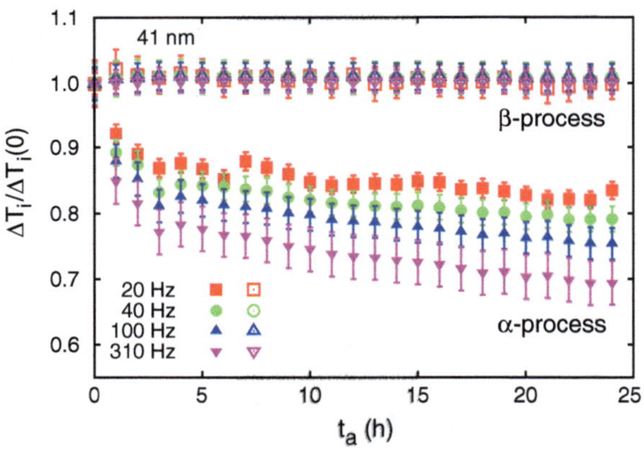

Fig. 16 The annealing time dependence of the relative width of the loss peak due to the α- and β-processes for the stacked PMMA thin films consisting of 41 nm thick layers. *Filled symbols* are for the α-process and *open ones* are for the β-process. The width in the temperature domain at the annealing time t_a for a given frequency is normalized with respect to the value at $t_a = 0$. The average values between ΔT_β^l and ΔT_β^r are used as the width for the β-process

4.2.3 Distribution of Relaxation Times and Relaxation Strengths of the α- and β-Processes

Figure 16 shows the annealing time dependence of the relative width in the temperature domain of the dielectric loss peaks due to the α- and β-processes for the stacked PMMA thin films consisting of 41 nm thick layers at various frequencies. The width is normalized with respect to the value at the initial time $t_a = 0$. This relative width in the temperature domain can be directly associated with the width of the distribution

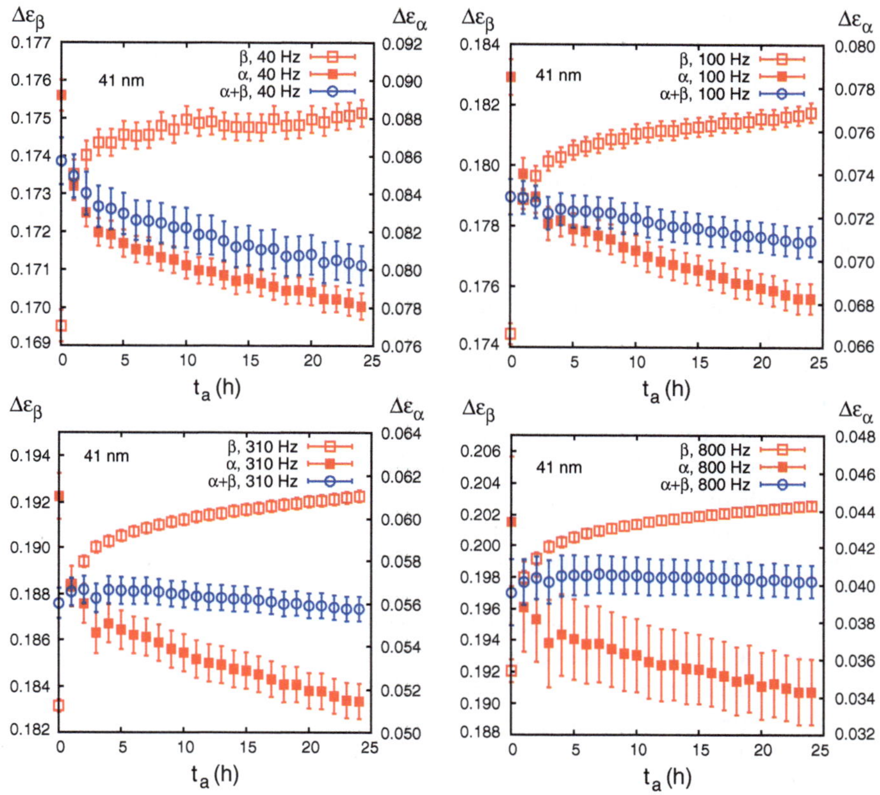

Fig. 17 The annealing time t_a dependence of the relaxation strength of the α- and β-processes at various frequencies from 40 to 800 Hz, for stacked PMMA thin films consisting of 41 nm thick layers. The *open* and *filled squares* stand for the values of the β- and α-processes, respectively. The *open circle* stands for the sum of the relaxation strength of both the β- and α-processes. The values of the sum $\Delta\varepsilon_\alpha + \Delta\varepsilon_\beta$ are shifted along the vertical axis by an arbitrary amount for ease of comparison

of relaxation times of the α-process, τ_α (β-process, τ_β) in the time domain. In Fig. 16, the width ΔT_β remains almost unchanged during the annealing process, while the relative width $\Delta T_\alpha/\Delta T_\alpha(0)$ decreases down to about 0.7 with increasing annealing time t_a. In other words, the distribution of the relaxation times of the β-process does not change with t_a, while that of the α-process becomes narrower with increasing t_a during the annealing process of the stacked PMMA thin films. This suggests that the distribution of the relaxation times of the α-process for the as-stacked PMMA thin films is broader than that of the bulk system. This correspondence between the distribution of τ_α of the as-stacked thin films and of the bulk is quite similar to the one observed for the distribution of τ_α of the single thin films and of the bulk [52]. However, the distribution changes from a thin-film-like one to a narrower one, usually observed in the bulk system, as the time elapses during the annealing process.

Figure 17 shows the annealing time t_a dependence of the relaxation strengths of the α- and β-processes, $\Delta\varepsilon_\alpha$ and $\Delta\varepsilon_\beta$, for the stacked PMMA thin films consisting of 41 nm thick layers during the annealing process at 409 K for various frequencies from 40 to 800 Hz. The sum of the relaxation strength of the α- and β-processes, $\Delta\varepsilon_\alpha + \Delta\varepsilon_\beta$, is also shown in Fig. 17. The relaxation strength of the α-process decreases with annealing time t_a, while that of the β-process increases during the annealing process. However, the sum of the relaxation strengths of both the processes remains almost constant, except for 40 Hz. Assuming that the orientational polarization of the dipole moment is the origin of the dielectric strength $\Delta\varepsilon$, and that the interaction between the orientational motions of the dipole moments can be neglected, the same relation as Eq. (8) is also valid for $\Delta\varepsilon$ in this case. Therefore, the present experimental result suggests that the number density of the dipole moments, which play a crucial role in the α-process of as-stacked PMMA thin films, decreases with t_a, while the number density of the dipole moments associated with the β-process increases at the expense of the number density of the dipoles associated with the α-process. This means that there is a strong coupling between the microscopic origins of the α- and β-processes, although the temperatures T_α and T_β show no correlation with respect to the annealing time dependence. For bulk PMMA, the coupling between the α- and β-processes has been investigated by multidimensional NMR [53] and DRS [54, 55]. Spiess et al. show that in PMMA, the β-process predominantly influences the time scale of the α-process, leading to particularly high mobility of the main chain itself. This coupling observed in bulk PMMA may be related to conversion from the α-process to the β-process.

In this section, the majority of the experimental results discussed are for stacked PMMA thin films consisting of 41 nm thick layers, as these serve as an example. For stacked PMMA thin films consisting of 65 nm and/or 15 nm thick layers, it should be noted that almost similar results are observed, and thus the above results are common amongst stacked thin films of PMMA layers of three different thicknesses.

4.2.4 Dynamics of Stacked PMMA Thin Films

In Sect. 4.2.1–4.2.3, the dielectric relaxation behavior of stacked PMMA thin films was discussed on the basis of the results observed in the temperature domain. Here, further dielectric results are shown for more insightful discussion. Figure 18 shows dispersion maps of stacked PMMA thin films of 41 nm thick layers. The temperature dependence of the relaxation rate of the β- and α-processes is shown in Fig. 18a and b, respectively. Here, the α- and β-peak temperatures T_α and T_β are evaluated from the temperature dispersion of the dielectric loss spectra at a given frequency f, so that the datasets (T_α, f) and (T_β, f) are obtained. These datasets are plotted as the datasets (T, f_α) and (T, f_β) in Fig. 18b, a, respectively.

As shown in Fig. 18a, the relaxation rates of the β-process for various annealing time t_a from 0 to 24 h are sufficiently reduced to a single straight line. This suggests that the dynamics of the β-process of stacked PMMA thin films remain unchanged during the annealing process at 409 K. From the slope of the straight line in Fig. 18a,

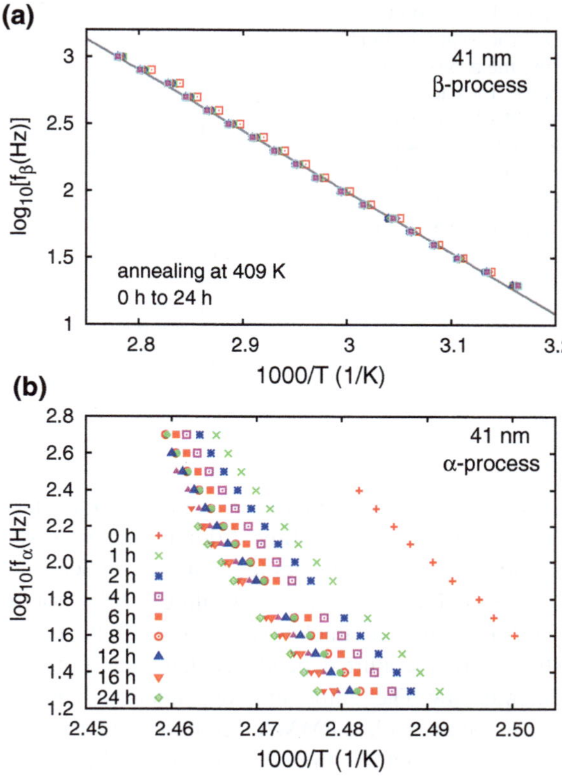

Fig. 18 Dispersion map of α- and β-processes at various annealing times from 0 to 24 h for the stacked PMMA thin films consisting of 41 nm thick layers during the annealing process at 409 K. **a** β-process and **b** α-process. The *straight line* is given by the Arrhenius law with an activation energy of 20.81 $\pm$ 0.04 kcal/mol

we can conclude that the microscopic origin of the β-process is controlled by a thermally activated process with an activation energy of 20.81 $\pm$ 0.04 kcal/mol. This value agrees well with that reported in the literature [37]. Contrary to the β-process, Fig. 18b shows that the temperature dependence of the relaxation rate of the α-process, f_α, changes significantly with increasing annealing time t_a during the annealing process. At a given temperature, f_α decreases with increasing annealing time, which suggests that the dynamics of the α-process become slower with annealing.

In order to discuss in more detail, the temperature dependence of f_α for stacked PMMA thin films, the results obtained by DSC are combined with the results obtained by DRS. The dataset (T_g, f_g) obtained by DSC is plotted in the dispersion map, as shown in Fig. 19. Here, f_g is the relaxation rate of the α-process at T_g, and is defined by $f_g = 1/2\pi\tau_g$, where $\tau_g = 10^3$ s. The DSC results for stacked thin films of PMMA layers with three different thicknesses were obtained in Sect. 4.1. Here, the datasets (T_g, f_g) obtained by DSC for as-stacked thin films consisting of PMMA layers of 80, 48, and 15 nm thickness are coupled with the datasets (T_α, f) observed at $t_a = 0$ h for stacked thin films of PMMA layers of 65, 41 and 15 nm thickness, respectively. Furthermore, the dataset (T_g, f_g) obtained by DSC for the bulk PMMA is coupled

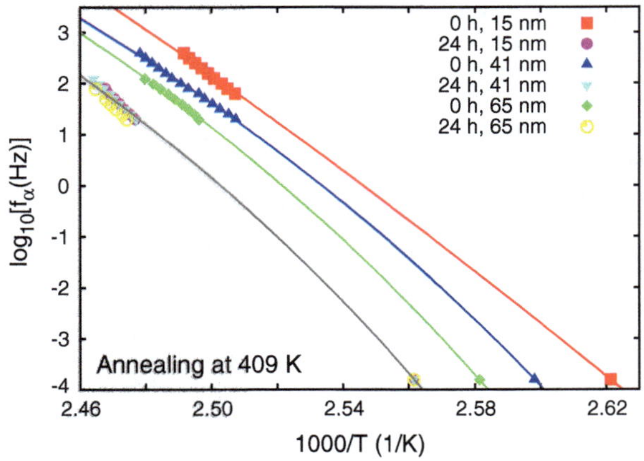

Fig. 19 Dispersion map of the α-process at $t_a = 0$ and 24 h during the annealing process at 409 K, for stacked thin films of PMMA layers with various film thicknesses [44]. The curves are evaluated using Eq. (14) with the best fitting parameters listed in Table 3

with all the data observed at $t_a = 24$ h during the annealing process for stacked thin films of PMMA layers of three different thicknesses.

Figure 19 shows the temperature dependence of the relaxation rate of the α-process before and after the annealing at 409 K for the stacked thin films of PMMA layers of various film thicknesses. In Fig. 19, the relaxation rate of the α-process has a stronger temperature dependence than that of the β-process, which is given by the Arrhenius law. The temperature dependence changes with the thickness of a single layer in the as-stacked PMMA thin films, and also changes with annealing. In order to evaluate this temperature dependence of the relaxation rate of the α-process, the Vogel-Fulcher-Tammann (VFT) law is adopted [56–59]:

$$\tau_\alpha(T) = \tau_{\alpha,0} \exp\left(\frac{U}{T - T_V}\right), \tag{14}$$

where τ_α is given by the relation $\tau_\alpha = 1/2\pi f_\alpha$, $\tau_{\alpha,0}$ is a positive constant, U is the apparent activation energy, and T_V is the Vogel temperature. The best-fitting parameters of VFT law are listed in Table 3. Using the values of U, T_V, and T_g, the fragility index m is also evaluated by the following definition [60]:

$$m = \left(\frac{d \log_{10} \tau_\alpha(T)}{d(T_g/T)}\right)_{T=T_g}. \tag{15}$$

As shown in Fig. 19, the temperature dependence of the relaxation rate of the α-process can be well-reproduced by the VFT law. Furthermore, Table 3 shows that the Vogel temperature T_V increases from 238 to 341 K as the thickness of the single layer in the as-stacked PMMA thin films increases from 15 nm to the bulk. Accordingly, the fragility index m increases from 139 to 192. This suggests that the temperature dependence of τ_α changes from an Arrhenius-like one to a VFT one. In other words, the glassy dynamics of the as-stacked thin films change from strong to

Table 3 The thickness of the single layer d, the glass transition temperature T_g, the Vogel temperature T_V, the apparent activation energy U, and the fragility index m [44]

d (nm)	T_g (K)	T_V (K)	U (K)	m
bulk	390.4	341.0	2756	191.5
65, 80	387.4	335.1	3068	188.4
41, 48	384.9	329.4	3236	175.4
15, 15	381.5	237.9	17283	138.8

fragile, as the thickness of each single layer increases. This thickness dependence of the α-process of stacked PMMA thin films is consistent with that of stacked P2CS thin films [30], and is also consistent with that observed for single thin films of PS [61]. Therefore, it can be concluded that the fragile to strong transition, which is observed by confinement [62–67], is present for stacked thin films of PMMA (and P2CS) as the thickness of each single layer decreases from the bulk to a few nanometer in thickness.

5 Concluding Remarks

In this review, experimental results made on two different types of multilayered thin polymer films using DSC and DRS were shown in Sects. 3 and 4. First, results and discussions on the relaxation behavior of multilayered thin films of P2CS and PS layers with five different geometries observed using DRS were given in Sect. 3. The positional dependence of the annealing effects on the α-process dynamics for the P2CS thin layer in multilayered thin polymer films was successfully observed. The results obtained in Sect. 3 are summarized as follows:

1. The effect of annealing on the relaxation rate of the α-process, f_{max}, for the P2CS layer in multilayered thin films is dependent on the position of the P2CS layer, although the relaxation rate f_{max} of the P2CS layer in the early stage of annealing is almost equal to that of 20 nm thick single P2CS thin films. The relaxation rate of the α-process for the P2CS layer near the upper electrode increases with annealing, while that near the bottom electrode remains almost constant or slightly decreases with annealing.
2. The relaxation strength of the α-process for the P2CS layer in multilayered thin films increases with annealing near the upper electrode and in the middle position of the films, whereas it decreases with annealing near the bottom electrode.

Second, results and discussions on the T_g and the dynamics of the α- and β-processes measured using DSC and DRS for stacked PMMA thin films during the annealing process were given in Sect. 4. The results can be summarized as follows:

1. The T_g and the dynamics of the α-process of as-stacked PMMA thin films exhibit thin-film-like behavior, in that the glass transition temperature is depressed and the dynamics of the α-process are faster than those of the bulk system.
2. Annealing at high temperatures causes T_g and the relaxation time of the α-process to increase to that of the bulk. The fragility index of as-stacked PMMA thin films increases from 138 to 192 with increasing thickness of the single layer from 15 nm to the bulk value.
3. The relaxation time of the β-process of stacked PMMA thin films is almost equal to that of the bulk PMMA, and is unaffected by annealing.
4. The relaxation strengths of the α- and β-processes show a strong correlation between each other, and the sum of the relaxation strengths remains almost unchanged, while the individual relaxation strengths change during the annealing process.

In the study of the first type of multilayered thin films, we have measured the relaxation rate of the α-process of the P2CS layer in multilayered thin films, where molecular transport is promoted at the interfaces between two polymer layers during annealing process. If the molecular mobility related to the translational diffusion is coupled with the orientational motion, as usually observed in the bulk system, the present results would suggest that the molecular mobility related to the translational diffusion near the upper electrode region is higher and that near the lower electrode is lower for the late stage of the annealing process. On the other hand, recent experimental results show that there is a decoupling between translational diffusion and orientational relaxation (α-process) at the nanoscale level under confinement. In the case of freely standing films, diffusion coefficient of small molecules in the polymer films shows no deviation from the value of the bulk [68], while the T_g and the relaxation time of the α-process of this type of thin films are strongly depressed from the values of the bulk [63, 69]. In the case of multilayered thin films of PS and labeled PS, it has been reported that diffusion of small molecules is decoupled from the structural relaxation process [70]. Therefore, if there is such a decoupling between translational and orientational motion also in the present system, it would be difficult to judge how diffusion motion is distributed in the thin films, from the observed results on the distribution of the dynamics of the α-process in the stack. Another type of experimental measurements such as neutron reflectivity measurements or dielectric measurements specially designed for the measurements of diffusivity [70] will be required.

In multilayered thin films of P2CS and PS, mutual diffusion occurs at the interface during annealing process, and finally the interface between two polymers disappears. Especially in the late stage of annealing, there is a blending effect of two different polymers. In this sense, we might not directly measure the glass transition dynamics and/or the relaxation time of the α-process of neat polymeric systems as a function of the distance from the surface or the substrate. In this study, nevertheless, we have successfully observed a positional dependence of the molecular motion related to the blending process. This positional dependence of the molecular motions should be associated with the positional dependence of the α-process of PS or P2CS in single thin films of homopolymers. The latter positional dependence is the one which Torkelson et al. and Inoue et al. have observed in their measurements using labeling methods. Therefore, in a next step, we have to try to extract this positional dependence of the glass transition dynamics from the present observed results quantitatively, although this is a difficult task and requires an appropriate theoretical modeling.

In our sample geometry, there are Al layers on both sides of polymeric layers. This is a usual geometry for DRS, unless one uses the recently devised thin film sample holders [71]. The lower Al layer is supported on glass substrate, while the upper Al layer is put on the polymeric layer by vacuum evaporation. Wrinkles on the surface of the upper Al electrode were often observed after measurements above T_g, which suggests that the upper electrode is not rigid enough not to move during annealing above T_g. This means that the upper and lower Al layers are not symmetric, in other words, the upper and bottom type multilayered films in this study does not have the same geometry related to the dynamics of the α-process. Therefore, it is not surprising that there is the difference of the relaxation rate and dielectric relaxation strength of the α-process between the types of multilayered thin films. In our previous study [32], a depression of T_g could be observed and its thickness dependence of T_g is almost same as the one observed for uncapped thin polymer films. Therefore, it is reasonable to assume that there is a mobile region within thin films. The present result in this chapter might be an experimental evidence of the validity of this assumption in an indirect sense.

In the study of the second type of multilayered thin films, the results obtained by DSC clearly showed that the T_g of as-stacked PMMA thin films is depressed from the bulk T_g, and is returned to the bulk T_g after annealing at high temperatures. This strongly suggests the existence of mobile interfacial regions between two PMMA thin layers. Hence, a thin PMMA layer in as-stacked PMMA films can be regarded as an ideal thin polymer film covered with mobile regions on both sides, and this might be similar to freely standing thin films.

In our previous study [52], both T_α and T_β for PMMA single thin films supported on an Al-deposited glass substrate, and covered with an upper Al layer, decreased with decreasing film thickness. These results are different from the results of the present measurements for stacked PMMA thin films. In the case of stacked PMMA thin films, most of the thin PMMA layers are covered with other thin layers of the same polymer on both sides, and only the thin PMMA layers located on the top or bottom of the stack are covered with a thin gold layer on one side. Thus, the difference between the previous and present results should come from the difference in the interactions between the substrate and polymer. As shown in Ref. [51], the T_g of PMMA thin films strongly depends on the substrate material, and hence we can judge that the T_g of the PMMA thin films is very sensitive to interactions between the substrate and polymer. As the thin layers in stacked PMMA thin films can be an ideal system, as mentioned earlier in this review, we believe that the present thickness dependence of T_α and T_β is intrinsic to the nature of the dynamics of PMMA thin films.

Our experimental results in Sect. 4 clearly show that there is a strong correlation between the depression of T_g and the decrease in the fragility index. This suggests that fragility can play a crucial role in determining the glass transition temperature and the related dynamics in thin polymer films. Recently, Evans et al. reported that there is a strong correlation between the fragility index and the depression of T_g in thin polymer films on the basis of the fluorescence measurements on polymeric systems with various fragility indices [72]. Related to the results in Sect. 3, it is expected that dielectric measurements made over a wider frequency range will reveal the dynamics of the α-process for P2CS layers inserted into arbitrary positions within multilayered thin films near T_g, and then the fragility index can be evaluated as a function of the position along the direction normal to the surface of the substrate. Such an investigation will be the next development of these studies, which will contribute to the elucidation of the glass transition dynamics in thin polymer films.

In this study, the observed experimental results suggest that the interfacial interaction can play a crucial role in determining the depression of T_g and the speed-up of the dynamics of the α-process, while leaving unaffected the properties of the β-process, except for its dielectric relaxation strength. Furthermore, there should be a possibility of controlling the glass transition dynamics of thin polymer films using an appropriate annealing condition. In order to understand the intricacies of the relationship between the interface and the dynamics of thin polymer layers, a direct investigation of the structure of the interface and the effect of annealing should be studied using X-ray or neutron reflectivity measurements, as the next step in this study.

Acknowledgments This work was supported by a Grant-in-Aid for Scientific Research (B) (No. 25287108) and Exploratory Research (No. 25610127, No. 23654154) from the Japan Society for the Promotion of Science. The authors would like to thank Prof. Friedrich Kremer for giving them an opportunity to publish the present article in this book. One of the authors (KF) would like to express his cordial thanks to Dr. Didier Long for giving him an opportunity to stay in his laboratory in Lyon as a visiting professor of CNRS, where half of the present article has been written.

References

1. Ngai KL (2004) In: Mark J (ed) Physical properties of polymers. Cambridge University Press, pp 72–152. http://dx.doi.org/10.1017/CBO9781139165167
2. Sillescu H (1999) J Non-Cryst Solids 243:81
3. Berthier L, Biroli G, Bouchaud J-P, Cipelletti L, van Saarloos W (2011) Dynamical heterogeneities in glasses, colloids, and granular media. Oxford University Press, New York
4. Muranaka T, Hiwatari Y (1995) Phys Rev E 51:R2735
5. Yamamoto R, Onuki A (1998) Phys Rev E 58:3515
6. Weeks ER, Crocker JC, Levitt AC, Schofield A, Weitz DA (2000) Science 287:627
7. Adam G, Gibbs JH (1965) J Chem Phys 43:139
8. Keddie JL, Jones RAL, Cory RA (1994) EPL Europhys Lett 27: 59
9. Alcoutlabi M, McKenna GB (2005) J Phys: Condens Matter 17:R461
10. Napolitano S, Capponi S, Vanroy B (2013) Eur Phys J E 36:61
11. Efremov MY, Kiyanova AV, Last J, Soofi SS, Thode C, Nealey PF (2012) Phys Rev E 86:021501
12. Tress M, Erber M, Mapesa EU, Huth H, Müller J, Serghei A, Schick C, Eichhorn K-J, Voit B, Kremer F (2010) Macromolecules 43:9937
13. Jérôme B, Commandeur J (1997) Nature 386:589

14. Fakhraai Z, Forrest JA (2008) Science 319:600
15. Yang Z, Fujii Y, Lee FK, Lam C-H, Tsui OKC (2010) Science 328:1676
16. Schwab AD, Agra DMG, Kim J-H, Kumar S, Dhinojwala A (2000) Macromolecules 33:4903
17. Tanaka K, Hashimoto K, Kajiyama T, Takahara A (2003) Langmuir 19:6573
18. Kawaguchi D, Tanaka K, Kajiyama T, Takahara A, Tasaki S (2003) Macromolecules 36:1235
19. Kajiyama T, Tanaka K, Takahara A (1997) Macromolecules 30:280
20. Ellison CJ, Torkelson JM (2003) Nat Mater 2:695
21. Priestley RD, Ellison CJ, Broadbelt LJ, Torkelson JM (2005) Science 309:456
22. Roth CB, McNerny KL, Jager WF, Torkelson JM (2007) Macromolecules 40:2568
23. Inoue R, Kawashima K, Matsui K, Kanaya T, Nishida K, Matsuba G, Hino M (2011) Phys Rev E 83:021801
24. Inoue R, Kawashima K, Matsui K, Nakamura M, Nishida K, Kanaya T, Yamada NL (2011) Phys Rev E 84:031802
25. Inoue R, Nakamura M, Matsui K, Kanaya T, Nishida K, Hino M (2013) Phys Rev E 88:032601
26. Rotella C, Napolitano S, De Cremer L, Koeckelberghs G, Wübbenhorst M (2010) Macromolecules 43:8686
27. Koh YP, McKenna GB, Simon SL (2006) J Polym Sci Part B: Polym Phys 44:3518
28. Koh YP, Simon SL (2008) J Polym Sci Part B: Polym Phys 46:2741
29. Fukao K, Oda Y, Nakamura K, Tahara D (2010) Eur Phys J Special Topics 189:165
30. Fukao K, Terasawa T, Oda Y, Nakamura K, Tahara D (2011) Phys Rev E 84:041808
31. Fukao K, Terasawa T, Nakamura K, Tahara D (2013) In: Kanaya T (ed) Glass transition, dynamics and heterogeneity of polymer thin films, Advances in polymer science, vol 252. Springer, Berlin. pp 65–106, ISBN 978-3-642-34338-4
32. Fukao K, Miyamoto Y (2000) Phys Rev E 61:1743
33. Priestley RD, Broadbelt LJ, Torkelson JM, Fukao K (2007) Phys Rev E 75:061806
34. Schmidt-Rohr K, Kulik AS, Beckham HW, Ohlemacher A, Pawelzik U, Boeffel C, Spiess HW (1994) Macromolecules 27:4733
35. Fytas G, Wang CH, Fischer EW, Mehler K (1986) J Polym Sci Part B: Polym Phys 24:1859
36. Sasabe H, Saito S (1968) J Polym Sci Part A-2: Polym Phys 6: 1401
37. McCrum NG, Read BE, Williams G (1967) Anelastic and dielectric effects in polymeric solids. Wiley, London
38. Ardi M, Dick W, Kubát J (1993) Colloid Polym Sci 271:739
39. Ribelles JLG, Calleja RD (1985) J Polym Sci: Polym Phys Ed 23: 1297
40. Bergman R, Alvarez F, Alegria A, Colmenero J (1998) J Chem Phys 109:7546
41. Paeng K, Ediger MD (2011) Macromolecules 44:7034
42. Paeng K, Richert R, Ediger MD (2012) Soft Matter 8:819
43. Takaki H, Fukao K Phys Rev E (submitted)
44. Hayashi T, Fukao K (2014) Phys Rev E 89:022602
45. Fukao K, Miyamoto Y (1999) EPL Europhys Lett 46: 649
46. Havriliak S, Negami S (1967) Polymer 8:161
47. Alvarez F, Alegría A, Colmenero J (1991) Phys Rev B 44:7306
48. Schönhals A, Kremer F (2003) In: Kremer F, Schönhals A (eds) Broadband dielectric spectroscopy. Springer, Berlin, pp 59–98. ISBN 978-3-642-62809-2
49. Chu B, Linliu K, Ying Q, Nose T, Okada M (1992) Phys Rev Lett 68:3184
50. Alexandrovich PS, Karasz FE, Macknight WJ (1980) J Macromol Sci Part B 17:501
51. Keddie JL, Jones RAL, Cory RA (1994) Faraday Discuss 98:219
52. Fukao K, Uno S, Miyamoto Y, Hoshino A, Miyaji H (2001) Phys Rev E 64:051807
53. Kuebler SC, Schaefer DJ, Boeffel C, Pawelzik U, Spiess HW (1997) Macromolecules 30:6597
54. Ménégotto J, Demont P, Lacabanne C (2001) Polymer 42:4375
55. Doulut S, Bacharan C, Demont P, Bernès A, Lacabanne C, Non-Cryst J (1998) Solids 235–237:645
56. Vogel H (1921) Phys Z 22:645
57. Fulcher GS (1925) J Am Ceram Soc 8:339
58. Fulcher GS (1925) J Am Ceram Soc 8:789

59. Tammann G, Hesse W, Anorg Z (1926) Allg Chem 156:245
60. Böhmer R, Angell CA (1992) Phys Rev B 45:10091
61. Fukao K, Miyamoto Y (2001) Phys Rev E 64:011803
62. Zhang C, Guo Y, Shepard KB, Priestley RD (2013) J Phys Chem Lett 4:431
63. Napolitano S, Wübbenhorst M (2010) Polymer 51:5309
64. Yin H, Napolitano S, Schönhals A (2012) Macromolecules 45:1652
65. Huwe A, Kremer F, Behrens P, Schwieger W (1999) Phys Rev Lett 82:2338
66. Riggleman RA, Yoshimoto K, Douglas JF, de Pablo JJ (2006) Phys Rev Lett 97:045502
67. Jain TS, de Pablo JJ (2004) J Chem Phys 120:9371
68. Pu Y, White H, Rafailovich MH, Sokolov J, Patel A, White C, Wu W-L, Zaitsev V, Schwarz SA (2001) Macromolecules 34:8518
69. Forrest JA, Dalnoki-Veress K, Stevens JR, Dutcher JR (1996) Phys Rev Lett 77:2002
70. Napolitano S, Rotella C, Wübbenhorst M (2011) Macromol Rapid Commun 32:844
71. Serghei A, Kremer F (2008) Rev Sci Inst 79:026101
72. Evans CM, Deng H, Jager WF, Torkelson JM (2013) Macromolecules 46:6091

Molecular Mobility and Phase Transformations of Several Low Molecular Weight Glass Formers Confined to Nanoporous Silica Matrices

M. Dionísio, N. T. Correia and A. R. Brás

Abstract The dynamical behavior of three low molecular weight glass formers confined to nanostructured mesoporous silica (100 % Si) with pore sizes ranging from 2.8 to 8.6 nm, is probed by Dielectric Relaxation Spectroscopy (DRS). The confined guests embrace different classes of materials: a surfactant, Triton X-100, a liquid crystal, E7 nematic mixture and the pharmaceutical drug, Ibuprofen. All three glass formers show two distinct dynamical domains inside the pores, as revealed by the detection of both bulk-like and surface processes. The latter is characterized by a slower mobility than the bulk-like process and the temperature dependence follows the characteristic Vogel-Fulcher-Tammann-Hesse (VFTH) law indicating a glassy dynamics of the molecules anchored to the pore surface. In the case of E7, the Vogel temperature (T_0) of this process is size dependent, decreasing with increasing pore size, which is taken as a finite size effect. Concerning the bulk-like process, assigned to the glassy dynamics of the molecules in the middle of the pore, the confinement effect becomes stronger depending on the material as follows: (1) Triton X-100, undergoing almost no change in the glass transition temperature (T_g), only a slight increase ~3 K is observed; (2) E7, with a maximum decrease of 10 K in T_g compared to the bulk temperature for a pore size of 6.8 nm; and (3) Ibuprofen, which shows not only a higher decrease in the glass transition temperature, ~30 K when confined to a pore size of 3.6 nm (MCM-41), but also its temperature dependence of relaxation times varies from VFTH to Arrhenius like, which is interpreted as 3.6 nm being

M. Dionísio (✉) · N. T. Correia · A. R. Brás
REQUIMTE/CQFB, Departamento de Química, Faculdade de Ciências e Tecnologia, Universidade Nova de Lisboa, 2829-516 Caparica, Portugal
e-mail: madalena.dionisio@fct.unl.pt

N. T. Correia
Unité Matériaux Et Transformations (UMET), UMR CNRS 8207, UFR de Physique, BAT P5, Université Lille Nord de France, 59655 Villeneuve d'Ascq Cedex, France

A. R. Brás
Jülich Centre for Neutron Science (JCNS-1) and Institute for Complex Systems (ICS-1), Forschungszentrum Jülich, 52425 Jülich, Germany

F. Kremer (ed.), *Dynamics in Geometrical Confinement*, Advances in Dielectrics, 213
DOI: 10.1007/978-3-319-06100-9_9,

a dimension that interferes with the length scale of cooperativity. Moreover, two secondary relaxations are detected in the pharmaceutical drug, the more local one being insensitive to confinement while the second process, taken as the precursor of the glass transition, becomes more mobile relatively to the respective process in bulk Ibuprofen. The results confirm that molecular dynamics of the probed low molecular weight guests confined into nanostructured mesoporous hosts is controlled by a counterbalance between confinement and surface effects.

Keywords Triton X-100 · E7 · Ibuprofen · Nanostructured mesoporous · Glass transition · Confinement · Surface process · Cooperativity · Debye process

Abbreviations

BJH	Barrett-Joyner-Halenda
CM	Coupling model
CN	Cyano group
CPG	controlled pore glasses
DSC	Differential scanning calorimetry
DRS	Dielectric relaxation spectroscopy
FTIR	Fourier transform infrared spectroscopy
HN	Havriliak- Negami
JG	Johari-Goldstein
LCs	Liquid-crystals
MWS	Maxwell-Wagner-Sillars
NMR	Nuclear magnetic resonance
M_W	Molecular weight
T_g	Glass transition temperature
T_{IN}	Isotropic to nematic transition
T_0	Vogel Temperature
TEM	Transmission electron microscopy
TEOS	Tetraethoxysilane
TGA	Thermogravimetric analysis
VFTH	Vogel-Fulcher-Tammann-Hesse
XRD	X-ray diffraction
$\Delta\varepsilon$	Dielectric strength

1 Introduction

Confinement studies emerged in the last few years as an attractive approach to explore fundamental concepts concerning the physics of condensed matter. A particular attention has been given to the influence of confinement on the glass transition, which physical origin, in spite of an intensive research, is far from being fully understood, namely if it is thermodynamic, kinetic, structural, or simply dynamical in nature [1–4].

In this context, confinement studies are related to the relevance of a length scale responsible for glassy dynamics [5], phase transitions or finite size effects. Additionally, it offers a mean to test theoretical models as the Adam Gibbs theory [6] that conceives cooperative rearranging regions with a length scale, ξ, that increases upon temperature decrease being related with the dynamic glass transition [1]. For a confined guest, if the pore dimensions of the confining host interfere with the spatial scale ξ of the molecular guest cooperative motion (in the scale of few nanometers), its molecular mobility can be dramatically changed from bulk state [6–13]. This could even lead to the suppression of the dynamically correlated regions [14], allowing estimating a minimal length scale for the cooperative motion underlying the glass transition, which manifests as a deviation from the typical curvature in the activation plot to Arrhenius dependence, the latter due to localized dipolar fluctuations [7].

Also, the coupling model predicts that the process associated with the dynamic glass transition (α-process) is transformed into the Johari-Goldstein (JG) process, taken as the precursor of the α-process, in the limit of extreme confinement when the pore size becomes smaller than the length scale of the α-process [15, 16].

Additionally, nanoscopic confinement provides a better understanding of the crystallization behavior in general, since it allows to access early stages of crystallization [17] and to stabilize crystalline forms that are metastable, transient, or inaccessible in bulk samples as demonstrated for acetaminophen inside controlled pore glasses (CPG) [18, 19]. The occurrence of unstable metastable forms in polymorphic materials as pharmaceutical drugs under nanoconfinement, can be rationalized by the stability dependence of certain crystalline forms on the size, which is controlled by both thermodynamic and kinetic effects [20, 21]. Therefore, crystallization under confinement could bring in-depth knowledge on the thermodynamic and kinetic properties of crystalline metastable forms difficult to access under bulk conditions.

Liquid crystals (LCs) allow going further on the fundamental investigation of phase transformations, since it exhibits additional transitions. It consist of anisotropic molecules that self-assemble into phases with orientational order, but often no positional order, in their simplest form, having structures between the crystalline and the isotropic liquid state [22, 23]. In LCs, the properties of both states are combined and mesomorphic phases are formed in which order and mobility compete [24]. They are formed for instance by molecules, which have a stiff, rod-like mesogenic unit in their structure [25], showing a high degree of shape anisotropy, which manifests itself in many ways, such as dielectric and optical anisotropies. The most important liquid crystalline structures are the nematic and the smectic mesophases, the former being the liquid crystalline state with the lowest order.

The various mesophases and phase transitions, which are sensitive to perturbations, make LCs suitable target materials when confined in porous media for the study of the quenched random disorder effects upon phase transitions and critical phenomena [26, 27]. Thus, many experimental and theoretical studies focused on these issues were carried out during the last decade [26, 28–31].

Besides this theoretical interest, also practical aspects are involved in these studies which are particularly relevant for the pharmaceutical science and industry. Indeed, nanoconfinement emerged in the past few years as a strategy to manipulate and stabilize unstable polymorphs as mentioned or even to prevent crystallization as shown for Ibuprofen by solid-state nuclear magnetic resonance (NMR) measurements [32] and X-ray diffraction (XRD) [33]. The stabilization of the amorphous form by nanoconfinement was demonstrated also for acetaminophen [34] and for other glass formers [18, 21, 32, 35–37] as well, since in certain geometrical confinements the guest molecules are unable to form a crystalline structure.The spontaneous amorphization of solid drugs by using mesoporous magnesium aluminum silicate was reported since 1986 [38]. In the 1990s, it was observed the amorphization of lead when confined to carbon nanotubes [39], which was rationalized as the pore size being smaller than the critical nucleus size [40].

Therefore, confinement to nanoscaled geometries allows the production of usable amorphous materials stabilized in a highly disordered state being crucial for controllable and efficient drug delivery [41, 42]. The molecularly disordered amorphous form is particularly desirable in the field of pharmaceutics since its dissolution and bioavailability is usually improved relative to the usually water poorly soluble crystalline state [43, 44]. But not only pharmaceutical drugs as happen more recently are the preferred guests. Also LCs [45] confined to different porous host systems such as porous glasses [46–50], molecular sieves [51–53] and also membranes with larger pore sizes [54] like Anopore membranes [55–59] have received particular interest. When crystallization is avoided, the guest enters in a supercooled regime and an acceleration of the guest dynamics could be induced by the confinement with a downshift of the T_g with decreasing pore size [7, 14, 60–64].

However, for rather low pore sizes, the guest molecules can undergo specific interactions with the pore wall as have been shown for systems forming hydrogen bonds between the guest molecules and the inner pore, leading to the formation of an interfacial layer, having a dynamics, which is slowed down compared to the bulk liquid [65, 66].

In drug delivery systems when specific host-guest interactions are present, the drug release kinetic profiles are influenced being demonstrated that the delivery rate can be modulated through modification of the surface pore walls. This was verified for alendronate confined in amino-functionalized MCM-41 and SBA-15 [67]. The same effect was also observed for captropril [68] and Ibuprofen incorporated in the same porous hosts modified by silylation [69]. Therefore, the dynamics of molecules confined to nanoporous hosts is controlled by a counterbalance between confinement and surface effects [7, 66]. Their relative influence is determined by the size of the pores and by the strength of the interaction between the guest molecules with the surface: finite size effects should dominate if the pore size is small [70].

If both effects manifest simultaneously, an additional relaxation processes can be detected, which is not characteristic for the bulk system as probed by dielectric relaxation spectroscopy (DRS) for Ibuprofen confined to SBA-15 [71] and MCM-41 [72]: two families of molecules with different molecular mobilities were observed, one due to molecules within the core of the pores with a higher mobility compared to the bulk at low temperatures and another with slower dynamics originated from molecules interacting with the pore walls. This was previously shown by solid-state NMR studies for pyridine [73] and Ibuprofen [74] confined to mesoporous silica; for the latter NMR [75] and fourier transform infrared spectroscopy (FTIR) [76] showed that weak hydrogen bonds are formed between Ibuprofen carboxylic acid group and the silanol groups. Also from DRS measurements, two relaxation processes were detected for phenyl salicylate (salol) [13, 77] and propyleneglycol [78] in CPG; the same heterogeneous dynamical behavior was observed for salol inside Vycor and Gelsil pores as probed by dynamical mechanical analysis [14]. In particular, for LCs, it was observed that the interaction of the LCs molecules with surfaces can induce the formation of a strongly adsorbed surface layer with a partial ordering, which is essentially slower relative to bulk molecules as observed for several cyanobiphenyl molecules [28, 29, 79, 80] and for a nematic mixture confined to Anopore membranes [81] and silica mesopores [82].

The existence of distinct mobilities could be calorimetric identified by the detection of two glass transitions as found for polystyrene/o-terphenyl solutions in CPG [83]; a few examples for low molecular weight glass formers are salol in silica mesoporos [84], propylene glycol in CPG [85], ortho-terphenyl in SBA-15 [86], acetaminophen in nanoporous Vycor glass [87].

Among the variety of host systems, this chapter will concern to mesoporous 100 % silica materials, which attract a great interest due to their particular physical properties, such as its chemical and mechanical stability, biocompatibility, high internal surface area, silanol containing surface, and ordered pore network [88–90]. A transmission electron microscopy (TEM) micrograph is shown in Fig. 1a evidencing the uniform honeycomb structure formed by parallel cylindrical pores organized in a hexagonal array [82]. The long-range ordering is reflected in the small angle XRD patterns ($2\theta < 10^0$) depicted in Fig. 1b in spite of the amorphous silica structure of the pore walls; the amorphous halo emerges for higher 2θ values (inset of Fig. 1b). Additionally, these materials are characterized by an uniform array of pores with very narrow pore diameter distribution that can be tuned from ~1.5–2 nm to several tens of nanometers by changing the chain length of the surfactant, employing polymeric structure-directing agents, or solubilizing auxiliary substances into micelles [91–93]; a representative example of the narrow pore width distribution is shown in Fig. 1c for SBA-15 of 5.7 nm average pore diameter, which was synthesized according to Gao et al. [94] employing the triblock copolymer, Pluronic P123 as template, and tetraethoxysilane (TEOS) as silica source (additional details in reference [95]).

These unique properties make silica matrices suitable carriers for drug delivery, which can exhibit dissolution enhanced properties irrespective to the physical chemical profile of the encapsulated drug [42, 96]. In the last decade, different pharmaceutics have been confined into these ordered mesoporous silica, such as Ibuprofen [97–99], erythromycin [100], alendronate [67], aspirin [101], captopril [102], sertraline [103], fenofibrate [104], in MCM-41, Ibuprofen and erythromycin [105] in MCM-48, Ibuprofen [106, 107], indomethacin [96], alendronate [67, 97] erythromycin, [105] gentamicin [108], amoxicillin [109], captopril, [102] fenofibrate [104] in SBA-15, flurbiprofen[110], salicylamide [111] in FSM-16; for a review on

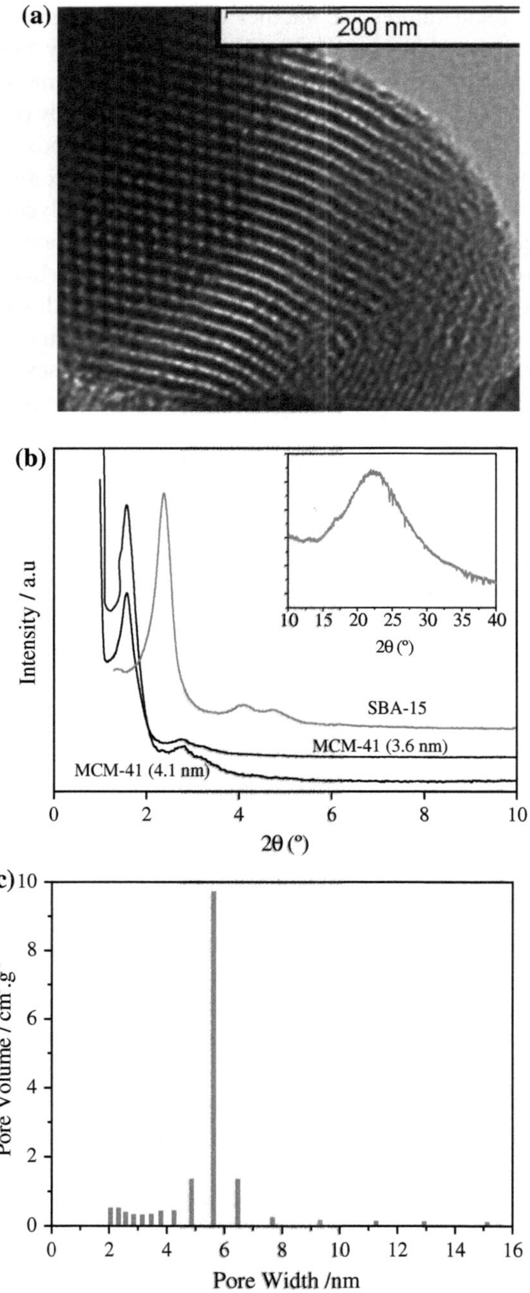

Fig. 1 **a** TEM microphotograph of SBA-15 evidencing the ordered pore structure, **b** Powder XRD patterns of MCM-41 and SBA-15 silica matrices of different pore sizes, exhibiting Bragg peak diffractions at small 2θ values due to long-range order (the inset depicts the amorphous hallo at higher 2θ values, adapted with permission from [95]. Copyright 2013 American Chemical Society), **c** Pore width distribution for SBA-15 (average diameter of 5.7 nm) as determined by the Barrett-Joyner-Halenda (BJH) method (details in reference [95])

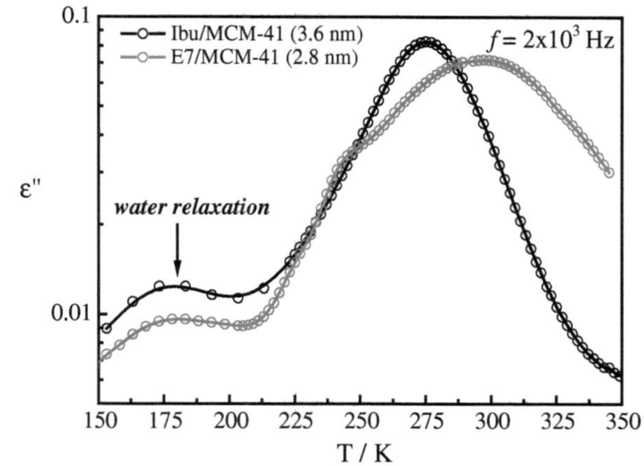

Fig. 2 Isochronal plots at 2 kHz for E7 (*gray circles*) and Ibuprofen (*black circles*) confined to MCM-41 of, respectively, 2.8 and 3.6 nm of pore diameter, before water removal evidencing at the lowest temperatures the additional process due to water relaxation

the use of mesoporous silica matrices for amorphization of pharmaceutics as drug delivery systems, see [41, 42].

In the following, three low molecular weight glass formers, Triton X-100, Ibruprofen, and the nematic mixture E7 confined into these silica-based mesoporous matrices (MCM-41 and SBA-15) will be described as case studies.

It must be emphasized that, due to the hydrophilic character of the silica matrices, an additional process is detected by DRS at cryogenic temperatures, below 200 K [82], irrespective to the molecular guest. This is exemplified in Fig. 2 for E7 and Ibuprofen confined in MCM-41 for water amounts between 1 and 3 %. This process is absent in Triton X-100 confined to SBA-15 for which the water content is 0.95 % (wt.). The unloaded dry matrix has a negligible small calorimetric [95] or dielectric [71] response as found for related hosts [53], compared to that observed for the loaded matrix; so the detected response for the composite is dominated by the response of guest molecules.

In the following sections, the reported results refer to data after water removal. It will be presented in order of increasing complexity regarding the multi-character of the processes as detected by DRS: the surfactant Triton X-100 with the emergence of a strong interfacial process allowing to detect two glass transitions, then a cyanobiphenyl-based nematic mixture, E7, with accelerated glassy dynamics and two surface processes, one with slower mobility due to molecules adsorbed inside pores and another with intermediate dynamical behavior due to molecules adsorbed on the outer pore surface and inter-grains and finally the pharmaceutical drug ibuprofen for which besides true confinement effects over the bulk-like glassy dynamics and surface process also secondary relaxations and a Debye-like process are detected.

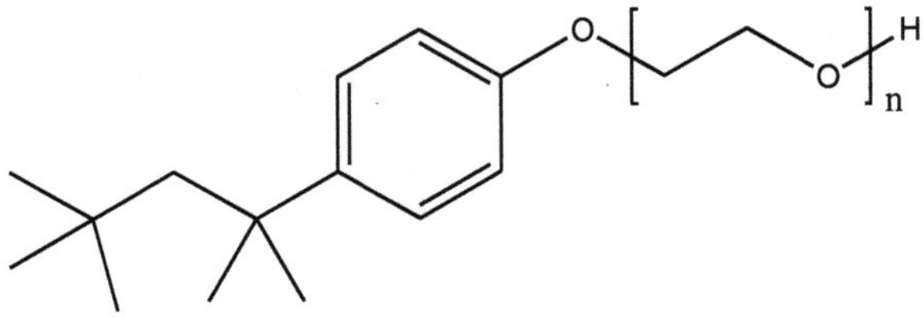

Fig. 3 Molecular structure of Triton X-100 (n ~ 9–10)

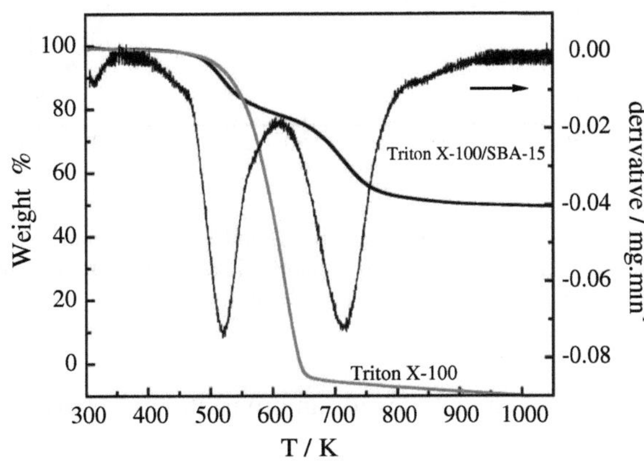

Fig. 4 Thermograms obtained by TGA analysis on heating at 10 K min^{-1} for Triton X-100 confined to SBA-15 (loading degree ~50 % (wt.)); the derivative plot is shown in the right-hand vertical axis, evidencing the two peaks corresponding to the bimodal decomposition of the guest; the small peak ~300 K is due to water removal (the estimated water content is negligible (<1 % (wt.)). The thermogram for bulk Triton X-100 is included for comparison. Adapted with permission from [95]. Copyright 2013 American Chemical Society

1.1 A surfactant: Triton X-100

The water-soluble, liquid surfactant Triton X-100, polyethylene glycol tertoctyl phenyl ether, with molecular formula $C_{14}H_{22}O(C_2H_4O)_n$ with an average number, n ~ 9–10 of oxyethylene units per molecule ($M_W \sim 625$) (see Fig. 3) is a glass former exhibiting a glass transition around 213 K.

In spite of Triton X-100 strong tendency to crystallize [112, 113], when confined into mesoporous silica (SBA-15), it is stabilized in the amorphous form over large periods of time (at least 2 years). This was confirmed by *X*-ray diffraction (XRD), differential scanning calorimetry (DSC) and DRS [95]. Moreover, it reveals a bimodal behavior concerning the molecular population distribution inside pores: a fraction of molecules in the pores center and another forming an interfacial layer due to molecules adsorbed in the inner pores wall. The two molecular populations give rise to two-step weight loss decomposition in the thermogravimetric analysis (TGA), seen as two strong peaks in the derivative plot while bulk Triton decomposes in a single weight loss (Fig. 4). The fraction due to the interfacial layer is comparable to

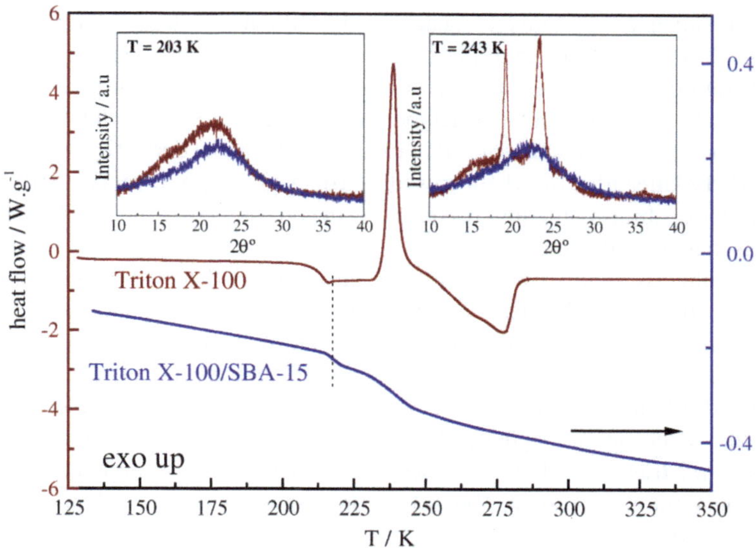

Fig. 5 Thermograms (heat flow vs. temperature) collected on heating at 30 K min^{-1} of confined Triton X-100 in SBA-15 (*dark blue line*) compared with bulk Triton X-100 (*brown line*). At the low temperatures, a bulk-like glass transition is detected for confined Triton (indicated by *dashed line*) and ~20 K above, a second heat flow step emerges due to the fraction of adsorbed molecules (adapted with permission from [95]. Copyright 2013 American Chemical Society). The inset shows the XRD patterns at the indicated temperatures for the composite (*dark blue curves*) compared with bulk triton X-100 (*brown curves*) evidencing that Triton X-100 remains amorphous inside pores

the fraction due to bulk-like molecules, being more thermal resistant decomposing at higher temperatures.

By DSC, the two molecular fractions give rise to two glass transitions as shown in Fig. 5: one emerging at only ~3 K above the T_g of bulk triton X-100 originated from the glass transition of bulk-like Triton X-100 molecules in the composite, and a second one shifted ~20 K toward higher temperatures relatively to the bulk Triton X-100 due to the glass transition of the molecules in the interfacial layer.

While a crystallization exotherm is observed at ~238 K for bulk Triton X-100 (Fig. 5), which is followed by an endotherm due to melting, the thermogram for the confined system, Triton X-100/SBA-15, does not exhibit either crystallization or melting. This means that the guest did not undergo crystallization either during inclusion or upon the thermal treatment as also confirmed by XRD illustrated in Fig. 5. In the upper left inset, the *X*-ray pattern taken in the subglass region at 203 K, shows the typical halo of the amorphous form for both bulk and confined Triton; in the upper right inset collected at 243 K, two Bragg peaks are detected for bulk Triton X-100 due to the emergence of crystallinity, whereas for confined Triton X-100 the *X*-ray pattern confirms that crystallization was suppressed and that the guest remains in the amorphous state inside pores.

The in-situ guest mobility as accessed by DRS in the frequency range from 10^{-1} to 10^{6} Hz, discloses multiple processes going from the subglass region up to temperatures well above the glass transition. Figure 6 shows the dielectric loss isotherms at some representative temperatures. The multimodal character is also evident from the $\varepsilon''(T)$ trace depicted in Fig. 7a and from the derivative of $\varepsilon'(T)$ trace [95, 114], the latter allowing a better resolution of the overlapping processes (Fig. 7b). The absence of abrupt depletion/discontinuity in both isothermal and isochronal plots confirms that crystallization is avoided.

In the subglass region, two secondary relaxations are detected, γ and β-process, which are not commonly observed in low molecular weight glass formers under confinement [7, 16]. The γ and β-processes are due to localized mobility, the former being assigned to dipolar fluctuations within the ethylene glycol moiety and the latter to hindered rotations of the octylphenyl ether group [113]. Those are followed by a bulk-like α-relaxation originated by the cooperative motion associated with the dynamic glass transition. Two processes are detected at higher temperatures, which are not found for bulk Triton X-100: (a) a S-process attributed to the dynamical glass transition associated to the guest molecules interacting with the pore surface of the host and (b) a Maxwell-Wagner-Sillars (MWS) process [115, 116] originated by the blocking of charge carriers at internal phase boundaries, which for the present case are the Triton guest/SBA matrix interfaces.

The relaxation times of all detected processes were extracted by simulating the complex permittivity data by a sum of Havriliak-Negami (HN) functions according to:

$$\varepsilon^*(f) = \varepsilon_\infty + \sum_j \frac{\Delta\varepsilon_j}{\left[1+\left(i\,\omega\,\tau_{HN_j}\right)^{\alpha_{HN_j}}\right]^{\beta_{HN_j}}} \tag{1}$$

where j is the index over which the relaxation processes are summed, $\Delta\varepsilon$ is the dielectric strength τ_{HN} is the characteristic HN relaxation time, and α_{HN} and β_{HN} are fractional parameters ($0 < \alpha_{HN} < 1$ and $0 < \alpha_{HN}.\beta_{HN} < 1$) describing, respectively, the symmetric and asymmetric broadening of the complex dielectric function [115]. From the estimated values of τ_{HN}, α_{HN} and β_{HN} parameters, a model-independent relaxation time, $\tau = 1/(2\pi f_{max})$, was calculated according to [115, 117, 118]:

$$\tau = \tau_{HN} \times \left[\frac{\sin\left(\frac{\alpha_{HN}\beta_{HN}\pi}{2+2\beta_{HN}}\right)}{\sin\left(\frac{\alpha_{HN}\pi}{2+2\beta_{HN}}\right)}\right]^{1/a_{HN}} \tag{2}$$

The extracted relaxation times are plotted in Fig. 8 after conversion to τ according to Eq. 2.

Figure 8 includes the relaxation times estimated from the analysis of the isochronal plot of the dielectric loss as measured and obtained from the derivative of $\varepsilon'(T)$ [114], both representations described by a superposition of k Gaussians functions; for details see reference [95]. The derivative plot allows distinguishing the secondary $\beta_{TritonX-100/SBA-15}$ process (Fig. 7b) but only for a few frequencies and so,

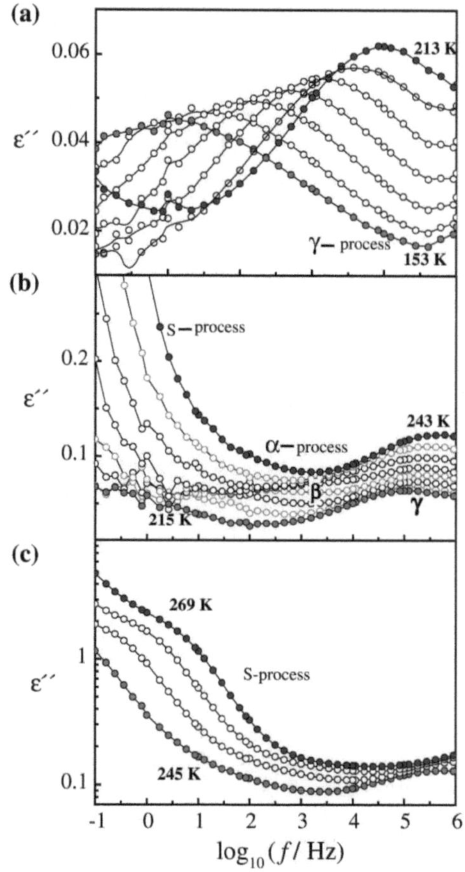

Fig. 6 Loss curves for confined Triton X-100 at some representative temperatures from the sub-glass to the supercooled regime **a** in steps of 10 K, **b** in steps of 4 K and **c** in steps of 8 K allowing distinguishing the different relaxation processes; at higher temperatures, a MWS process emerges. The *lines* are guides for the eyes

the respective relaxation times are not included in the relaxation map. As will be described later on this chapter, the resolution of secondary processes over a wider frequency and temperature range was possible for Ibuprofen confined to the same mesoporous matrix, allowing drawing the complete relaxation map.

The temperature dependence of the well-resolved γ-process for confined Triton X-100 is relatively similar to the homologous process observed in bulk Triton X-100, both obeying to an Arrhenius law: $\tau(T) = \tau_\infty \exp(E_a/RT)$, where τ_∞ is the pre-exponential factor, E_a is the activation energy, R is the ideal gas constant, and T the temperature. The estimated activation energy, E_a, is 36.3 ± 0.1 and 33.4 ± 0.4 kJ mol^{-1} for, respectively, bulk and confined Triton X-100 [95].

Both α (either bulk or confined) and S processes show nonlinear temperature dependence obeying to the well-known Vogel-Fulcher-Tammann-Hesse (VFTH) equation [119–121] that reads,

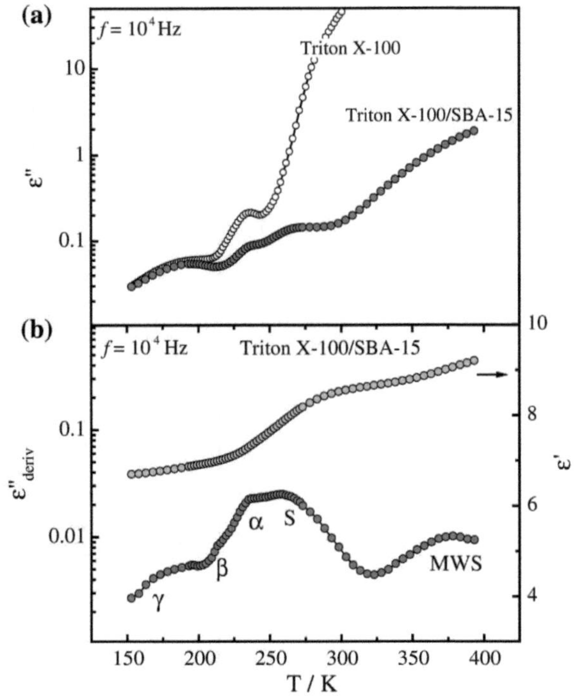

Fig. 7 **a** Dielectric loss (ε″) of confined Triton (Triton X-100/SBA-15) (*filled circles*) versus temperature at a frequency of 10^4 Hz compared with bulk Triton X-100 (*open circles*), **b** ε′ isochronal plot (*gray filled circles*) and its derivative, ε″ deriv(T) (*black filled circles*) that enhances the multimodal profile. The labeling assigns the different relaxation processes detected for confined Triton. Adapted with permission from [95]. Copyright 2013 American Chemical Society

$$\tau(\mathrm{T}) = \tau_\infty \exp\left(\frac{\mathrm{B}}{\mathrm{T} - \mathrm{T}_0}\right) \tag{3}$$

where τ_∞ and B are constants and T_0 is the so-called Vogel temperature.

The glass transition temperature obtained for Triton/SBA from extrapolation of the VFTH equation to $\tau = 100\,\mathrm{s}$ [122, 123], is $\mathrm{T}_{g,\alpha\, Tr/SBA} = 211.7\,\mathrm{K}$, 3.4 K higher than $\mathrm{T}_{g,\alpha\,\mathrm{bulk}}$ (208.3 K). This increase of $\mathrm{T_g}$ is in good agreement with the increment determined from DSC results (~4 K) [95], meaning that the molecular dynamics of the guest under confinement is slowed down compared to the bulk. This positive shift of the bulk-like $\mathrm{T_g}$ is not unusual being observed for a variety of glass formers [7, 66] being of the order of the increase observed for propylene glycol in CPG (4.5 K) [78], for which two distinct processes are detected. Indeed, the relatively slower Triton X-100 molecular dynamics seen as modest positive deviation in the glass transition of the bulk-like relaxation process, must be distinguished from the strong dynamical hindrance revealed by the S-process due to the dynamical domain of molecules interacting with the pore surface. For the latter, the glass transition

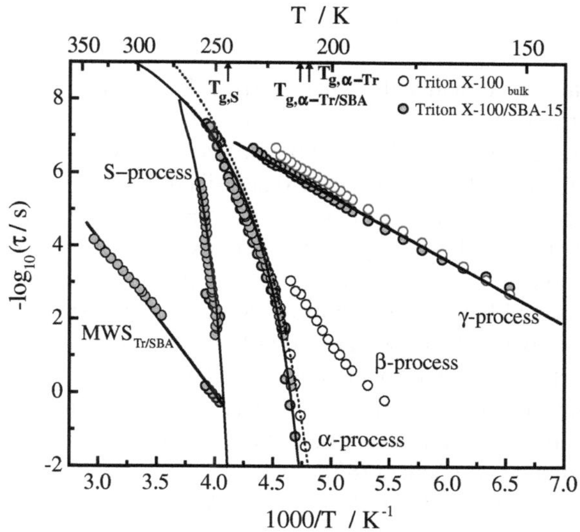

Fig. 8 Activation plot of all detected processes for Triton X-100/SBA-15 (*filled circles*) and bulk Triton X-100. The *lines* are fits of the Arrhenius and VFTH equations to the corresponding data. The *upper arrows* indicate the dielectric glass transition temperatures, $T_g(\tau = 100\ s)$, of the α-process for bulk and Triton X-100/SBA-15 and for the additional surface S-process detected in the composite. Adapted with permission from [95]. Copyright 2013 American Chemical Society

temperature estimated according the same criterion ($\tau = 100\,s$), $T_{g,S} = 243.5\,K$, is shifted ~35 K toward higher temperatures relative to bulk Triton X-100 as determined from dielectric measurements.

Therefore, Triton X-100 provides a further example of a glass former which behavior when confined to SBA-15 (5.7 nm) exhibits dynamically distinct domains, which calorimetrically show up as two glass transitions compatible with a two T_g scenario as described and sketched in [66] (see also references therein).

1.2 A Liquid Crystalline Material: Nematic E7

As mentioned in the Introduction of this chapter, additional effects like ordering and/or phase transitions can take place for molecules forming LCs in the bulk as guests. Generally, confinement induces disorder, decreases transition temperatures, and may change the order of the transition. It is also expected that confinement and disorder have a considerable influence on the molecular mobility and therefore on the relaxation processes as well, yielding to a change of the bulk-like behavior [7].

In this section, the behavior of E7 is under analysis. It is a nematic liquid crystalline mixture of alkylcyanobiphenyls, with the following composition: 4-cyano-4'-pentyl-1,1'-biphenyl (51 %), 4-n-heptyl-4'-cianobiphenyl (25 %), 4,4'-

n-octyloxycyanobiphenyl (16 %) and 4'-n-pentyl-4-cyanoterphenyl (8 %) w/w [124, 125].

Its bulk state was dynamically characterized by means of DRS in the frequency range from 20 to 1 MHz [126] and updated by broadband DRS over the frequency range 10^{-2}–10^{9} Hz, complemented by specific heat spectroscopy (10^{-3} Hz–2×10^{3} Hz) [127], which will next described shortly. In the isotropic phase, above the clearing temperature where occurs the transition from nematic to isotropic state, $T_{NI} = 331$ K [128–130], only one broad relaxation process is detected, which splits into two processes in the nematic phase region (Fig. 9a). These modes are theoretically expected being assigned to the tensorial components of the complex dielectric function parallel and perpendicular to the director [24]: the δ-relaxation, which corresponds to rotational fluctuations around its short axis, determining mainly the parallel component and the tumbling mode at higher frequencies than the former one, due to librational fluctuations around the long molecular axis mostly related with the perpendicular component [82].

Figure 9b displays the $\varepsilon'(T)$ at 10^4 Hz for bulk E7, which is a sensitive tool to monitor phase transformations offering here a mean to clearly identify the frequency independent nematic to isotropic transition in excellent accordance with calorimetric measurements [128]. This is almost imperceptible in the $\varepsilon''(T)$ trace since it is masked by conductivity effects, which does not affect the temperature dependence of the real part of the complex permittivity. By other side, the isochronal plot of ε'' allows distinguishing the two processes detected in the nematic phase whose location is frequency dependent due to the underlying kinetic nature and which relative magnitude depends on the orientation of the molecules relatively to the outer electric field, the δ-relaxation being the more intense; this means that the sample exhibits a preferential parallel orientation of the nematic director.

The respective temperature dependencies of relaxation times of the two processes detected in the nematic phase, obey to the VFTH law (see Fig. 10). The combination of specific heat spectroscopy with DRS measurements allowed to assign the tumbling mode as the responsible by the dynamic glass transition, corresponding to the α-process found in conventional glass formers [127]. The calorimetric T_g of bulk E7 is 210 K [125, 128]. An additional process, slower than the δ-relaxation is found for the bulk [126, 128, 131], which molecular assignment is still under discussion so, its analysis will not be considered in this section for the confined guest.

E7 was loaded in MCM-41 with 2.8, 3.6 and 4.1 nm of pore diameter and SBA-15 with 6.8 nm under vacuum at 343 K in its isotropic state; feeling degrees between 50 and 60 % as determined in [82] were obtained allowing direct comparison of results.

A multimodal dynamical behavior is found for E7 inside pores as investigated by dielectric spectroscopy exhibiting, in general, three bulk-like and two additional processes. The correspondence with the already described behavior for other guests under confinement, allows assuming that the bulk-like processes are originated by E7 molecules lying in the pores core, while the additional processes are associated with molecules interacting with the surface of the confining porous host by hydrogen bonding. For E7 the hydrogen bonding interaction was demonstrated to be via the cyano (CN) group of the E7 constituents and the silica hydroxyl groups by FTIR

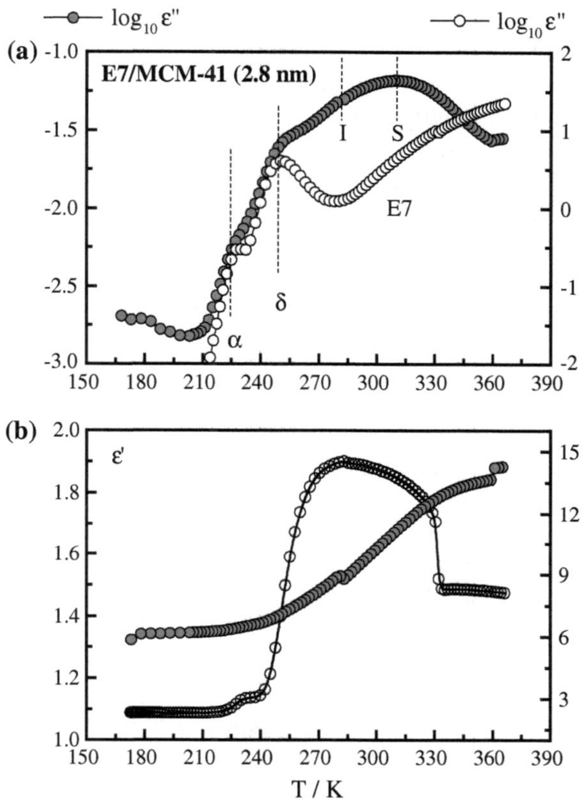

Fig. 9 **a** Dielectric loss of E7 versus temperature at a frequency of 10 kHz confined to the molecular sieve MCM-41 2.8 nm (*filled circles*) during cooling in comparison to the bulk (*open circles*). The dielectric loss of bulk E7 was scaled in order to match the values of the confined material, **b** ε' (T) trace at 10 kHz allowing to identify the nematic to isotropic transition for bulk, which is not perceptible for confined E7 (same symbols used)

analysis [82]. The bulk–like processes involve the isotropic process that is detected above the clearing point, which also splits in the two α and δ processes as found for the bulk in the nematic range, although the region of the transition cannot be identified due to the low dielectric strength of the bulk-like processes and the overlapping of the two strong additional processes. These two intense relaxation processes appear at frequencies lower (higher temperatures) than the processes observed for the bulk, being attributed to E7 molecules adsorbed at the pores wall, both in the inner surface (S-process) and at the outer surfaces of the microcrystals of the silica host and/or at its inter grains (I-process); this will become clearer in the text below.

All the processes are illustrated in Fig. 9a by the isochronal plot at 10 kHz for E7 confined in MCM-41 of 2.8 nm of pore size together with the respective plot for bulk E7. As mentioned above, in the temperature range of the nematic phase, the relative intensity of the dielectric response depends on the orientation of the molecules relatively to the applied electric field, which can differ for each pore size, meaning that not all bulk-like relaxation processes are observed for each sample. The isochronal plot for the real component of the dielectric permittivity at 10 kHz shown in Fig. 9b do not allows identifying the isotropic to nematic transition T_{IN}, although the detection of only one broad process at temperature above the bulk transition T_{IN} and of two in the bulk nematic region, lead to conclude that the confined guest keeps its liquid crystalline properties.

The analysis of the dielectric spectra by a sum of HN equations (Eq. 1), provides the deconvolution in the individual processes, which relaxation times, after conversion to a model independent τ value (Eq. 2), are plotted against 1/T in the relaxation map (Fig. 10).

Figure 11 is a scale-up of relaxation map low temperature region allowing comparing the $\tau_\alpha(1/T)$ and $\tau_\delta(1/T)$ dependencies for the confined guest in the different matrices with bulk E7. It reveals that the temperature dependencies on approaching the glass transition are less abrupt for the confined guest than the ones obeyed by the bulk, the effect being quite significant for the α-process of E7 included in SBA-15 (6.8 nm) (triangles in the figure). This means that the glass transition temperature of the bulk-like E7 molecules decreases, in general, under confinement relative to bulk, which is shown in the inset of the figure versus the inverse of the pore diameter. The higher decrease in T_g estimated by extrapolation of $\tau(1/T)$ to $\tau = 100$ s [122, 123] is close to 10 K for E7 in MCM-41 with 6.8 nm of pore diameter. A decrease of ~30 K is found for the respective Vogel temperature T_0.

Figure 10 also displays the temperature dependence of relaxation times for the two additional processes. The dynamic behavior of the *I*-process, which is located between the bulk-like and the S-process, is consistent with the one observed for a surface layer of 5CB molecules, the main component of E7, adsorbed on nonporous silica spheres [80], and therefore it is assigned to E7 molecules anchored at the outer surface of the micro crystals of the mesoporous silica host. Its dynamical behavior does not show a dependence on the pore size [82].

It is interesting to observe that the $\tau(1/T)$ traces of the S and I processes, which strongly deviate from each other at the lowest temperatures, merge in a single process in the isotropic liquid; this is a further evidence that a nematic transition occurs for the confined guest although is not clear if it coincides with bulk. A similar behavior is observed for the S-process and the process associated with molecules absorbed nearly as a monolayer over silica spheres of aerosil of 8OCB [29].

The temperature dependence of the slowest S-process, follows a VFTH law (Eq. 3) taken as a signature of the glassy dynamics and therefore this process is attributed to the glass transition of the guest molecules anchored to the inner pore surface; a scale up of the relaxation map in the region of the surface process is displayed in Fig. 12. The steepest temperature dependence of molecules adsorbed inside pores relative to molecules adsorbed in the outer surface, is attributed to a stronger interaction in multiple inner surfaces within the three-dimensional pore structure relative to the single outer surface; the possibility of different alignment inside pores relative to the outer surface, mostly parallel in the former and possibly perpendicular in the latter,

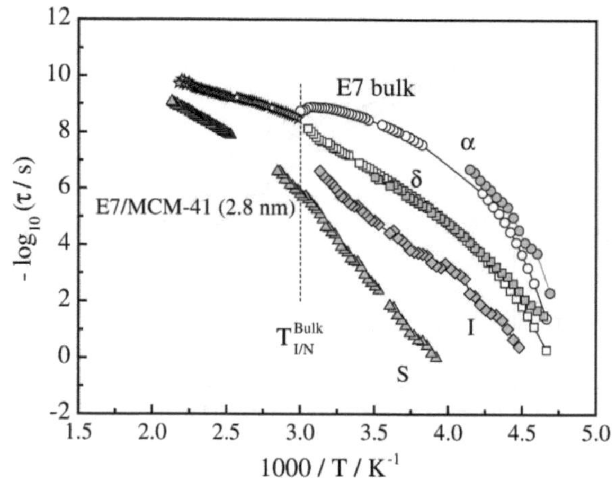

Fig. 10 Activation plot for E7 embedded in the molecular sieve MCM-41 2.8 nm (*filled symbols*) in comparison to the bulk (*open symbols*); the temperature dependence of the two additional I and S processes is included. The *lines* are guides for the eyes. (Reprinted with permission from [82] Copyright (2010), AIP Publishing LLC)

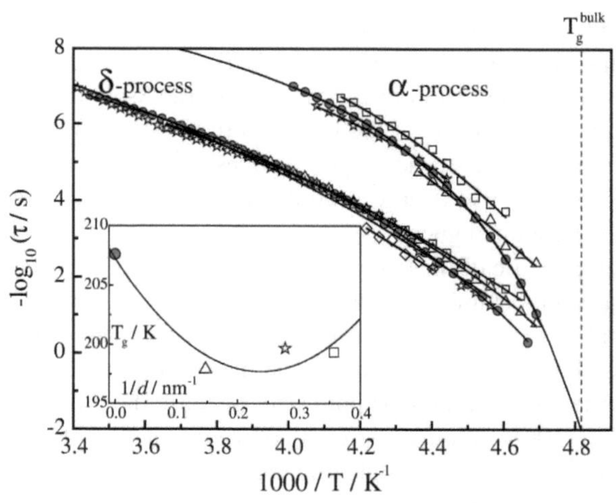

Fig. 11 Activation plot of the temperature dependencies of the α and δ processes in the low temperature region of the relaxation map for E7 confined in the different hosts: *squares*-MCM-41 2.8 nm; *stars*-MCM-41 3.6 nm; *diamonds*-MCM-41 4.1 nm; *triangles*- SBA-15 6.8 nm; no α-process is detected for E7 inside MCM-41 4.1 nm. *Filled gray circles*: data for bulk E7. The *inset* displays the glass transition temperatures, T_g, obtained by extrapolation of $\tau_\alpha(1/T)$ to $\tau = 100\,s$. (Reprinted with permission from [82]. Copyright (2010), AIP Publishing LLC)

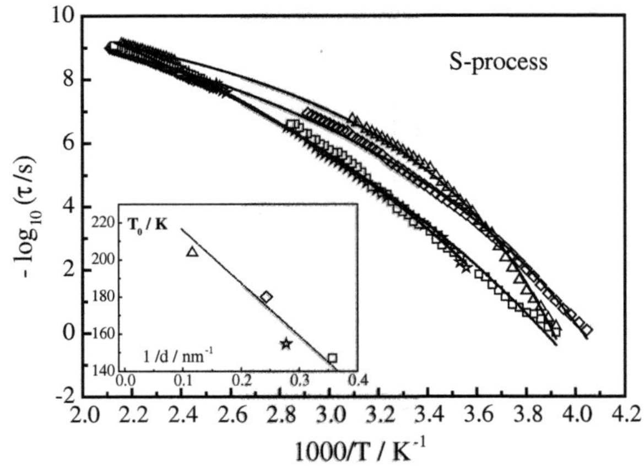

Fig. 12 Activation plot of the relaxation map in the region of the surface process for E7 confined in the different hosts: *squares*-MCM-41 2.8 nm; *stars*-MCM-41 3.6 nm; *diamonds*-MCM-41 4.1 nm; *triangles*-SBA-15 6.8 nm. The inset displays the decrease of the Vogel temperature, T_0 (Eq. 3), with the inverse of the pore diameter, *d*. (reprinted with permission from [82]. Copyright (2010), AIP Publishing LLC)

could also provide different points of interaction between the guest molecules and the silanol groups as suggested for 5CB and 8OCB confined to Al and Si pores [80].

The steepness of the temperature dependence for the S-process of confined E7 is discussed in [82], being higher for the higher pore size (6.8 nm). For the latter, the Vogel temperature T_0 of the surface S-process lyes nearly 50 K above the respective temperature for the bulk-like α-process, reflecting the interaction between the wall of the pore and the confined guest.

The Vogel temperature of these surface interacting molecules decreases with the pore size decrease, almost linearly with 1/d as seen by the plot in the inset of Fig. 12.

It is important to have in mind that the glass transition and Vogel temperatures of the S process should be due to two effects: the strength of the interaction of the molecules with the pore wall and the pore size. Because the different hosts have the same composition (100 % silica) the strength of the interaction of the molecules with the pore wall is the same for all samples, so the decrease of T_0 with decreasing pore size seems to be a finite size effect. This is the opposite that is observed for 8OCB in Al-MCM-Si pores for which an increase of T_0 with the decrease of pore size is found but which is due to a different concentration of Al centers, which provide stronger interaction relative to Si centers, being highly concentrated close to the pores wall for lower pore size [29].

1.3 A Pharmaceutical Drug: Ibuprofen

Ibuprofen is a nonsteroidal worldwide used pharmaceutical compound, which belongs to the category of 2-arylpropanoic acid (see inset of Fig. 13) showing analgesic, antipyretic and anti-inflammatory properties [132]. It is a crystalline solid at ambient temperature (melting temperature $T_m = 347\,K$), which easily fails to crystallize upon cooling from the molten state. It exhibits a glass transition with an onset at 228 K as calorimetrically measured on heating [133, 134], undergoing cold-crystallization to two different polymorphs depending on the specific experimental conditions [134].

When confined to mesoporous SBA-15 (8.6 nm) and MCM-41 (3.6 nm) ibuprofen remains in a long-term stable amorphous form avoiding crystallization under different thermal treatments. This was evaluated 2 and 3 years after drug loading by, respectively, calorimetry and dielectric spectroscopy.

The molecular mobility of the guest inside pores of either SBA-15 [71] or MCM-41 [72] with, respectively 27 wt-% and 32 wt-% feeling degrees (details on its calculation described in reference [71]), was probed by DRS revealing a multimodal character: three bulk-like relaxation processes and two additional ones (see example of Fig. 13 for Ibuprofen confined to MCM-41). The bulk-like processes are the ones detected at the lowest temperatures: the γ-process, which is simply thermally activated, followed by a relaxation assigned to a Johari-Goldstein β_{JG}–process [72, 133], which, in the framework of the coupling model (CM) is taken as the precursor of the α-process associated with the dynamic glass transition, the latter also found for the confined guest being related with the glassy dynamics of the molecules in the core of the pore. The assignment to a β_{JG} process was done by the agreement between the relaxation time predicted by the CM and the τ_β at T_g of this secondary process in both bulk [133] and confined Ibuprofen [72]. For confined Ibuprofen, this was also confirmed by the ratio of $E_{a,\beta}$ (E_a, is the activation energy) by RT_g [72], which is 21 close to 24 as proposed by Kudlik et al. [135] for a JG process.

At higher temperatures than this bulk-like α-process, an intense process emerges becoming the dominant one in the dielectric spectra, which is due to dynamical behavior of molecules interacting with the walls of the pores host, the S-process. Since Ibuprofen is a hydrogen bonding associating liquid, a Debye-like shape D-process associated with the dynamics of the hydrogen-bonded aggregates is detected for the bulk at lower frequencies relative to the α-process [133]. Under confinement, this Debye bulk-like process is not perceived, possibly because it becomes submerged under the strong S-process which also appears in the low frequency flank of the bulk-like α relaxation. Instead, in the confined system, a new Debye-like process is found at higher temperatures and lower frequencies relative to the additional S process; the dynamics of these two additional processes seem to be correlated.

The isochronal plot of the imaginary part of the complex permittivity at 0.1 Hz, Fig. 14a provides a clear picture of the multi-character of the dynamical behavior allowing identifying the different processes and offering a first mean to evaluate the

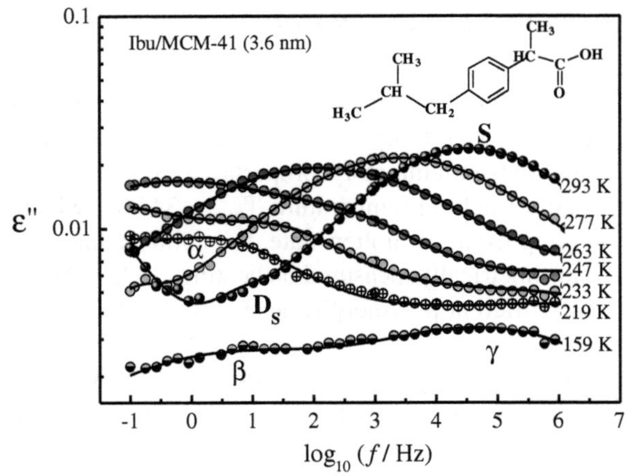

Fig. 13 Dielectric loss versus frequency for Ibu/MCM-41 at the indicated temperatures; the *solid lines* are the overall HN fitting curves (Eq. 1) to the data considering the superposition of multiple processes. The chemical structure of Ibuprofen is included as inset

effect of confinement over the bulk—like processes, namely the shift toward lower temperatures of the β_{JG} and the α-process.

Therefore, two distinct dynamical domains are found for confined Ibuprofen.

Figure 14b puts in evidence, by the isochronal ε' (T) plot, the effect of confinement deviating the temperature location of the α-bulk-like relaxation around 20 and 30 K to lower temperatures for Ibuprofen confined in, respectively, SBA-15 and MCM-41, also leading to the emergence of the S-process. The latter keeps almost the same location irrespective to the confining media since the kind of guest-host interaction between Ibuprofen molecules and the silanol groups is the same in both silica matrices.

Figure 15 provides the dynamical fingerprint of the different systems: (a) the bulk-like processes and (b) the additional S and D-process. It is important to note that secondary relaxations were detected under confinement, which is not usual for conventional glass formers [7, 16] as already mentioned, and able to be analyzed through the HN equation (Eq. 1) over a significant temperature range. The extracted relaxation times for the γ, β_{JG} and α-process nicely reflect the influence of finite size effects when compared with bulk behavior: the more local γ-process almost remains unchanged meaning that length scale motions below 3.6 nm are involved in this secondary relaxation mode; deviations are felt for the β_{JG} and α-process, which become stronger the higher the length scale behind the process that is under analysis. Indeed, the Johari–Goldstein β_{JG}–process, which is not as local as the γ relaxation, becomes more mobile relative to bulk and its trace for the confined systems deviates for a smaller slope in the glass, corresponding to a lower activation energy. Nevertheless, the difference between the temperature dependencies of this process for both matrices is not obvious in the relaxation map. The more dramatic

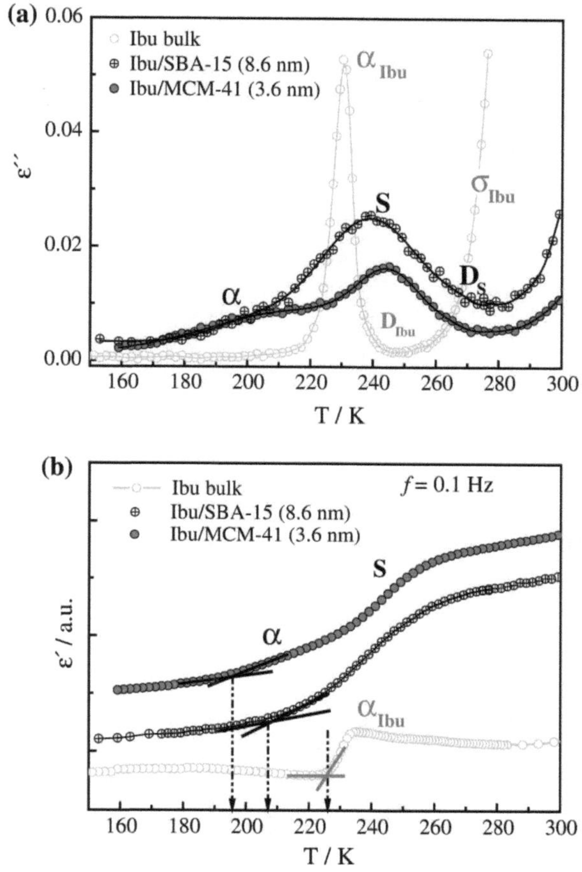

Fig. 14 **a** Isochronal plots at a frequency of 0.1 Hz of the dielectric loss for Ibuprofen confined to the MCM-41 and SBA-15 mesoporous hosts in comparison to the bulk, **b** The same representation is shown for the real part of the complex permittivity evidencing the shift to lower temperatures of the glassy dynamics α-relaxation relative to molecules in the middle of the pore and the emergence at higher temperatures of the surface S-process

changes occur for the α process, which dynamics is associated with motions over a large length. The respective traces change from VFTH dependence in bulk, which curvature reflects the underlying cooperative nature of the α-process, to Arrhenius-like inside pores, and all the α-traces split from each other.

The mobility becomes more enhanced the lower the pore size, while the respective glass transition temperature estimated at $\tau = 100\,\mathrm{s}$ [122, 123] is reduced: $T_{g,\tau=100\,s}$ is 226 K for bulk, 204 K for Ibu/SBA-15 (8.6 nm) and 194 K for Ibu/MCM-41 (3.6 nm). This effect is illustrated in the inset of Fig. 15a. This means that the scale of the confining media is interfering with the length scale of the cooperative motion underlying the glassy dynamics. However, the limit for which the glass transition is fully

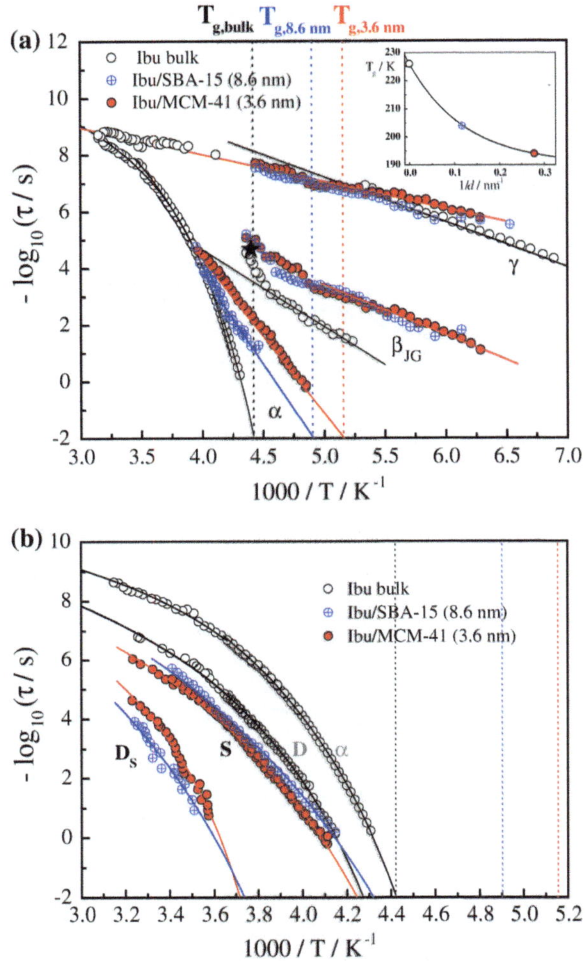

Fig. 15 Activation plots for the relaxation time, τ, versus 1/T for **a** bulk-like processes and **b** additional S and D-processes. *Open black symbols*—bulk Ibuprofen (adapted with permission from [133]. Copyright (2008) American Chemical Society); *blue crossed circles*—Ibu/SBA-15 (adapted with permission from [71]. Copyright (2011) American Chemical Society) and *red filled symbols*—Ibu/MCM-41 (data from [72]). The plot includes relaxation times estimated by the HN fit to raw data and by the Gaussian fit of the isochronal ε'' (T) plot for all studied frequencies f ($\tau = 1/(2\pi f)$, $1/T_{max}$). Lines are fits of the Arrhenius and VFTH formulas to the corresponding data. *Black star* indicates the JG relaxation time, τ_{JG}, estimated from the coupling model for the bulk Ibuprofen. *Vertical dashed lines* indicate the dielectric glass transition temperature T_g, $\tau = 100$ s for bulk (226 K), Ibu/SBA-15 (204 K) and Ibu/MCM-41 (194 K). The decrease of T_g with the inverse of the pore diameter, *d*, is depicted in the inset of **a** ($d = \infty$ and $1/d = 0$ for bulk)

suppressed is not reached yet. The decrease in cooperativity on going from the bulk to the confined guest can be evaluated by this change in the relaxation times temperature dependence and by comparing the linear τ(T) dependencies for the two confined systems: the pre-exponential factor, τ_∞, of the Arrhenius law increases from $\sim 10^{-30}$ s for Ibuprofen in SBA-15 (8.6 nm) to $\sim 10^{-26}$ s for Ibuprofen in MCM-41 (3.6 nm). The obtained τ_∞values, are unphysical low compared with the typical 10^{-12}–10^{-14} s of local orientational fluctuations, which is rationalized as an indication that some degree of cooperativity still affects the glass transition dynamics, however being lower for the smaller pore size. The relatively high activation energy also supports the cooperative nature that persists in the dynamics of the α-process but denounces a significant decrease from 400 kJ mol^{-1} for bulk Ibuprofen to 122 kJ mol^{-1} for Ibu/SBA-15, decreasing even further to 104 kJ mol^{-1} for Ibu/MCM-41.

The coupling model predicts that under the conditions of extreme confinement, the α-process is transformed into the β_{JG} [16] which is not observed so far, but the separation between the α and β_{JG} traces is going smaller from bulk to 8.6 and 3.6 nm; this points out as Ibuprofen being a suitable system to test CM prediction.

An additional parameter derived from the HN simulation of the dielectric spectra is the dielectric strength, which is plotted for each process in Fig. 16 which analysis complements the discussion based on the relaxation map.

The plot offers also a mean to confirm the assignment of the different processes. Interesting, for the bulk is the strong increase of the dielectric strength for the γ-process above T_g, an effect that already happen for the $\tau_{\gamma,Ibu}$–trace that bents off toward the $\tau_{\alpha,Ibu}$-trace in the relaxation map (Fig. 15a), the respective relaxation times being longer than the ones obtained by extrapolating the glassy state Arrhenius dependence [133]. A rationalization of this behavior is given in [16] as being an hybridization of the γ with the β_{JG} relaxation, which reflects a transference of the β_{JG}-process properties to the γ-process. For the respective γ-process under confinement no such bent occurs in the relaxation time T dependence, however its dielectric strength slightly increases on crossing the glass transition of the confined guest. The same for $\Delta\varepsilon_{\beta JG}$, supporting the attribution to a JG process as found for bulk [133]. In what concerns the α-process, the dielectric strength temperature dependence decreases with the temperature increase, recognized as a characteristic sign of the α-relaxation detected in conventional glass formers and polymers [136]. However, in the temperature region near the T_g of the respective system, the rate at which $\Delta\varepsilon_\alpha$ decreases with the temperature increase, is higher for the bulk which can be interpreted as a higher degree of cooperativity of the bulk dynamics relative to the guest inside pores. Indeed, the Onsager-Kirkwood-Fröhlich equation [137–139] predicts a decrease of $\Delta\varepsilon$ with the temperature decrease but less accentuated as occurs for the α-process of low glass formers and polymers, being considered as a general feature of this process [139] and originated by dipole-dipole cross-correlations with neighboring dipolar units becoming more significant when cooperative interactions are important [140].

Interesting here, is the identical magnitude of the dielectric strength of the bulk-like α-process, for Ibuprofen confined in MCM-41 and SBA-15 mesopores. Since the dielectric strength is proportional to the volume density of dipoles [137–139], the mobility enhancement and the reduction on glass transition temperature of Ibu /MCM-41 relative to Ibu/SBA-15 is originated by real finite size effects on decreasing

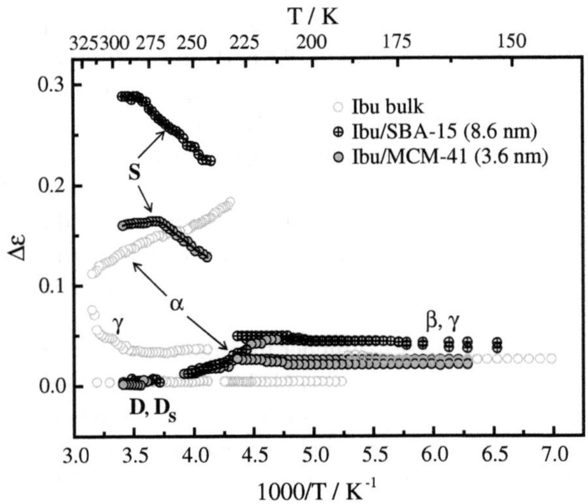

Fig. 16 Dielectric strength, $\Delta\,\varepsilon$, *versus* 1/T for all the processes detected for bulk Ibuprofen, Ibu/SBA-15 and Ibu/MCM-41

the pore size instead of being caused by a less dense packing of Ibuprofen molecules in the middle of the MCM-41 pores, as observed previously for E7 in Anopore membrane [81].

The nonmonotonous variation of the dielectric strength for the S-process is identical for Ibuprofen confined to both matrices and explained as a counterbalance of two effects: at the lowest temperatures, the increase of $\Delta\,\varepsilon(T)$ it is caused by a weakening with the temperature increase of the interaction of Ibuprofen molecules with the pore wall, which leads to the reorientation of greater parts of the molecular dipole vector (increased fluctuation angle) and/or an increased number of fluctuating dipoles as observed for the S-process detected in E7 confined in the same kind of hosts, described in previous section and reported in [82]. At the highest temperatures, $\Delta\,\varepsilon$, follows the decrease with the temperature increasing due to thermal energy predicted by the Onsager-Kirkwood-Fröhlich equation [137–139].

The Debye-like D-process for confined Ibuprofen follows identical temperature dependence of the respective dielectric strength in all the three systems. However, it must be emphasized that this process for the confined guest has a relatively different origin relative to bulk. By means of Molecular Dynamics simulation studies [141], it was demonstrated that the dynamics of the D-process observed for bulk Ibuprofen, which belongs to the class of hydrogen bonded associating liquids, was governed by the internal cis-trans conversion of the O=C–O–H group coupled to the change of the inter-molecular linear/cyclic hydrogen-bonded structures; furthermore, since the cis-trans conversion is much slower (1–2 decades) than the overall rotation of the molecules, all the O=C–O–H group environments are equal on average. This means that the effective rotational potential energy barriers of the O=C–O–H groups due to the surroundings are averaged and the dipolar relaxation follows a simple Debye law, which is in the origin of the Debye-like nature of this process. Moreover, its dynamics is highly correlated with the dynamics of the α-process as found by broadband DRS for bulk Ibuprofen [133]. It is not clear if this process is also observed for confined Ibuprofen due to the overlapping of the strong surface process. However, the Debye-like process detected for Ibuprofen under confinement either in SBA-15 or MCM-41, which emerges in the low frequency tail of the S-process, has a temperature dependency of the relaxation times that evolves quite in parallel with the one of the surface process (remember Fig. 15b) pointing out for a correlation between the dynamic of the two processes; besides the identical temperature dependencies of $\Delta \varepsilon_{Dbulk}$ and $\Delta \varepsilon_{DS}$ as above mentioned. Therefore, this D_S process is taken as being governed by the cis-trans conversion of the O=C–O–H groups of the molecules that belong to the interfacial layer.

2 Conclusions

A review of the dynamical behavior of three quite different low molecular weight glass formers confined to nanostructured mesoporous silica (100 % Si), with pore sizes between 2.8 and 8.6 nm, is provided in this chapter. The confined guests are a surfactant, Triton X-100, a nematic mixture, E7, and the pharmaceutical drug, Ibuprofen. DRS was explored to enlighten both phase transformations and molecular mobility of the confined guests. So, besides the fundamental importance on clarifying theoretical aspects of the glass transition phenomenon, these studies have also a practical interest most relevant to pharmaceutical science and industry confirming the use of confinement as a mean to stabilize the amorphous form of the guest. The long-term stability of the respective amorphous forms inside pores was established for these particular guests.

A common characteristic in the three glass formers is the existence of two distinct dynamical domains inside pores as revealed by the detection of a surface process besides a bulk-like one, the former always with much slower mobility than the bulk-like process, exhibiting a curved plot in the relaxation map taken as a signature of the glassy dynamics. Therefore, this process was assigned to the glassy dynamics of guest molecules anchored at the inner pore walls. For Triton X-100, the molecular population in this interfacial layer is highly significant giving rise to a strong peak in the derivative plot of the mass loss thermogram and allowing its calorimetric detection. Hence, a two-step profile was observed in the heat flux by DSC corresponding to the bulk-like and the interfacial glass transitions of Triton X-100 confined to SBA-15 (8.6 nm). For E7, this surface process presents a dependence on the pore size, the respective Vogel temperature decreasing with increasing pore size, which is taken as a finite size effect. An additional process was detected for E7, for which no dependence on the pore size was found that it is related with the interaction of the molecules anchored at the outer surface of the micro crystals of the mesoporous. Since Ibuprofen belongs to the class of associating materials, a Debye-like process is

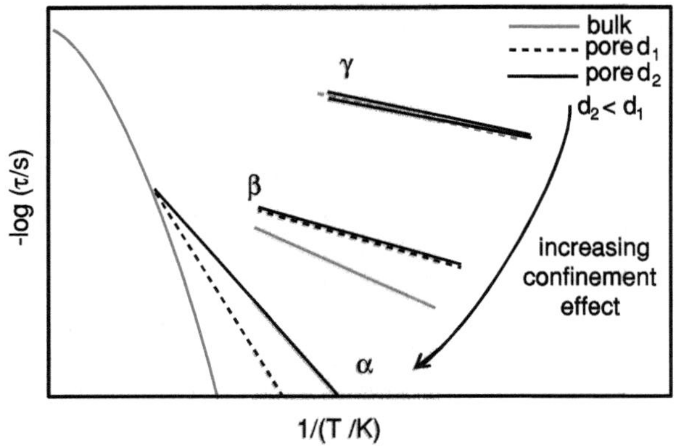

Fig. 17 Schematic representation/scenario of the effect of confinement on the different relaxation processes of Ibuprofen loaded in mesoporous silica, allowing splitting its temperature dependence the higher the length scale of the motional process that is in its origin

found, having a temperature dependence which is correlated with the surface process, being associated to the dynamics of hydrogen bonded aggregates.

Concerning the bulk-like process while no mobility enhancement was observed for Triton X-100, instead a slight hindrance that reflects in an increase between 3 and 4 K in the glass transition temperature, an acceleration is found for E7 inside pores which T_g of the bulk-like process is ~10 K inferior to bulk for a pore size of 6.8 nm (SBA-15); this effect is even higher for Ibuprofen confined to 3.6 nm (MCM-41) for which the glass transition temperature reduces more than 30 K. The temperature dependence of the α-process taken as the responsible for the glassy dynamics, is less accentuated near T_g, for confined E7 and Ibuprofen than for the bulk guest. For Ibuprofen the temperature dependence shows a transition from VFTH to Arrhenius which is interpreted as 3.6 nm being a dimension that interferes with the length-scale of cooperativity. The detection, in this glass former, of the γ and β secondary relaxations, the latter assigned to the Johari-Golstein process, allows to observe confinement effects over different processes which take place over different time and length scales. A schematic representation of the confinement effects is given in the sketch of Fig. 17. In short, confinement almost leaves unaffected the very local γ-process, allows distinguishing the JG process for the confined guest from bulk, a process that owns some degree of cooperativity above the glass transition; nevertheless, the pore dimensions do not allow differentiating the behavior between each matrix to which Ibuprofen is confined to 3.6 and 8.6 nm. The stronger effect occurs over the cooperative α-relaxation, which involves a higher length scale of the underlying mobility, for which confinement allows to split the temperature dependences in the relaxation map for the bulk and the two pore sizes.

Acknowledgments M. Dionísio dedicates this chapter to J. J. Moura Ramos and G. Williams who introduce her in the dielectrics world. The authors deeply acknowledge the fruitful collaboration with Prof. Dr. Andreas Schönhals by the use of broadband dielectric equipment to investigate confined Ibuprofen and E7, being grateful by his deeply knowledge in this matter whose papers supports a significant part of the work here presented. Our collaboration started in 2006 in the framework of the PhD of A. R. Brás, which grant SFRH/BD/23829/2005 is also acknowledged. The authors also acknowledged the careful reading and revision of this chapter by Professor Friedrich Kremer. Financial support for the work here reported was provided through the projects PTDC/CTM//64288/2006 and PTDC/CTM/098979/2008 implemented within the framework of the Programme "Promover a Produção Científica, o Desenvolvimento Tecnológico e a Inovação 002: Investigação Científica e Tecnológica (3599-PPCDTI)" financed by Fundação para a Ciência e Tecnologia (FCT).

References

1. Donth E (2001) Relaxation dynamics in liquids and disordered materials, vol 48. Springer Series in Materials Science, Berlin
2. Leocmach M, Tanaka H (2012) Roles of icosahedral and crystal-like order in the hard spheres glass transition. Nat Commun 3:974–982
3. Cavagna A (2009) Supercooled liquids for pedestrians. Phys Rep 476:51–124
4. Berthier L, Biroli G (2011) Theoretical perspective on the glass transition and amorphous materials. Rev Moder Phys 83(2):587–645
5. Schönhals A, Goering H, Schick Ch, Frick B, Zorn R (2005) Polymers in nanoconfinement: what can be learned from relaxation and scattering experiments? J Non-Cryst Solids 351:2668
6. Adam G, Gibbs JH (1965) On the temperature dependence of cooperative relaxation properties in glass-forming liquids. J Chem Phys 43:139–146
7. Kremer F, Huwe A, Schönhals A, Różański SA (2003) Molecular dynamics in confining space in broadband dielectric spectroscopy; Schönhals, A.; Kremer F., Eds.; Springer, Berlin. Chap. 6.
8. Ngai KL (1993) In: Richert R, Blumen A (eds) Disorder effects on relaxational processes. Springer, Berlin
9. Roth CB, Dutcher JR (2005) Glass transition and chain mobility in thin polymer films. J Electroanal Chem 584:13–22
10. Sappelt D, Jäckle J (1993) The cooperativity length in models for the glass transition. J Phys A: Math Gen 26:7325–7341
11. Fischer EW, Donth E, Steffen W (1992) Temperature dependence of characteristic length for glass transition. Phys Rev Lett 68:2344–2346
12. Gorbatschow W, Arndt M, Stannarius R, Kremer F (1996) Dynamics of h-bonded liquids confined to nanopores. Europhys Lett 35:719–724
13. Arndt M, Stannarius R, Gorbatschow W, Kremer F (1996) Dielectric investigations of the dynamic glass transition in nanopores. Phys Rev E 54:5377–5390
14. Koppensteiner J, Schranz W, Puica MR (2008) Confinement effects on glass forming liquids probed by dma. Phys Rev B 78:054203–054214
15. Ngai KL (2007) Predicting the changes of relaxation dynamics with various modifications of the chemical and physical structures of glass-formers. J Non-Cryst 353:4237–4245
16. Ngai KL (2011) Relaxation and diffusion in complex systems. Springer, New York, Chap 2
17. Beiner M (2008) Nanoconfinement as a tool to study early stages of polymer crystallization. J Polym Sci: Part B, Polym Phys 46:1556
18. Beiner M, Rengarajan GT, Pankaj S, Enke D, Steinhart M (2007) Manipulating the crystalline state of pharmaceuticals by nanoconfinement. Nano Lett 7:1381–1385
19. Rengarajan GT, Enke D, Steinhart M, Beiner M (2011) Size-dependent growth of polymorphs in nanopores and ostwald's step rule of stages. Phys Chem Chem Phys 13:21367–21374
20. Johari GP (2005) Water's size-dependent freezing to cubic ice. J Chem Phys 122:194504–194509
21. Rengarajan GT, Enke D, Beiner M (2007) Crystallization behavior of acetaminophen in nanopores. Open Phys Chem J 1:18–24

22. de Gennes PG (1975) The physics of liquid crystals. Clarendon press, Oxford
23. Chandrasekhar S (1992) Liquid crystals. Cambridge University Press, Cambridge
24. Kremer F, Schönhals A (2003) Molecular and collective dynamics of (polymeric) liquid crystals. In: Kremer F, Schönhals A (eds) Broadband dielectric spectroscopy. Springer, Berlin, Chap. 10
25. Demus D, Goodby J, Gray GW, Spiess HW, Vill V (eds) (1998) Handbook of liquid crystals. Wiley-VCH Weinheim, Weinheim
26. Bellini T, Radzihovsky L, Toner J, Clark NA (2001) Universality and scaling in the disordering of a smectic liquid crystal. Science 294:1074–1079
27. Iannacchione GS (2004) Review of liquid-crystal phase transitions with quenched random disorder. Fluid Phase Equilib 177:222–223
28. Sinha GP, Aliev FM (1998) Dielectric spectroscopy of liquid crystals in smectic, nematic, and isotropic phases confined in random porous media. Phys Rev E 58:2001–2010
29. Frunza L, Frunza S, Kosslick H, Schönhals A (2008) Phase behavior and molecular mobility of n-octylcyanobiphenyl confined to molecular sieves: dependence on the pore size. Phys Rev E 78:051701–051712
30. Iannacchione GS, Crawford GP, Žumer S, Doane JW, Finotello D (1993) Randomly constrained orientational order in porous glass. Phys Rev Lett 71:2595–2598
31. Krause C, Schönhals A (2013) Phase transitions and molecular mobility of a discotic liquid crystal under nanoscale confinement. J Phys Chem C 117:19712–19720
32. Azais T, Tourne-Peteilh C, Aussenac F, Baccile N, Coelho C, Devoisselle JM, Babonneau F (2006) Solid-state nmr study of ibuprofen confined in mcm-41 material. Chem Mater 18:6382–6390
33. Charnay C, Begu S, Tourne-Peteilh C, Nicole L, Lerner DA (2004) Devoisselle j.m., inclusion of ibuprofen in mesoporous templated silica:drug loading and release property. Eur J Pharm Biopharm 57:533–540
34. Rengarajan GT, Enke D, Steinhart M, Beiner M (2008) Stabilization of the amorphous state of pharmaceuticals in nanopores. J Mater Chem 18:2537–2539
35. Beiner M (2008) Nanoconfinement as a tool to study early stages of polymer crystallization. J Polym Sci: Part B Polym Phys 46:1556–1561
36. Jackson CL, McKenna GB (1996) Vitrification and crystallization of organic liquids confined to nanoscale pores. Chem Mater 8:2128–2137
37. Bergman R, Swenson J (2000) Dynamics of supercooled water in confined geometry. Nature 403:283–286
38. Konno T, Kinuno K, Kataoka K (1986) Physical and chemical changes of medicinals in mixtures with adsorbents in the solid state. i. effect of vapor pressure of the medicinals on changes in crystalline properties. Chem Pharm Bull (Tokyo) 34:301–307
39. Ajayan PM, Iijima S (1993) Capillarity-induced filling of carbon nanotubes. Nature 361:333–334
40. Prasad R, Lele S (1994) Stabilization of the amorphous phase inside carbon nanotubes: solidification in a constrained geometry. Philos Mag Lett 70:357–361
41. Qian KK, Bogner RH (2012) Application of mesoporous silicon dioxide and silicate in oral amorphous drug delivery systems. J Pharm Sci 101(2):444–463
42. Laitinen R, Lobmann K, Strachan CJ, Grohganz H, Rades T (2013) Emerging trends in the stabilization of amorphous drugs. Int J Pharm 453:65–79
43. Hancock BC, Shamblin SL, Zografi G (1995) Molecular mobility of amorphous pharmaceutical solids below their glass transition temperatures. Pharm Res 12(6):799–806
44. Yoshioka M, Hancock BC, Zografi G (1995) Inhibition of indomethacin crystallization in poly(vinylpyrrolidone) coprecipiatates. J Pharmaceu Sci 84(8):983–986
45. Crawford GF, Žumer S (eds) (1996) Liquid crystals in complex geometries. formed by polymer and porous networks. Taylor and Francis, London.
46. Aliev F, Sinha G (2001) Non-debye relaxation and glass-like behavior of confined liquid crystals. Mol Cryst Liq Crys 364:435–442

47. Sinha G, Leys J, Glorieux C, Thoen J (2005) Dielectric spectroscopy of aerosil-dispersed liquid crystal embedded in anopore membranes phys. Rev E 72:051710
48. Cramer Ch, Cramer Th, Kremer F, Stannarius R (1997) Measurement of orientational order and mobility of a nematic liquid crystal in random nanometer confinement. J Chem Phys 106:3730–3742
49. Massalska-Arodź M, Krawczyka J, Procyk B, Kremer F (2007) Dielectric relaxation studies of 4-(2-hexyloxyethoxy)4'-cyanobiphenyl (6o2ocb) enclosed in sio2 nanopores. Phase Transitions 80:687–695
50. Werner J, Otto K, Enke D, Pelzl G, Janowski F, Kresse H (2000) Dielectric investigations of the n-smb transition in a porous glass. Liq Cryst 10:1295–1300
51. Frunza S, Frunza L, Schönhals A (2000) Dielectric measurements of liquid crystals confined to molecular sieves. J Phys IV France 10:115–118
52. Frunza L, Kosslick H, Frunza S, Schönhals A (2006) Molecular dynamics of 4-n-octyl-4'-cyanobiphenyl in partially filled nanoporous sba-type molecular sievesmicropor. Mesopor Mater 90:259–270
53. Frunza S, Frunza L, Schönhals A, Zubowa HL, Kosslick H, Carius HE, Frick R (1999) On the confinement of liquid crystals in molecular sieves:dielectric measurements. Chem Phys Lett 307:167–176
54. Rózanski SA, Stannarius R, Kremer F, Diele S (2001) Structure and dynamics of ferroelectric liquid crystals under random geometrical restrictions liq. Cryst 28:1071–1083
55. Rózanski SA, Kremer F, Groothues H, Stannarius R (1997) The dielectric properties of nematic liquid crystal, 5cb confined to treated and untreated anopore membranes. Mol Cryst Liq Crys 303:319–324
56. Nazario Z, Sinha GP, Aliev FM (2001) Dynamics of librational mode of nematic liquid crystal confined in cylindrical pores. Mol Cryst Liq Crys 367:333–340
57. Diez S, Pérez-Jubindo MA, De la Fuente MR, López DO, Salud J, Tamarit JL (2006) On the in?uence of cylindrical sub-micrometer con?nement on heptyloxycyanobiphenyl (7ocb). a dynamic dielectric study. Chem Phys Lett 423:463–469
58. Bengoechea MR, Aliev FM (2005) Dielectric relaxation in thin liquid crystal layers formed on cylindrical pore walls. J Non-Cryst Solids 351:2685–2689
59. Brás AR, Dionísio M, Schönhals A (2008) Confinement and surface effects on the molecular dynamics of a nematic mixture investigated by dielectric relaxation spectroscopy. J Phys Chem B 112:8227–8235
60. Leys J, Sinha G, Glorieux C, Thoen J (2005), Influence of nanosized confinements on 4-n-decyl-4'-cyanobiphenyl (10CB): a broadband dielectric study. Phys Rev E 71:051709 (13 pages).
61. Koppensteiner J, Schranz W, Carpenter MA (2010), Revealing the pure confinement effect in glass-forming liquids by dynamic mechanical analysis. Phys Rev B 81:024202 (8 pages).
62. Kranbuehl D, Knowles R, Hossain A, Hurt M (2003) Modelling the effects of confinement on the glass transition temperature and segmental mobility. J Phys: Condens Matter 15:S1019–S1029
63. Aliev FM (1996) Liquid crystals and polymers in pores, the influence of confinement on dynamic and interfacial properties. In: Crawford GF, Žumer S (eds) Liquid crystals in complex geometries, formed by polymer and porous networks. Taylor and Francis, London
64. Pissis P, Daoukaki-Diamanti D, Apekis L, Christodoulides C (1994) The glass transition in confined liquids. J Phys Condens Matter 6:L325–L328
65. Iacob C, Sangoro JR, Papadopoulos P, Schubert T, Naumov S, Valiullin R, Kärger J, Kremer F (2010) Charge transport and diffusion of ionic liquids in nanoporous silica membranes. Phys Chem Chem Phys 12:13798–13803
66. Richert R (2011) Dynamics of nanoconfined supercooled liquids. Ann Rev Phys Chem 62:65–84
67. Balas F, Manzano M, Horcajada P, Vallet-Regí M (2006) Confinement and controlled release of bisphosphonates on ordered mesoporous silica-based materials. J Am Chem Soc 128:8116

68. Qu F, Zhu G, Huang S, Li S, Qiu S (2006) Effective controlled release of captopril by silylation of mesoporous mcm-41. ChemPhysChem 7:400–406
69. Tang Q, Xu Y, Wu D, Sun Y (2006) Hydrophobicity-controlled drug delivery system from organic modified mesoporous silica. Chem Lett 35:474–475
70. Dadmun MD, Muthukumar M (1993) The nematic to isotropic transition of a liquid crystal in porous media. J Chem Phys 98:4850–4852
71. Brás AR, Merino EG, Neves PD, Fonseca IM, Dionísio M, Schönhals A, Correia NT (2011) Amorphous ibuprofen confined in nanostructured silica materials: a dynamical approach. J Phys Chem C 115:4616–4623
72. Brás AR, Fonseca IM, Dionísio M, Schönhals A, Affouard F, Correia NT Influence of nanoscale confinement on the molecular mobility of ibuprofen. submitted.
73. Buntkowsky G, Breitzke HH, Adamczyk A, Roelofs F, Emmler T, Gedat E, Grünberg B, Xu Y, Limbach H-H, Shenderovich I et al (2007) Structural and dynamical properties of guest molecules confined in mesoporous silica materials revealed by NMR. Phys Chem Chem Phys 9:4843–4853
74. Izquierdo-Barba I, Sousa E, Doadrio JC, Doadrio AL, Pariente JP, Martínez A, Babonneau F, Vallet-Regí M (2009) In?uence of mesoporous structure type on the controlled delivery of drugs: release of ibuprofen from mcm-48, sba-15 and functionalized sba-15. J Sol-Gel Sci Technol 50:421–429
75. Azaïs T, Hartmeyer G, Quignard S, Laurent G, Tourné-Péteilh C, Devoisselle JM, Babonneau F (2009) Solid-state nmr characterization of drug-model molecules encapsulated in mcm-41 silica. Pure Appl Chem 81(8):1345–1355
76. MuWoz Rámila A, Pérez-Pariente J, Díaz I, Vallet-Regí M (2003) MCM-41 organic modification as drug delivery rate regulator. Chem Mater 15:500–503
77. Arndt M, Stannarius R, Groothues H, Hempel E, Kremer F (1997) Length scale of cooperativity in the dynamic glass transition. Phys Rev Lett 79:2077–2080
78. Schüller J, Mel'nichenko YB, Richert R, Fischer EW (1994) Dielectric studies of the glass transition in porous media. Phys Rev Lett 73:2224–2227
79. Cramer Ch, Cramer Th, Kremer F, Stannarius R (1997) Measurement of orientational order and mobility of a nematic liquid crystal in random nanometer confinement. J Chem Phys 106(9):3730–3742
80. Frunza S, Frunza L, Tintaru M, Enache I, Beica T, Schönhals A (2004) Dynamics of the surface layer in cyanobiphenyl-aerosil nanocomposites with a high silica density. Liq Cryst 31:913–932
81. Brás AR, Dionísio M, Schönhals A (2008) Confinement and surface effects on the molecular dynamics of a nematic mixture investigated by dielectric relaxation spectroscopy. J Phys Chem B 112(28):8227–8235
82. Brás AR, Frunza S, Guerreiro L, Fonseca IM, Corma A, Frunza L, Dionísio M, Schönhals A (2010), Molecular mobility of nematic E7 confined to molecular sieves with a low filling degree. J Chem Phys 132:224508 (1–12).
83. Park J-Y, Mckenna GB (2000) Size and con?nement effects on the glass transition behavior of polystyrene/o-terphenyl polymer solutions. Phys Rev B 61:6667–6676
84. Trofymluk O, Levchenko AA, Navrotsky A (2005), Interfacial effects on vitrification of confined glass-forming liquids. J Chem Phys 123:194509 (1–7).
85. Zheng W, Simon SL (2007), Confinement effects on the glass transition of hydrogen bonded liquids. J Chem Phys 127:194501 (1–11).
86. Le Quellec C, Dosseh G, Audonnet F, Brodie-Linder N, Alba-Simionesco C, Haüssler WB (2007) Frick influence of surface interactions on the dynamics of the glass former ortho-terphenyl confined in nanoporous silica. Eur Phys J Spec Top 141:11–18
87. Rengarajan GT, Enke D, Steinhart M, Beiner M (2008) Stabilization of the amorphous state of pharmaceuticals in nanopores. J Mater Sci 18:2537–2539 (ESI; Fig. S3).
88. Van Speybroeck M, Mellaerts R, Martens J, Annaert P, Van Den Mooter G, Augustijns P (2011) Ordered mesoporous silica for the delivery of poorly soluble drugs. In: Wilson CJ, Crowley PJ (eds) Controlled release in oral drug delivery: advances in delivery science and technology. Springer New York, Chap, p 10

89. Ruiz-Hitzky E, Ariga K, Lvov YM (eds) (2008) Bio-inorganic hybrid nanomaterials: strategies, syntheses, characterization and applications. Wiley-VCH, Weinheim
90. Salonen J, Kaukonen AM, Hirvonen J, Lehto VP (2008) J Pharm Sci 97:632–653
91. Beck S, Vartuli JC, Roth WJ, Loenowicz ME, Kresge CT, Schmitt KD, Chu CTW, Olson DH, Sheppard EW, McCullen SB, Higgins JB, Schlenker JL (1992) J Am Chem Soc 114:10834
92. Fan J, Yu C, Gao F, Lei J, Tian B, Wang L, Luo Q, Tu B, Zhou W, Zhao D (2003) Angew Chem 115:3254
93. Fan J, Yu C, Gao F, Lei J, Tian B, Wang L, Luo Q, Tu B, Zhou W, Zhao D (2003) Angew Chem Int Ed 42:3146
94. Gao L, Wang Y, Wang J, Huang L, Shi L, Fan X, Zou Z, Yu T, Zhu M, Li Z (2006) A novel znii-sensitive fluorescent chemosensor assembled within aminopropyl-functionalized mesoporous sba-15. Inorg Chem 45:6844–6850
95. Merino EG, Neves PD, Fonseca IM, Danéde F, Idrissi A, Dias CJ, Dionísio M, Correia NT (2013) Detection of two glass transitions on triton x-100 under confinement. J Phys Chem B 117:21516–21528
96. van Speybroeck M, Barillaro V, DoThi T, Mellaerts R, Martens J, van Humbeeck J, Vermant J, Annaert P, van den Mooter G, Augustijns P (2009) Ordered mesoporous silica material sba-15: a broad-spectrum formulation platform for poorly soluble drugs. J Pharm Sci 98:2648–2658
97. Vallet-Regí M, Balas F, Arcos D (2007) Mesoporous materials for drug delivery. Angew Chem Int Ed 46:7548–7758
98. Vallet-Regí M, Rámila A, del Real RP (2001) Pérez-pariente, a new property of mcm-41:? Drug delivery system. J Chem Mater 13:308–311
99. Muñoz Rámila A, Pérez-pariente J, Díaz I, Vallet-Regí M (2003) MCM-41 organic modification as drug delivery rate regulator. Chem Mater 15:500–503
100. Doadrio C, Sousa EMB, Iizquierdo-Barba I, Doadrio AL, Pérez-Pariente J, Vallet-Regí M (2006) functionalization of mesoporous materials with long alkyl chains as a strategy for controlling drug delivery pattern. J Mater Chem 16:462–466
101. Zeng W, Qian XF, Zhang YB, Yin J, Zhu ZK (2005) organic modified mesoporous mcm-41 through solvothermal process as drug delivery system. Mater Res Bull 40:766–772
102. Qu F, Zhu G, Huang S, Li S, Sun J, Zhang D, Qiu S (2006) Controlled release of captopril by regulating the pore size and morphology of ordered mesoporous silica. Micropor Mesopor Mater 92:1–9
103. Nunes CD, Vaz PD, Fernandes AC, Ferreira P, Romão CC, Calhorda MJ (2007) Loading and delivery of sertraline using inorganic micro and mesoporous materials. Eur J Pharm Biopharm 66:357–365
104. van Speybroeck M, Mellaerts R, Mols R, DoThi T, Martens JA, van Humbeeck J, Annaert P, van den Mooter G, Augustijns P (2010) enhanced absorption of the poorly soluble drug feno?brate by tuning its release rate from ordered mesoporous silica. Eur J Pharm Sci 41:623–630
105. Izquierdo-Barba I, Martínez A, Doadrio AL, Pérez-Pariente J, Vallet-Regí M (2005) Release evaluation of drugs from ordered three-dimensional silica structures. Eur J Pharm Sci 26:365–373
106. Mellaerts R, Jammaer JAG, Van Speybroeck M, Chen H, Van Humbeeck J, Augustijns P, Van den Mooter G, Martens JA (2008) Physical state of poorly water soluble therapeutic molecules loaded into sba-15 ordered mesoporous silica carriers: a case study with itraconazole and ibuprofen. Langmuir 24:8651–8659
107. Shen SC, Ng WK, Chia L, Dong YC, Tan RB (2010) Stabilized amorphous state of ibuprofen by co-spray drying with mesoporous sba-15 to enhance dissolution properties. J Pharm Sci 99:1997–2007
108. Doadrio AL, Sousa EMB, Doadrio JC, Pérez-Pariente J, Izquierdo-Barba I, Vallet-Regí M (2004) Mesoporous sba-15 hplc evaluation for controlled gentamicin drug delivery. J Controlled Release 97:125–132
109. Vallet-Regí M, Doadrio JC, Doadrio I, Izquierdo-Barba I, Pérez-Pariente J (2004) hexagonal ordered mesoporous material as a matrix for the controlled release of amoxicillin. Solid State Ionics 172:435–439

110. Tozuka Y, Wongmekiat A, Kimura K, Moribe K, Yamamura S, Yamamoto K (2005) Effect of pore size of fsm-16 on the entrapment of flurbiprofen in mesoporous structures. Chem Pharm Bull 53:974–977
111. Tozuka Y, Oguchi T, Yamamoto K (2003) Adsorption and entrapment of salicylamide molecules into the mesoporous structure of folded sheets mesoporous material (fsm-16). Pharm Res 20:926–930
112. Merino EG, Rodrigues C, Teresa Viciosa M, Melo C, Sotomayor J, Dionísio M, Correia NT (2011) phase transformations undergone by triton x-100 probed by differential scanning calorimetry and dielectric relaxation spectroscopy. J Phys Chem B 115:12336–12347
113. Merino EG, Danéde F, Derrollez P, Dias CJ, Viciosa MT, Correia NT, Dionísio M (2013) Investigating the influence of morphology in the dynamical behavior of semicrystalline triton x-100: insights in the detection/nondetection of the α?-process. J Phys Chem B 117:9793–9805
114. van Turnhout J, Wübbenhorst M (2002) Analysis of complex dielectric spectra. II: evaluation of the activation energy landscape by differential sampling. J Non-Cryst Sol 305:50–58
115. Schönhals A, Kremer F (2003) Analysis of dielectric spectra. In: Kremer F, Schönhals A (eds) Broadband dielectric spectroscopy. Springer, Berlin, Chap. 3
116. Steeman PAM, van Turnhout J (2003) Dielectric properties of inhomogeneous media. In: Kremer F, Schönhals A (eds) Broadband dielectric spectroscopy. Springer, Berlin, Chap. 13
117. Boersma A, Van Turnhout J, Wübbenhorst M (1998) Dielectric characterization of a thermotropic liquid crystalline copolyesteramide:? 1 Relaxation peak assignment. Macromolecules 31:7453–7460
118. Schröter K, Unger R, Reissig S, Garwe F, Kahle S, Beiner M, Donth E (1998) Dielectric spectroscopy in the α β splitting region of glass transition in poly(ethyl methacrylate) and poly(n-butyl methacrylate):? different evaluation methods and experimental conditions. Macromolecules 31:8966–8972
119. Vogel H (1921) The temperature dependence law of the viscosity of fluids. Phys Zeit 22:645–646
120. Fulcher GS (1925) Analysis of recent measurements of the viscosity of glasses. J Am Ceram Soc 8:339–355
121. Tammann G, Hesse W (1926) The dependancy of viscosity on temperature in hypothermic liquids. Z Anorg Allg Chem 156:245–257
122. Böhmer R, Ngai KL, Angell CA, Plazek DJ (1993) Nonexponential relaxations in strong and fragile glass formers. J Chem Phys 99:4201–4209
123. Moynihan CT, Macebo PB, Montrose CJ, Gupta PK, DeBolt MA, Dill JF, Dom BE, Drake PW, Esteal AJ, Elterman PB et al (1976) Structural relaxation in vitreous materials. Ann NY Acad Sci 279:15–35
124. Brás ARE, Henriques S, Casimiro T, Aguiar-Ricardo A, Sotomayor J, Caldeira J, Santos C, Dionísio M (2007) Characterization of a nematic mixture by reversed-phase hplc and uv spectroscopy: application to phase behavior studies in liquid crystal-co2 systems. Liq Crys 34(5):591–597
125. Maschke U, Benmouna M, Coqueret X (2002) Electro-optical properties of polymer-dispersed liquid crystals macromol. Rapid Commun 23:159
126. Viciosa MT, Nunes AM, Fernandes A, Almeida PL, Godinho MH, Dionisio M (2002) Dielectric studies of the nematic mixture e7 on a hydroxypropylcellulose substrate liq. Cryst 29:429
127. Brás AR, Dionísio M, Huth H, Schick Ch, Schönhals A (2007) The origin of glassy dynamics in a liquid crystal studied by broadband dielectric and thermal spectroscopy. Phys Rev E 75:061708
128. Brás AR, Viciosa MT, Dias JC, Rodrigues C, Dionísio M (2006), Changes in molecular dynamics upon formation of a polymer dispersed liquid crystal. Phys Rev E 73 (6):061709 (1–11).
129. Zhong ZZ, Schuele DE, Gordon WL, Adamic KJ, Akins RB (1992) Dielectric properties of a pmma/e7 polymer-dispersed liquid crystal. J Polym Sci B: Polym Phys 30:1443–1449

130. Roussel F, Buisine JM, Mascchke U, Coqueret X (1998) Photopolymerization kinetics and phase behaviour of acrylate based polymer dispersed liquid crystals. Liq Cryst 24:555–561
131. Capaccioli S, Prevosto D, Bets A, Hanewald A, Pakula T (2007) Applications of the rheo-dielectric technique. J Non-Cryst Solids 353:4267–4272
132. Goodman LS, Gilman A (1990) The pharmacological bases of therapeutics, 8th edn. Pergamon Press, New York
133. Brás AR, Noronha JP, Antunes AMM, Cardoso MM, Schönhals A, Affouard F, Dionísio M, Correia NT (2008) Molecular motions in amorphous ibuprofen as studied by broadband dielectric spectroscopy. J Phys Chem B 112:11087–11099
134. Dudognon E, Danède F, Descamps M, Correia NT (2008) Evidence for a new crystalline phase of racemic ibuprofen. Pharm Res 25:2853–2858
135. Kudlik A, Tschirwitz C, Blochowicz T, Benkhof S, Rössler E (1998) Slow secondary relaxation in simple glass formers. J Non-Cryst Solids 235–237:406–411
136. Schönhals A (2003) Molecular dynamics in polymer model systems. In: Kremer F, Schönhals A (eds) Broadband dielectric spectroscopy. Springer, Berlin, Chap. 7
137. Böttcher CJF (1973) Theory of dielectric polarization, vol 1, 2nd edn. Elsevier, Amsterdam
138. Böttcher CJF, Bordewijk P (1978) Theory of dielectric polarization, vol 2. Elsevier, Amsterdam
139. Schönhals A, Kremer F (2003) Theory of dielectric relaxation. In: Kremer F, Schönhals A (eds) Broadband dielectric spectroscopy. Springer, Berlin, Chap. 1
140. Kremer F, Schönhals A (2003) The scaling of the dynamics of glasses and supercooled liquids. In: Kremer F, Schönhals A (eds) Broadband dielectric spectroscopy. Springer, Berlin, Chap. 4
141. Affouard F, Correia NT (2010) Debye process in ibuprofen glass-forming liquid: insights from molecular dynamics simulation. J Phys Chem B 114:11397–11402

Deviations from Bulk Glass Transition Dynamics of Small Molecule Glass Formers: Some Scenarios in Relation to the Dimensionality of the Confining Geometry

Michael Wübbenhorst and Simone Napolitano

Abstract Deviations from the bulk behavior of the temperature dependence of the structural relaxation time τ(T), typically well described by the Vogel-Fulcher-Tammann (VFT) relation, are the most obvious signature for "confinement effects," which comprise various scenarios ranging from slight changes in the VFT parameters to cases where the VFT behavior completely breaks down to an Arrhenius law. An attractive idea that has finally stimulated many studies on nanoconfined glass formers is the concept of cooperativity with a cooperative length scale ξ that dates back to Adams Gibbs and is also a subject of modern concepts for the description of glasses and supercooled liquids. In this chapter, we discuss the question in how far experiments on glass-forming liquids, confined to nanometer-sized geometries, can prove the cooperative nature and of the dynamic glass transition and provide quantitative information about its characteristic length scale ξ. Here, we have particularly focused on experimental evidence regarding deviations of τ(T) from the bulk behavior. To classify experimental findings from very diverse systems, we introduce three major "deviation scenarios," which range from a complete breakdown of the VFT-law (type I) via a VFT-Arrhenius cross-over scenario (type II) to a scenario (III) representing a perturbed VFT behavior in terms of an overall accelerated or retarded glass transition dynamics. Four cases are discussed in detail, all of them referring to H-bonding molecules or molecular groups. EG confined in various zeolitic hosts revealed either bulk dynamics or the scenario I, while EG mixed with amylopectine (starch) showed an evolution through the cross-over scenario (II) toward genuine Arrhenius behavior (I). A third system, representing a H-bonding "liquid" confined to a self-assembled (smectic) layer structure, yielded a clear-cut scenario II along

M. Wübbenhorst (✉)
Laboratory of Acoustics and Thermal Physics, Department of Physics and Astronomy, KU Leuven, Celestijnenlaan 200D, 3001 Leuven, Belgium
e-mail: wubbenhorst@fys.kuleuven.be

S. Napolitano
Laboratory of Polymer and Soft Matter Dynamics, Faculté des Sciences, Université Libre de Bruxelles (ULB), Bâtiment NO, Boulevard du Triomphe, 1050 Bruxelles, Belgium

F. Kremer (ed.), *Dynamics in Geometrical Confinement*, Advances in Dielectrics, 247
DOI: 10.1007/978-3-319-06100-9_10, © Springer International Publishing Switzerland 2014

with a linear relation between the layer thickness and the cross-over frequency. Subsequently, the dynamics of ultrathin films of glycerol having one free surface was discussed as an example for a perturbated VFT-scenario (type III). Apart from all molecular details and the diversity of confining host structures, it is evident that the strength of the confinement, expressed by the number of spatially restricted dimensions (1D = layer, 3D = cavities), should be linked to the severeness of changes in the dynamics. Inventarization of various experimental findings reported in the literature including them described above indeed confirm a correlation between the "deviation scenario" and the number of constrained axes. Finally, the effect of mechanisms other than cooperativity arguments, e.g., surface interactions, density, packing, and orientation effects, will briefly be discussed to rationalize obvious exceptions from the proposed correlation.

Keywords Glass transition · Nanometer confinement · Dielectric relaxations

Abbreviations

Al	Aluminum
α-	Structural relaxation process
AFM	Atomic force microscopy
AG	Adams-Gibbs (approach)
AP	Amylopectine
BDS	Broadband dielectric spectroscopy
CRR	Cooperatively rearranging regions
DRS	Dielectric relaxation spectroscopy
DSC	Differential scanning calorimetry
$\Delta\varepsilon$	Dielectric strength
s	Electrical conductivity
EG	Ethylene glycol
ε_∞	Instantaneous dielectric constant
g	Kirkwood factor
h	Film thickness
HN	Havriliak-Negami (function)
k_B	Boltzmann constant
MCT	Mode-coupling theory
min	Minute(s)
mol	Mole(s)
M_w	Weight average molecular weight
OMBD	Organic molecular beam deposition
ξ	Cooperativity length
τ_α	(Structural) relaxation time
τ	Relaxation time
t	Time

T	Temperature
T_g	Glass transition temperature
VFT	Vogel-Fulcher-Tammann (equation)
1D	One-dimensional

1 Introduction

The cooperative dynamics in nanoconfined glass formers is a topic of persisting interest in condensed matter physics for more than 20 years. When a supercooled liquid is subjected to a geometry restricted in size on the nanometer scale, deviations from its bulk dynamics are commonly observed, which particularly manifest themselves in changes in the characteristic temperature dependence as captured by the Vogel-Fulcher-Tammann (VFT) relation [1–3]

$$\tau(T) = \tau_\infty \exp\left[\frac{E_V}{R(T - T_V)}\right]. \tag{1}$$

Deviations from the bulk-VFT behavior are the most obvious signature for "confinement effects," which comprise various scenarios ranging from slight changes in the Vogel-parameters to cases where the VFT behavior completely breaks down to an Arrhenius law. A basic idea that has finally stimulated many studies on nanoconfined glass formers is the concept of cooperativity with a cooperative length scale ξ that dates back to Adams Gibbs [4]. Figure 1 sketches the temperature-dependent size of cooperatively rearranging regions (CRR) as proposed by the AG-theory, and the structural relaxation time τ_α. In this context, the size of the CRR (i.e., the volume $\sim\xi^3$) is taken as a measure for the degree of cooperativity, which increases upon cooling due to densification of the melt and thus causes the progressive increase in the structural relaxation time.

It is evident that the physical picture suggests by Fig. 1 is oversimplified as it ignores both the actual (diversity in) shape of dynamically correlated regions and the (dynamically) heterogeneous nature of supercooled liquids. Moreover, there are more recent theoretical approaches that predict the existence of intrinsic length scales of glassy dynamics such as the random first order transition (RFOT) model of Wolynes [5] and the two order parameter (TOP) model of Tanaka [6]. In the RFOT, an intrinsic length is defined as the correlation length of the dynamics within "entropic droplets," which are solid-like regions with aperiodic crystalline structure. Here, ξ also represents the characteristic length scale of the dynamic heterogeneity, as the droplets are spatially and morphologically distinct.

In contrast, the TOP model replaces the concept of CRR's by solid-like islands with nonzero medium range crystalline order, and the concept of cooperative motion is overcome introducing the idea of orientational correlation among neighboring molecules, established via intermolecular bonds. In the RFOT theory, the formation of the droplets is driven by configurational entropy alone, while in the TOP model

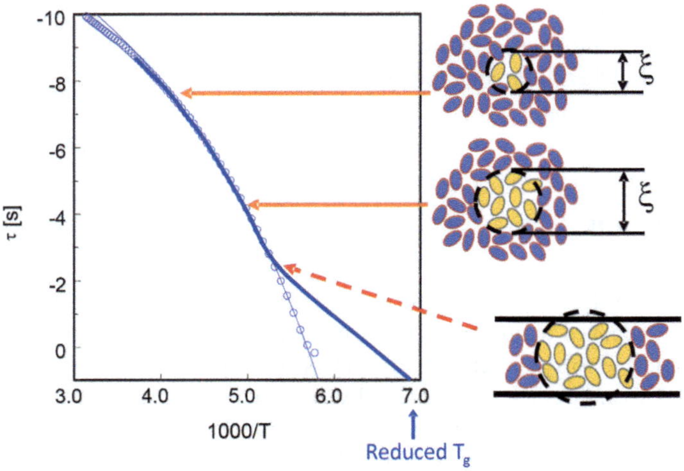

Fig. 1 Typical temperature dependence of the dielectric relaxation time of a small molecule glass former (1,2-propanediol, *blue symbols*) along with a fictive deviation scenario (*blue line*). On the *right*, a cartoon of molecular sub-ensembles (CRR's) is shown to give an (over) simplified clue about the CRR's shape and size in different situations

regions rich in medium range crystalline order are induced by the simultaneous maximization of the density and of intermolecular bonds.

Despite its shortcomings that have been discussed elsewhere, the simple CRR-picture helps to rationalize so-called confinement effects as they concern cooperativity as illustrated on the bottom of Fig. 1. When the intrinsic length scale of cooperativity approaches, upon cooling, the dimension of the confining geometry, we may expect a deviation from the bulk-VFT-law into a less progressive temperature dependence with true Arrhenius behavior being the limiting case. Indeed, transitions from VFT-behavior to a scenario of constant activation energy (Arrhenius-law) at a specific temperature/relaxation time have been reported by several authors for a large diversity of systems, sometimes denoted as strong-fragile transition [7] as particularly debated for confined water [8]. The widespread of this observation is a strong support for the relevance of a cooperativity length in studies of nanoconfined glass formers and has yielded estimates for ξ being in good agreement with results from other approaches based on c_p-measurements [9, 10] or a combination of the Adam-Gibbs theory with the self-concentration concept [11].

Given the validity of a relation between the temperature dependence of the cooperativity volume and the effective activation barrier, we might consider the apparent activation energy $E_{app}(T)$ as a measure of the (T-dependent) degree of cooperativity itself, which reads for the VFT-equation as follows:

$$E_{app}(T) = \frac{E_V}{(1 - T_V/T)^2} \tag{2}$$

Analyzing the local activation energy $E_{app}(T)$ for nanoconfined systems will thus provide a powerful approach to detect perturbations in the dynamics from bulk behavior [12, 13]; this sensitive tool will be discussed in the last section below.

As the apparent activation energy is related to the local slope of the relaxation time in an activation diagram $\log(\tau)$ versus $1/T$, another parameter, the dynamic fragility m corresponds to the limiting case of the slope at $T = T_g$ in a normalized "Angell-plot" according to Eq. 3:

$$m = \left. \frac{d \log\langle\tau\rangle}{d(T_g/T)} \right|_{T=T_g}, \tag{3}$$

which is linked to the VFT parameters by:

$$m = \frac{E_V}{2.303R} \frac{T_g}{(T_g - T_V)^2}. \tag{4}$$

This book chapter is organized as follows. In the next section, some systematics will be given regarding typical confining geometries including host materials, their dimensionality, and expected strength of the expected confinement effects. We will then define some basic "deviation scenarios" for the primary relaxation time as a working hypothesis to classify the large diversity of experimental results.

The following four sections will be devoted to specific glass forming systems that exemplify one or more of the *deviation scenarios* as well transitions between the three proposed basic scenarios.

Finally, we will establish a correlation between typical dynamic scenarios and the dimensionality of the confining host. In this context, we will also discuss other mechanisms that might affect the structural relaxation dynamics in nanometer geometries, e.g., the role of attractive or repulsive interactions [13–15] between the glass-forming liquid and the solid host (interfacial effects), density effects, the role of curvature, etc.

2 Dynamics in Confinement: Common Observations

In the search for intrinsic length scales of glass-forming liquids, numerous experiments on various glass-forming systems in nanometer scale geometric confinement have been performed so far, a representative summary about which was given recently by McKenna [16]. Typical confined glass-forming systems comprise H-bonding liquids (ethylene glycol, propylene glycol, water), liquid crystals, many Van der Waals-glass formers, and other small molecule glass formers. Owing to their low viscosity, those small molecules can easily be confined in nanoporous systems like glasses, porous membranes, various kinds of channel structures, or in a thin film geometry.

On the other hand, polymeric glass formers are typically studied as ultrathin polymer films, since the large molecule size and viscosity do not facilitate efficient

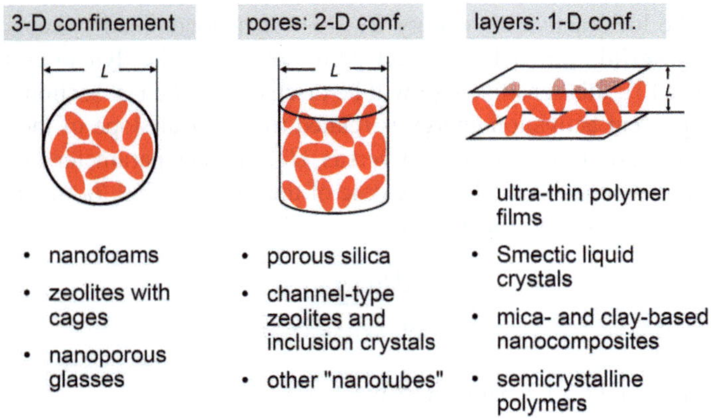

Fig. 2 Three basic geometries providing 3D (cavities), 2D (pores), or 1D (films) confinement

filling of nanoporous host systems. Nevertheless, polymeric systems might adopt other than layer-type geometries by nanophase separation as present in clay-based nanocomposites or polymeric liquid crystals.

In this chapter, however, we will largely restrict ourselves to small molecule glass formers since polymer thin films are widely discussed elsewhere in this book.

A common fashion to classify the several options to confine glass formers on nanometer scale is based on the dimension of confining host structure, cf. Fig. 2. Here, cavities that restrict the size and thus limiting length of cooperativity in all three spatial dimensions, represent the ultimately strongest geometry. Typical systems that realize such 3D confinement are nanoporous glasses, (polymeric) nanofoams, or zeolites with closed cages.

In contrast, channel-type host geometries facilitate molecular packing and dynamics in one "infinite" direction and thus allow to study glass transition dynamics in a two-dimensional confinement. Well-known examples are zeolites with channel structures (e.g., $AlPO_4$-5) [17], anodic aluminum oxides (AAO) containing straight channels with adjustable pore diameter [18] or various types of porous silica.

Finally, any type of ultrathin layer structure can be considered as an one-dimensional (1D) confinement. Here, it is useful to distinguish between genuine thin films as prepared by, e.g., spin- or drop casting of a glass former onto a flat substrate, or multilayer structures that are formed by self-assembly of the molecules into layer structures as found in smectic liquid crystals (cf. Sect. 2.3), semi-crystalline polymers [19], and clay-based polymer nanocomposites [20]. A specific case concerns liquid films, the thickness of which can be varied during *in-situ* characterization in an organic molecular beam deposition (OMBD) chamber [21] (cf. Sect. 2.4). Regardless the specific type of 1D-geometry we intuitively expect the least strong deviation scenario in all layer structures.

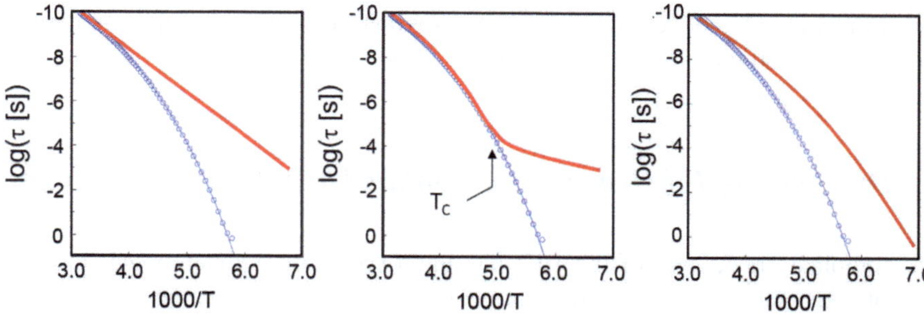

Fig. 3 Three main scenarios for the deviation of the mean relaxation time (*red lines* assumed characteristics) from the bulk VFT-behavior (*blue symbols* real data from 1,3-propane diol)

Despite the tremendous diversity in glass-forming systems, confining hosts and experimental techniques, a couple of recurrent signatures of "confined dynamics" can be identified:

i. The structural relaxation response typically broadens and eventually splits into subprocesses having their specific temperature dependence and strength.
ii. Upon reduction of the pore or cavity size, the strength of the α-relaxation decreases, a phenomenon that typically goes along with a change from a negative slope in $d\Delta\varepsilon/dT$ (bulk case) to a positive slope.
iii. The temperature dependence of the (mean) structural relaxation time shows deviations from the bulk (VFT) behavior in a more or less pronounced manner.

The latter observation is often discussed as the main "observable" since (peak) relaxation times are readily available from many experimental techniques and thus form a large body of experimental findings. For this reason, we will further restrict ourselves to experimental evidence concerning effects in $\tau_\alpha(T)$.

Among the various effects on $\tau_\alpha(T)$ in glass forming systems under confinement, we can distinguish three common scenarios as depicted in Fig. 3.

The ultimate change in the temperature dependence of the relaxation time is represented by the left panel (type I) illustrating a complete transition from VFT- to Arrhenius behavior. Type II (middle panel) corresponds to a crossover from VFT- to Arrhenius behavior at a particular cross-over temperature T_c and cross-over relaxation time τ_c. Note that here T_c has a different meaning than other cross-over temperatures as discussed in the context of $\alpha - \beta$ merging scenarios [22] or the characteristic temperature predicted by the mode-coupling theory. The third deviation scenario (type III) covers all experimental evidences of a slowed-down or accelerated structural dynamics that maintains VFT-behavior. Consequently, this type might be regarded as the manifestation of "modest" confinement effects.

In the next four sections, different H-bonding glass-forming systems will be described that exemplify the above-mentioned deviation scenarios. While two systems show a single deviation scenario (type I and III), two other cases demonstrate transitions and links between deviation scenarios that can be provoked upon structural evolution or subtile molecular adjustments.

2.1 Dynamics of Ethylene Glycol Confined to Nanoporous Zeolites

The complete breakdown of the VFT-law upon confining a supercooled liquid is evidently the most severe manifestation of a dynamics that changes from being cooperative in the sense of the AG-theory to a molecular ensemble, which experiences a temperature independent energy barrier. The breakdown of VFT-type dynamics, on the other hand, does not necessarily imply that cooperative reorientation ceases, since Arrhenius-type relaxation processes are also known from many dense systems such as smectic or discotic liquid crystals, plastic crystals, and semi-crystalline polymers. A solid criterion to judge whether a simple activated relaxation process involves cooperativity and thus an entropic contribution to the overall (apparent) barrier, is the pre-exponential factor τ_∞ being in the order of 10^{-13}s for noncooperative motions. A more quantitative analysis of Arrhenius-type processes is provided by the Starkweather approach [23], based on the Eyring theory for reaction kinetics [24], that allows to separate the activation enthalpy from the entropic contribution.

One of the most comprehensive studies of glass formers in nanometer confinement was done by Huwe and coworkers, who incorporated ethylene glycol (EG) guest molecules into zeolitic host systems of different topology [25]. Depending on the type and size of the zeolite channel system, the authors revealed either bulk-like VFT dynamics or a completely changed, Arrhenius-type behavior.

Three zeolite host structures were studied. Silica sodalite, a clathrasil compound contains a system of isolated cavities, so-called β cages, with an inner diameter of 0.6 nm. Ethylene glycol is one of the few organic molecules that act as template during the hydrothermal synthesis of silica sodalite [26]. As the result, EG molecules become trapped as single molecules during synthesis and represent a perfect system where molecular interaction between EG molecules are practically impaired. Besides silica sodalite, silicalite-I, and zeolite beta were used as zeolitic host systems with three-dimensional pore systems. Silicalite, a pure silica material, has two different types of elliptical channels with cross-sections of 0.56 nm $\times$ 0.53 nm and 0.55 nm $\times$ 0.51 nm [27]. The third host, zeolite beta, an aluminosilicate with a Si:Al ratio of 40, exhibits even larger channels in [100] and [010] directions with a diameter of 0.76 nm $\times$ 0.64 nm, whereas the channels in the [001] direction are smaller (0.55 nm $\times$ 0.55 nm). Since silicalite and zeolite beta have an open pore system they were filled with guest molecules after synthesis and calcination. A detailed description of the sample preparation and instrumentation for the dielectric spectroscopy measurements can be found in [25].

Dielectric measurements were performed in the frequency range from 10^{-2} to 10^9 Hz on the three EG/zeolite systems as well as on EG in bulk. To extract relaxation parameters like the mean relaxation time τ, relaxation strength $\Delta\varepsilon$, one or more (n)

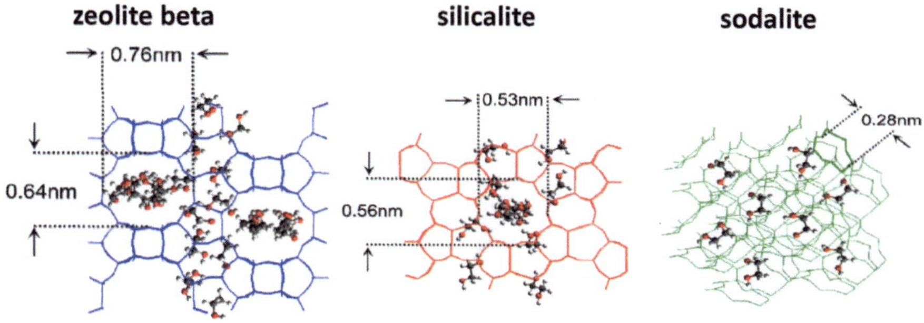

Fig. 4 Zeolitic frameworks of zeolite beta, silicalite, and sodalite, together with the confined EG molecules according to molecular simulations. Published with permission by Friedrich Kremer

generalized relaxation functions according to Havriliak and Negami [28] were used in combination with a second term accounting for electrical conductivity.

$$\varepsilon'' = -\sum_{k-1}^{n} \mathrm{Im}\left\{\frac{\Delta\varepsilon_k}{(1+(i\omega\tau_k)^{a_k})^{b_k}}\right\} + \frac{\sigma}{\varepsilon_0\omega} \tag{5}$$

In this notation, ε_0 , σ and $\Delta\varepsilon$ denote the vacuum permittivity, dc conductivity, and the dielectric strength, respectively. The parameters a and b describe the symmetric and asymmetric broadening of the relaxation time distribution function. From the fits according to Eq. 5, the mean relaxation rate $1/\tau_{\max}$ was evaluated which is displayed in Fig. 4 for all systems.

Figure 5 shows the logarithmic mean relaxation rate as a function of the inverse temperature for EG as bulk liquid and confined in zeolites. Ethylene glycol in zeolite beta has a mean relaxation rate very close to that of bulk EG following the characteristic temperature dependence according to Vogel-Fulcher-Tammann (Eq. 1). In contrast, the isolated molecules in sodalite show an Arrhenius-type temperature dependence indicative for a single molecule relaxation which is, for $T \approx 155$ K, about 6 decades faster compared to the bulk liquid. The activation energy of this relaxation is about 26 kJ/mol and corresponds to the value for bulk EG at high frequencies and high temperatures $E_V = 29$ kJ/mol [29]. For silicalite, the relaxation of EG shows an intermediate behavior. Its activation energy ($\sim$35 kJ/mol) is in between those of EG/sodalite and the apparent activation energy (Eq. 2) of bulk EG.

In case of sodalite, Arrhenius-type temperature dependence evidently resembles the single molecule relaxation of isolated EG molecules. In an attempt to reveal the reason for the transition from a single molecule relaxation to liquid-like behavior, the authors performed computer simulations with the particular purpose to get information about the packing of EG in the various host systems. The result is partially shown in Fig. 4, the full details are given in [25].

As the main result of the simulation, no significant differences between the bulk and the confining geometry were found concerning the distance between molecules,

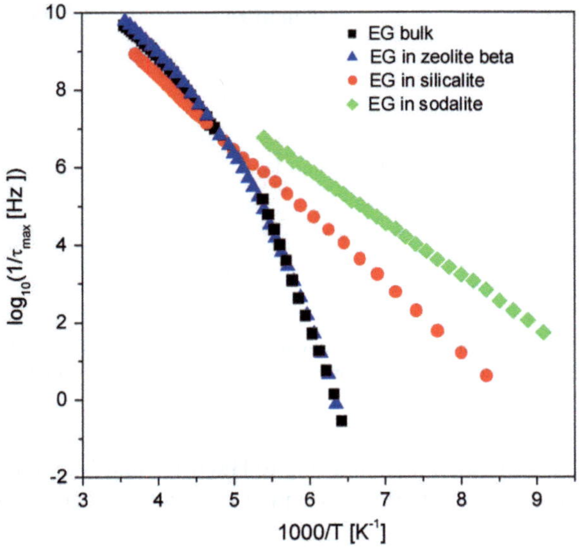

Fig. 5 Activation plot for ethylene glycol (EG) as a bulk liquid (*squares*) and confined in sodalite (*diamonds*), in silicalite (*circles*), and in zeolite beta (*triangles*). The errors are smaller than the size of the *symbols*. Published with permission by Friedrich Kremer

the average length of hydrogen bonds, and the density. However, for the number of neighboring molecules (coordination number), a pronounced difference was found. Within a radius of $r = 0.66\,\text{nm}$ a coordination number of 11 ± 1 was found for bulk EG, a number that corresponds to the maximum value in the case of the random close packing model [30]. Ethylene glycol in zeolite beta has only five neighboring molecules; hence, one has to conclude that an ensemble as small as six molecules is sufficient to facilitate liquid-like dynamics. A further reduction in the channel size as in the case of silicalite leads to a decrease of the average number of neighboring molecules from 5 to 4, a minor change that obviously results in a sharp transition from liquid-like dynamics to single molecule dynamics (Fig. 6).

2.2 A Continuous Change from Scenario 3 to 1: Curing Effects on the Dynamics of Starch/Ethylene Glycol Mixtures

While the previous example revealed a clear-cut "all-or-nothing" case—subtle changes in the confining geometry resulted in an abrupt change from bulk behavior to the scenario type I—we will, in the following, discuss a study showing an evolution between different deviation scenarios in a single system [31].

Starch and starch-based polymeric materials play an important role in both human diet and, increasingly, as biodegradable material. For processing reasons or to inhibit

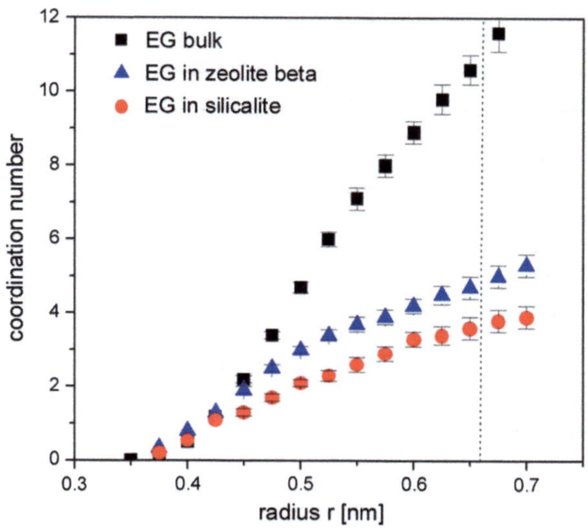

Fig. 6 Average number of neighboring molecules as a function of the radius of a surrounding sphere as calculated from the simulation for EG bulk liquid (*squares*), EG confined in zeolite beta (*triangles*), and confined in silicalite (*circles*). Adapted from Huwe et al. [25] with permission of the American Physical Society

retrogradation of starch-based products, monoglycerides are often used as "plasticizers," which interact strongly with the starch fraction (amylose and amylopectin) by building a H-bonded structure. In this context, the interactions of glycerol and ethylene glycol with dry starch polysaccharides have been investigated using DSC, solid-state NMR spectroscopy and dielectric spectroscopy (DRS) [31–33].

The aim of the original work [31] was to study the interaction between dry amylopectin and ethylene glycol. For this purpose, dry amylopectin/ethylene glycol samples were prepared by mixing ethylene glycol with dried potato amylopectin (amylopectin UG, Avebe, NL). The potato amylopectin was examined with X-ray diffraction confirming a crystallinity similar to that of native potato starch (~25 %). Amylopectin was dried for 48 h in a vacuum oven at 70 °C. The dried material (<3 % H_2O) was mixed under a nitrogen gas flow with 21 wt % ethylene glycol and subsequently pressed between circular brass electrodes to obtain disk-shaped samples of a thickness in the range of 300–600 μm.

Dielectric experiments on the AP/EG mixtures were performed using a conventional broadband dielectric spectroscopy setup in the frequency range from 10^{-1} to 10^6 Hz [34]. Three freshly prepared samples were subjected to specific annealing schemas, the thermal history of which is given in Table 1. Here, the temperatures T_{ann} have the meaning of either the highest temperature (25 °C) or the temperature at which the sample was annealed just prior to a subsequent cooling run to −120 °C by steps of 5 °C. Such a cooling program was chosen to ensure that kinetic changes, if

Table 1 Maximum annealing temperature and VFT parameters for the dielectric relaxation process α_{EG} of the EG-fraction

Sample	T_{ann} [°C]	T_V [K]	E_V [kJ/mol]	log(τ_∞ [s])
1	25	131	13.5	−12.8
1	40	110	25.2	−15.5
2	60	133	15.4	−12.8
2	80	74	36.6	−16.8
2	100	7	54.5	−17.7
2	120	0	57.9	−18.2
3	100	132	13.4	−20.0
3	120	0	57.1	−18.0

Typical errors margins $\Delta T_V = 2$ K, $\Delta E_V = 0.5$ kJ/mol, Δ log(τ_∞ [s]) = 0.5

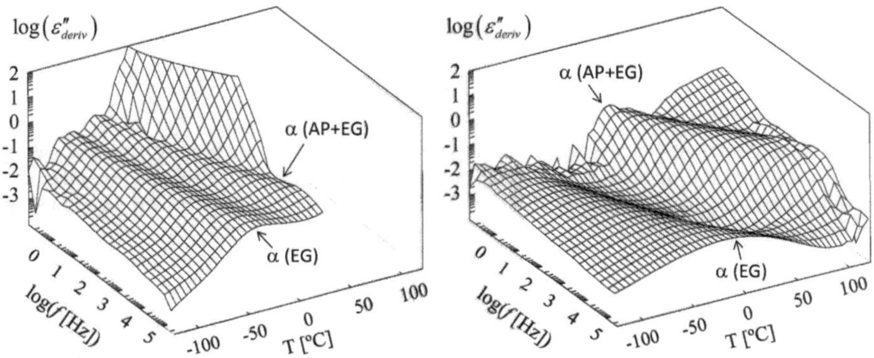

Fig. 7 *Left panel* Three-dimensional representation of the "conduction-free" loss $\varepsilon''_{deriv}(f, T)$ for a freshly mixed AP/EG sample (sample 1) showing two distinct non-Arrhenius relaxation processes. *Right panel* Relaxation pattern of sample 2 after annealing at 120 °C. Reproduced from Smits et al. [31] with permission of the American Chemical Society

present, were restricted to the upper (start) temperature interval. The average cooling rate was about 1 °C/min.

A typical three-dimensional loss spectrum of sample 1, which was immediately cooled from room temperature to −120 °C after sample preparation, is shown in Fig. 7 (left). To eliminate strong contributions to the loss due to Ohmic conduction we have computed the "conduction-free" dielectric loss from the permittivity spectra according to Eq. 6 [35]

$$\varepsilon''_{deriv} = -\frac{\pi}{2}\frac{\partial \varepsilon'(\omega)}{\partial \ln \omega}. \tag{6}$$

By using this derivative technique, two strong and well-resolved relaxation processes were revealed (cf. Fig. 7), which could be assigned to the dynamic glass transitions of phase separated ethylene glycol (α_{EG})—the fast or low-temperature process—and of "plasticized" amylopectin (α_{AP}). Relaxation time data and strength were obtained

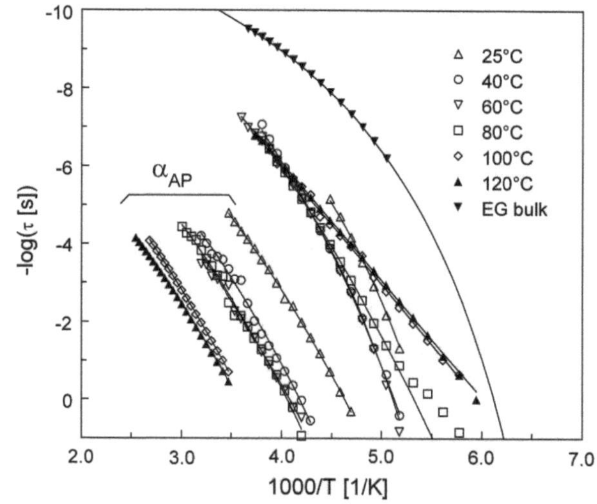

Fig. 8 Relaxation time versus inverse temperature for samples 1 and 2: sample 1 fresh (25 °C), and after annealing at 40 °C, and sample 2 after annealing at 60, 80, 100, and 120 °C. The relaxation time data for bulk EG are given for comparison. Reproduced from Smits et al. [31] with permission of the American Chemical Society

by either fitting the raw spectra $\varepsilon''(f)$ to a set of HN-functions (Eq. 5) or, alternatively, by fitting the derivative-based loss spectra $\varepsilon''_{\mathrm{deriv}}(f)$ to the analytical derivative of the Havriliak-Negami function given in [35].

The relaxation times for both processes α_{AP} and α_{EG} are summarized in Fig. 8, the solid lines indicate fits to the VFT equation. One clearly sees that indeed most of the relaxation processes show a curvature according to the VFT law as a characteristic feature of dynamic glass transition processes. The VFT parameters for the α_{EG} – process are listed in Table 1, the corresponding data for the α_{AP}.-relaxation can be found in [31].

Figure 8 reveals two main features: (a) the glass transition dynamics of the AP/EG complexes shows a continuous up-shift upon annealing. This phenomenon was discussed in [31] in terms of chain stiffening of the amylopectin upon evolution toward a strongly H-bonded structure, in which EG molecules preferentially form intra-AP hydrogen-bonds. This picture is supported by the extremely low dynamic fragility (Eq. 3) of the α_{AP}-process in the order of $m = 25$, which corresponds to its weak curvature as seen in Fig. 8.

The second feature (b) concerns the relaxation process assigned to an excess fraction of ethylene glycol. That is, EG molecules that are not part of the supramolecular AP/EG structure. While the α relaxation process of the AP-rich fraction continuously slows down with increasing annealing temperature (25–120 °C), the (low-temperature) dynamics of the EG phase shows a more complex behavior as shown in Fig. 9 in more detail:

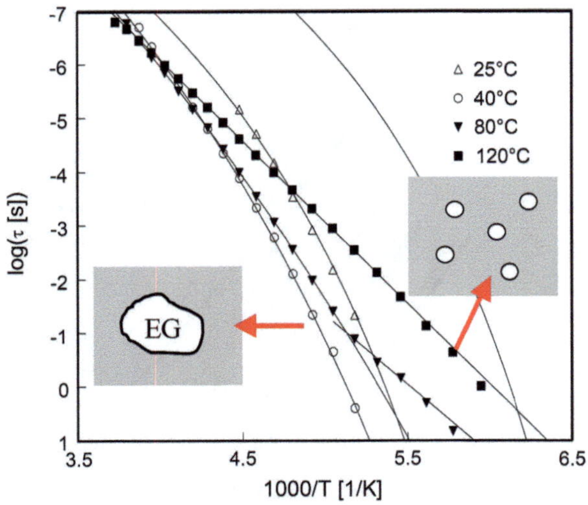

Fig. 9 Relaxation time versus inverse temperature for samples 1 and 2: sample 1 fresh (25 °C), and after annealing at 40 °C, and sample 2 after annealing at 60 (not shown here), 80, 100, and 120 °C. The relaxation time data for bulk EG are given for comparison. Adapted from Smits et al. [31] with permission of the American Chemical Society

i. Upon moderate annealing (40 and 60 °C [Fig. 8]), a general slowing down of the α_{EG} dynamics is observed (scenario III).
ii. After heat treatment at 80 °C, the relaxation time-temperature characteristics reveals an unchanged dynamics at the highest frequencies, followed (upon cooling) by an overall speed-up and flattening of the VFT curvature until it shows a cross-over to genuine Arrhenius behavior (deviation scenario II).
iii. Further annealing at 100 and 120 °C finally results in full Arrhenius behavior (deviation scenario I), which goes along with a tremendous speed-up of the α_{EG} process at lower temperatures.

This evolution of the relaxation behavior of the EG-rich phase, though it was not the main objective of the study, has clearly the signature of a nanometer confined glass-forming liquid. It is the strength of this work to demonstrate, how a single host/guest system evolves from (almost) bulk-like dynamics through the three “deviation scenarios” to the state of total breakdown of the VFT-dynamics as discussed in the previous section. To the best knowledge of the authors, this system is also the only sample system reported so far in the literature that revealed all three “deviation scenarios.”

Unfortunately, no extensive structural information regarding the size and topology of the nanophase separated material was reported in [31], information that would anyway be challenging to obtain by either imaging techniques or scattering experiments. The only direct clue about the length scale of heterogeneity comes from SEM measurements by the authors, which proved that the initial course structure (μm-scale) evolves to a material showing no heterogeneities in the limit of optical resolution of the scanning electron microscope being in the order of 100 nm.

Regarding topology of the heterogeneous system we can only speculate, however, the dynamic scenario allows to build a conclusive physical picture:

Let's start from a phase-separated system that evolves from a microporous to a nanoporous material upon "curing," which is opposed to systems obeying spinodal decomposition that typically leads to coarsening. The molecular dynamics of the EG-rich phase will be governed by the balance between surface and confinement effects: while interactions with the inner surfaces will delay the molecular dynamics, the reduction in pore size might speed up the glass transition dynamics by virtue of cooperativity-based finite size effects according to Fig. 1. In this sense, the initial retardation of the α_{EG} process in the AP/EG system compared to bulk EG implies a larger amount of less mobile EG located near the inner surfaces of EG droplets after annealing. However, the drastic difference in the dynamics between bulk EG and dispersed EG by 2–3 decades cannot solely be explained by surface anchoring alone, in particular not in the initial coarse morphology.

We therefore hypothesize that the overall delay of the EG dynamics in AP/EG hints to the existence of a dispersed phase constituted by EG with some fraction of dissolved AP molecules or segments. Such dilute solution of AP in EG would perfectly explain the up-shifted glass transition of EG without affecting the cooperative nature of the EG dynamics as manifested in the VFT behavior.

After annealing at 80 °C, the $\tau(T)$ dependence becomes flatter and shows a cross-over to Arrhenius behavior. This change in dynamics marks the point at which the size (geometrical confinement) of the glass-forming molecular ensemble starts to influence its cooperative dynamics. In other words, the departure of $\tau(T)$ from VFT behavior implies a characteristic size of the EG-rich fraction ("droplets") in the order of the length of cooperativity ξ being typically 2–3 nm near T_g [36]. Further annealing at 100 and 120 °C then results to scenario I according to Fig. 2, i.e., the dynamics of EG has changed from cooperative liquid dynamics to thermally activated, single molecule dynamics.

2.3 *Glass Transition Dynamics of H-bonding End Groups Phase Separated in a Smectic Liquid Crystal*

In the first and the previous sections, we have discussed the deviation scenario II, the cross-over from VFT to Arrhenius behavior, as the only confinement scenario that would give direct access to the intrinsic length scale ξ by interfering with an extrinsic length imposed by the confining host geometry. To prove this idea, one would need to study the dynamic cross-over in function of a known, adjustable host size. Unfortunately, the system starch/EG did neither provided control over nor quantitative information about the size of the confining geometry.

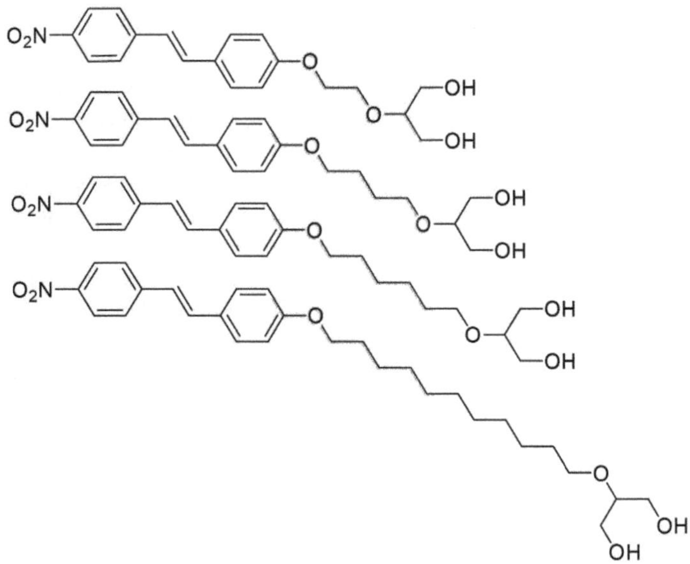

Fig. 10 Chemical structure of the nitrostilbene-based mesogenes **8a–d** having different alkyl spacers with 2, 4, 6, or 11 carbon atoms (from *top* to *bottom*) [37]

In this section, we will discuss a dielectric relaxation study on a liquid crystalline system that turned out to be an excellent model system for studying "confined dynamics" in a nanometer-spaced layer structure. Starting point of this study was the availability of a homologue series of nitrostilbene-based mesogenic compounds **8a–b** (cf. Fig. 10), which were originally prepared as precursors for the synthesis of side chain liquid polycarbonates [37]. Interestingly, both the lc monomers and polymers showed liquid crystalline behavior, predominantly smectic A phases, which were investigated thoroughly by DSC, dynamic mechanical spectroscopy (DMA), optical microscopy, WAXS [37] as well as by dielectric spectroscopy [38].

DRS measurements were performed in the wide frequency range from 10^{-1} to 10^9 Hz using the setup described in the previous section at lower frequencies (10^{-1} to 10^9 Hz) in combination with a Hewlett-Packard Rf-analyzer HP4291A for the frequency range from 10^6 to 1.8×10^9 Hz. Circular samples were prepared by melting the freeze-dried powder together with fused silica spacers between polished, circular brass disks, which resulted in samples of about 300 micrometer thickness. Samples were measured under isothermal conditions, starting above the clearing temperature, down to -140 °C in steps of 2.5 K, which lead to an effective cooling rate of ~0.5 K/min. A typical result is given in Fig. 11 showing the three-dimensional presentation of the "conduction-free" loss $\varepsilon''_{\text{deriv}}$ as function of temperature and frequency.

On the first glance, the dielectric spectra of compound **8c** show the typical features of a smectic A material: At high temperatures, in the S_A phase, two relaxation processes (I and II) show up, which are separated in frequency by less than two

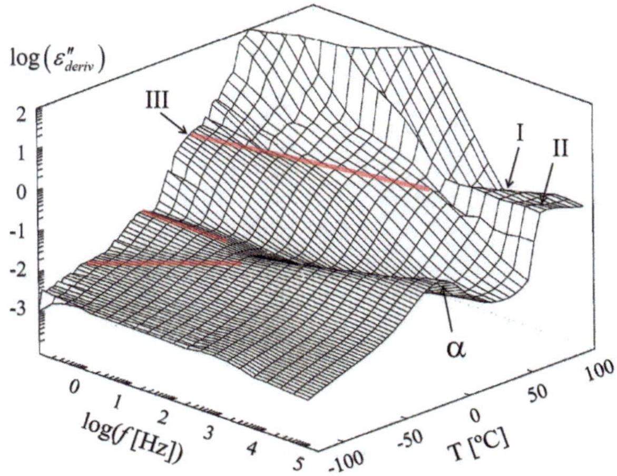

Fig. 11 "Conduction free" loss spectrum dε'/dlnf of **8c** (C_6-spacer) revealing two "mesogenic" relaxations (I, II) in the S_A-phase, a third process (III) in the S_X-phase and a bifurcated α-relaxation

decades and obey Arrhenius behavior (cf. Fig. 12). Both processes are the expression of rotational diffusion of the rod-like mesogens in an anisotropic potential around the short (I) and long (II) axis.

Upon cooling below 55 °C, the material undergoes a first order phase transition that was identified as crystallization into a smectic crystal phase S_X, based on WAXS results and the occurrence of a new, strong, and highly activated relaxation process (III). The crystal structure of all compounds was found to be monoclinic, however, the exact assignment to a smectic crystal type could not be done unambiguously.

Further cooling the sample yields another relaxation process (α), co-existing with the much slower process III, that is highly activated and bifurcates at a frequency around 100 Hz. The relaxation parameters of all processes were analyzed according to the above-mentioned procedures and are summarized in an activation plot (Fig. 12). More experimental details including DRS measurements, data analysis, and supporting characterization by DSC, XRD and optical microscopy can be found in [39].

The most striking feature of this low-temperature relaxation is its VFT-behavior, being the signature of the dynamic glass transition. Having in mind that the mesogenic cores are organized in a crystalline layer structure and are dynamically frozen at the time scale of the α-process, the authors concluded that this peculiar dynamics must originate from a mobile, disordered phase intercalated between the smectic core arrays. In this picture, which is illustrated in Fig. 13, the disordered phase is constituted by the H-bonding 1,3-propane diol motifs that are tethered via aliphatic spacers to the nitrostilbene mesogenic cores.

The observation of glassy dynamics in partially ordered systems is actually not unique; Glass transitions have been reported from many systems that combine

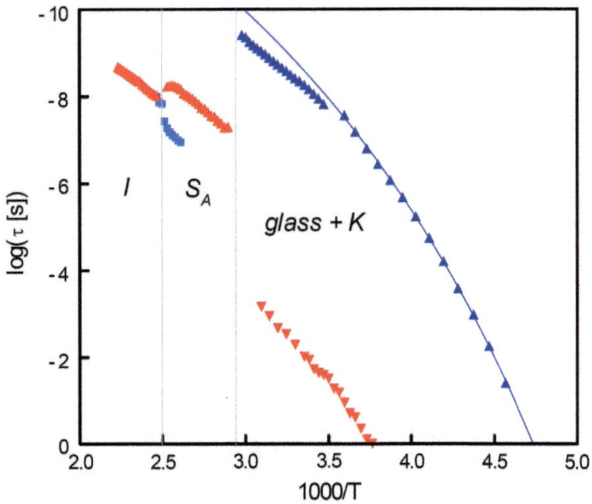

Fig. 12 Relaxation time versus inverse temperature for the liquid crystalline compound **8c** showing relaxation processes in the isotropic, smectic A, and crystalline/glassy state

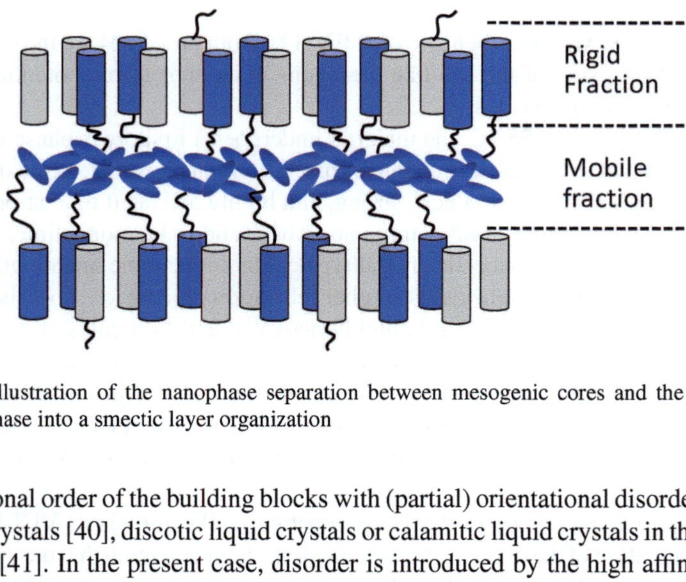

Fig. 13 Illustration of the nanophase separation between mesogenic cores and the disordered tail/diol phase into a smectic layer organization

translational order of the building blocks with (partial) orientational disorder, such as plastic crystals [40], discotic liquid crystals or calamitic liquid crystals in the smectic B phase [41]. In the present case, disorder is introduced by the high affinity of the diol-endgroups to form H-bonds, the mobility of which is facilitated by presence of an aliphatic fraction (alkyl spacers) and possible packing frustrations. To check the hypothesis of a H-bonded liquid layer in analogy to pure H-bonded liquids, the authors investigated two structurally similar diols, 1,2-propane diol and 1,3-propane diol. The relaxation times are given in Fig. 14 along with the $\tau(T)$ dependence for the compound **8d** (C_{11} spacer).

On a first glance, the relaxation time dependences show great similarity, differences in the glass transition temperature (indicated by the intersection of the VFT fit

Fig. 14 Relaxation time versus inverse temperature for the structural relaxation time τ_α for **8d** together with data from two model H-bonding liquids (1,2-PD and 1,3-PD)

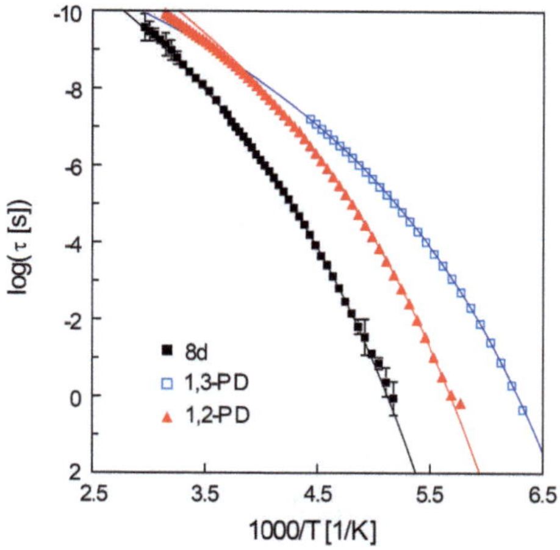

curve with the upper limit at log $\tau = 2$) between the two isomeres are in the same order of magnitude as the up-shift in T_g for **8d**. At the very high frequency limits, the dynamics of the model liquids seems to merge, while the relaxation time of 8d remains slower by less than one decade. Knowing that high temperature dynamics is governed by individual molecule rotations rather than cooperative reorientations, the remaining gap in the relaxation time can be explained by rotational restrictions of the diol endgroup in 8d due to the linkage to the spacer part. This linkage particularly inhibits the fastest rotational modes of the diol around the long molecule axis (C–C–C axis).

On the other hand, the general slowdown of the dynamics of **8d** over pure 1,3-PD is plausible as well: the anchoring of the diol-endgroup to the mesogenic "hard walls" causes a strong interaction of the glass-forming moieties that restricts both their rotational and translational diffusion capabilities. Hence, we might expect a surface-induced, retarded liquid dynamics as discussed above as one of the possible consequences of the deviation scenario III.

The authors then studied the α-relaxation in compounds **8a–d** systematically, which revealed a couple of other remarkable observation (cf. Fig. 15):

i. The high frequency dynamics ($f > 250$ kHz) appears to be independent from the spacer length, which proves that the cooperative dynamics at higher frequencies exclusively involves the diol endgroups and certainly not more than one adjacent methylene group.
ii. Upon reduction of the spacer length from $n = 11$ to $n = 2$, the relaxation time τ_α starts to depart from the VFT behavior toward faster dynamics, which obeys unambiguously an Arrheneus law for **8a**. The characteristic "cross-over" frequency f_c increases accordingly.
iii. The cross-over goes along with the emergence of a 2nd, slow relaxation mode, which occasionally joins the former VFT-trajectory and, in all cases where present takes over the overall relaxation strength [39].

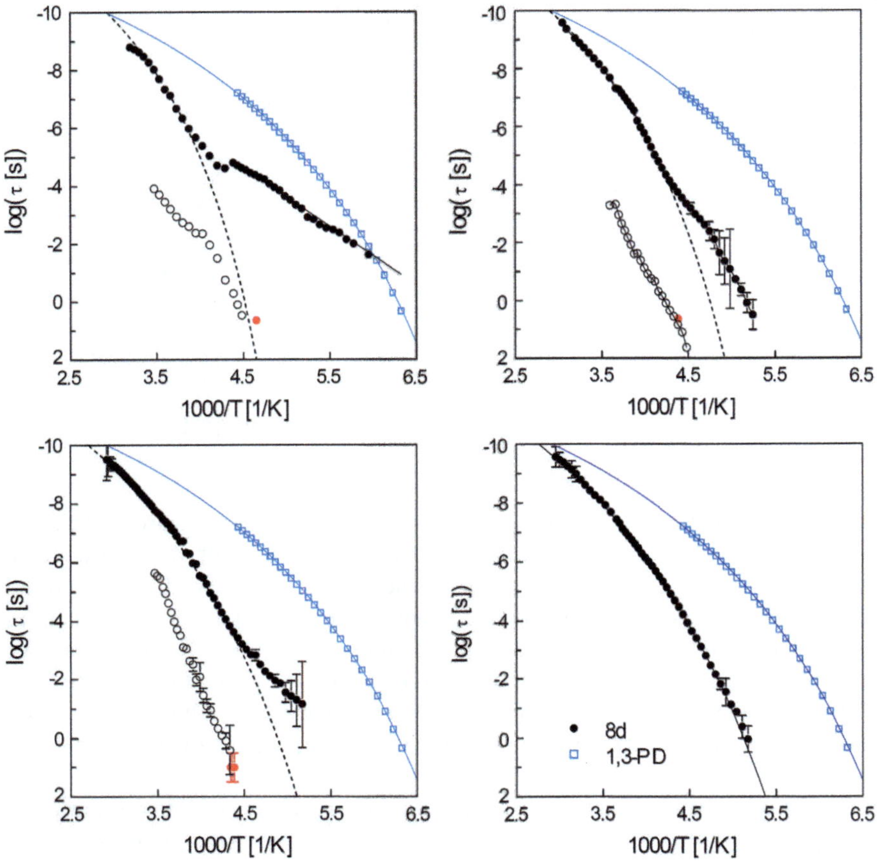

Fig. 15 Arrhenius presentation of the dynamic glass transition in compounds **8a** (*top left*), **8b** (*top right*), **8c** (*bottom left*) and **8d** (*bottom right*). *Black filled symbols* denote the main process, a second, slow process is observed for **8a–c** (*open circles*). The *blue symbols* and fit *curve* refer to 1,3-PD for comparison. The *black*, dashed VFT *curve* corresponds to the fit of $\tau_\alpha(T)$ of **8d**

The existence of two coupled relaxation modes provokes the question, which of the modes is the cooperative one, i.e., which of the processes has a calorimetric signature. To address this point, the authors performed additional DDSC (dynamic DSC) measurements at a fixed, low frequency. The results are shown in Fig. 15 as a red, filled circle, and give a clear answer: In all samples that showed the additional slow relaxation, the calorimetric relaxation time coincides with the slow dielectric mode in an excellent way. However, this calorimetric relaxation time is longer than the **8d** (C_{11}) reference for the two intermediate spacers.

Given the fact that the systems **8a–d** resembles a cross-over according to the scenario II (cf. Fig. 3), one is tempted to relate the thickness of the mobile fraction to the cross-over frequency.

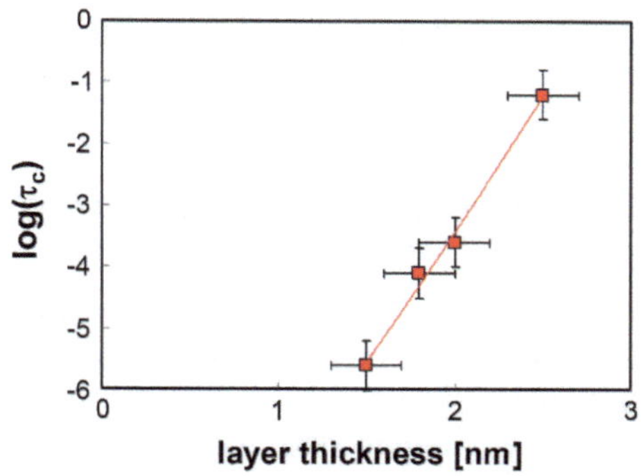

Fig. 16 Cross-over relaxation time τ_c versus thickness of the mobile layer for materials **8a–d**

The thickness of the mobile layer L_m was obtained from the periodicity of the smectic layer structure, determined from the WAXS results, from which the length of the rigid mesogenic core as calculated by molecular modeling software (ChemOffice 6.0) was substracted. As expected, L_m scales linearly with the number of carbon atoms in the alkyl-spacer, which is in line with the assumption of a spacer length independent density of both the rigid and the mobile layer.

Figure 16 displays the cross-over relaxation time $\tau_c = 1/2\pi f_c$ as a function of L_m showing a linear relation between log τ_c and L_m within the experimental errors. This finding is qualitatively in agreement with the AG picture that postulates an increase in cooperativity with a decreasing temperature with respect to T_g. Moreover, the length scale of cooperativity ξ, varying from 1.5 to 2.5 nm according to Fig. 16, even agrees quantitatively with ξ-values that were calculated in the framework of the fluctuation approach to the glass transition by Donth [42] from heat capacity spectroscopy data using the fluctuation formula [36]

$$N_\alpha = \frac{RT_\omega^2 \Delta(1/c_V)}{M_0 \delta T^2} \approx \frac{RT_\omega^2 \Delta c_p}{M_0 (\delta T)^2 \bar{c}_p^2} \tag{7}$$

Here, R is the gas constant, $\Delta(1/c_V) \approx \Delta(1/c_p) \approx \Delta c_p/\bar{c}_p^2$ the calorimetric α intensity, δT^2 the mean temperature fluctuation of CRRs as taken from the width, δT, of the α-peak in $c_p''(\omega, T)$, while M_0 and T_ω denote the molar mass per monomer unit and the dynamic glass temperature for a given frequency ω. A typical cooperativity near T_g is $N_\alpha\ (T_g) \approx 100$ corresponding to typical CRR sizes ξ of about 1.5–3.5 nm [43]. Equation 7 also predicts the temperature dependence of N_α above T_g, which can be approximated by $N_\alpha^{1/2}(T) \propto (T_{on} - T)$ with T_{on} being the onset temperature of the α-process [22].

Comparing the range and the temperature dependence of ξ-values obtained by the fluctuation approach with our cross-over length L_m brings us to the conclusion that the initial assumption (cf. Sect. 1) $\xi \approx L$(confinement) might indeed hold for specific glass-forming systems like the present smectic monomers. In view of that we can also rationalize the emergence of the second, slow relaxation mode that was observed for compounds **8a–c**. Though cooperative dynamics is restricted in the layer normal direction, there is no limit for growing cooperativity in the lateral dimension of the "infinite" mobile layer, which finally restores VFT behavior.

2.4 Ultrathin Layers of Glycerol on an Attractive Surface

Among the basic geometries used in studies of "confined dynamics" (cf. Fig. 2), nanometer-sized layers impose the least geometrical restriction on the cooperative dynamics in glass-forming liquids, since cooperativity can be established in two remaining "infinite" lateral directions. As a consequence, ultrathin films of glass-forming liquids show typically perturbations from their bulk-VFT behavior (scenario III) in the sense of a changed fragility and/or Vogel-temperature, which results in either an increase or decrease of the glass transition temperature.

Here, the vast majority of experimental work was done on polymeric systems, since polymers allow easy film formation and investigations even in the freely standing film geometry [44–48]. On the other hand, there exist only a very limited number of relaxation studies on small glass formers confined to a truly planar geometry as, e.g., provided by liquid film suspended in a surface force apparatus (SFA) [49]. A specific experimental situation is represented by studies on microporous host systems that are partially filled by small glass-forming molecules, which "coat" the pore surfaces and thus might be regarded as nanometer thin liquid films on a curved substrate [50, 51].

In this section, we will discuss results from a recent dielectric spectroscopy study on glycerol in a truly planar film geometry based on organic molecular beam deposition [21, 52]. The key idea of the authors was to combine a continuous film forming technique—OMBD—with an open electrode system that senses the relaxation dynamics in situ and in real-time. The technical principle of this approach is depicted in Fig. 17.

The glass-forming material is evaporated from an effusion cell (Knudsen cell) into an UHV vacuum chamber (pressure $p < 10^{-8}$ mbar), which results in a collision-free flux of molecules that strictly depends on the geometry and temperature of the Knudsen cell and the vapor pressure of the source. This molecular beam hits a target on which an inter-digitated comb electrode (IDE) is mounted, the temperature of which is accurately controlled by combined action of a resistance heating and a cooling flux of gaseous nitrogen. Owing to its open electrode configuration, the IDE sensor is accessible for the incoming molecules forming a stable liquid film between the IDE-fingers that is facilitated by H-bonds between the glass former and the oxide substrate of the IDE. Simultaneously, the thickness of the deposited layer is monitored by a quartz crystal microbalance (QCM), located close to the IDE sensor inside the chamber.

Dielectric spectra in the frequency range from 1 Hz and 1 MHz are continuously acquired by means of a high-resolution dielectric analyzer (ALPHA N, Novocontrol Technologies) as described above. In contrast to the parallel plate configuration, commonly used for dielectric spectroscopy measurements on ultrathin (polymer)

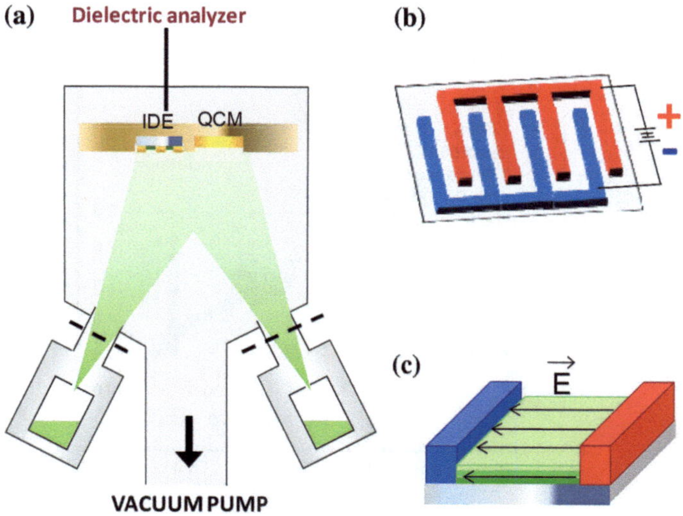

Fig. 17 Sketch of the experimental set up (**a**). The structure of the IDE is represented for an empty sensor (**b**) and in the presence of a film (**c**). Reproduced from Capponi et al. [52] with permission of the American Chemical Society

films, the direction of the probing electric field in the IDE is parallel to the surface of the film (see Fig. 17c). As a consequence, the total dielectric response represents the simple sum of all individual contributions of the molecular layers constituting the film [53]. The dielectric loss spectra $\varepsilon''(\omega)$ were analyzed in the frequency domain using the Havriliak-Negami (HN) function (cf. Eq. 5).

On the basis of this methodology, the authors investigated the structural relaxation of glycerol films in the thickness range from several tens of nanometers down to 0.7 nm, as a function of the temperature. Here, one has to realize that the actual thickness of the liquid film is subject to continuous changes driven by two counteracting processes: the presence of a molecular beam leads to an increase of the film thickness, while, at the same time, the liquid film undergoes continuous thinning due to evaporation at its free surface as the consequence of the low pressure in the vacuum chamber ($p \ll$ vapor pressure). Evaporation, however is a thermally activated process that can be readily neglected in the vicinity of T_g. Taking advantage of this interplay between deposition and desorption, the authors were using two different measurement protocols: (a) Fast evaporation of a glycerol film up to a desired thickness, followed by stepwise changes of the sample temperature under nearly constant film thickness conditions, or (b) evaporation and subsequent monitoring of the dielectric spectrum upon desorption driven thickness reduction under isothermal conditions. Both types of experiments will be discussed in the following.

Though no finite size effects according to scenario I or II have been observed in any of the measured samples, for films below 3 nm, a broadening in the loss peak was revealed that went along with a slowdown in the dynamics for samples of a thickness corresponding to about three molecular layers (1.6 nm). Both effects have

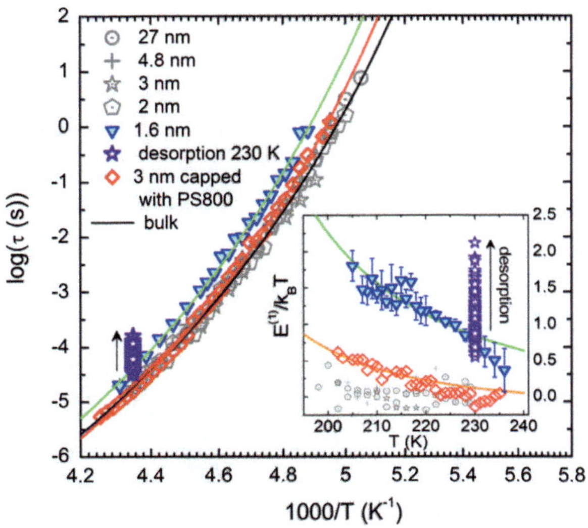

Fig. 18 Thermal and thickness evolution of the structural relaxation. *Solid lines* VFT fit *curves* using the same pre-exponential factor $\tau_\infty = 10^{-15}$ s for all films. $T_0 = 144$ K for 1.6 nm, $T_0 = 142$ K for the 3 nm capped film and $T_0 = 139$ K for all of the other thicknesses. *Inset* perturbation of the activation energy $E^{(1)}$ in units of $k_B T$. Reproduced from Capponi et al. [52] with permission of the American Chemical Society

been attributed to the presence of attractive interactions between glycerol and the quartz substrate of the IDE sensor.

The temperature dependence $\tau(T)$ for films in the thickness range between 27 and 1.6 nm, measured right after deposition, is shown in Fig. 18. The values of τ have been evaluated using the above-mentioned fit procedure using the HN-function. Subsequently, the values $\tau(T)$ for all measured films were fitted to the VFT-equation (cf. Eq. 1).

Films of thickness between 2 and 27 nm (gray symbols) revealed an almost identical temperature dependence $\tau(T)$ compared to bulk glycerol (black line). The value of the glass transition temperature T_g, considered as the temperature at which $\tau = 100$ s, was found to be constant within the experimental error. The value obtained by this approach (193.8 K) agrees with previous measurements in bulk [54–56]. However, the thinnest film with a mean thickness of 1.6 nm revealed an overall slowdown in the relaxation times with respect to the other films, which manifests in a higher Vogel temperature T_0 (up-shift by 5 K) and a corresponding increase of T_g by 3.5 K, while the other VFT parameters E_V and τ_∞ remained unchanged. The same trend was also observed during isothermal desorption at 230 K, where the film thickness decreased from initially 3 down to 0.7 nm at a constant rate of 0.4 nm/h. During this process, the structural relaxation time increased by almost 1 decade. Similar results have been reported from confinement experiments in Vycor glasses with pore sizes of 7 nm [57].

From the temperature-dependent experiments, a physical picture emerges that suggests the presence of a reduced mobility layer (RML) in close proximity of the substrate, characterized by slower dynamics [13]. The existence of a RML in the particular systems can be ascribed to the attractive H-bonding interactions between the adsorbed glycerol molecules and the substrate of the IDE.

Small variations of the Vogel temperature while leaving the values of E_V and τ_∞ unchanged, indicate the presence of a subset of molecules relaxing as in bulk (same dynamic fragility) but with a slightly modified activation energy. Particularly, in the framework of an energy landscape scenario, the specific case of increase of T_0 can be rationalized as a vertical shift of the potential wells reducing the probability for rotational jumps, without affecting the actual contour of the landscape. Consequently, the resulting increase in the relaxation times can be used to quantify the impact of interfacial interactions, via a perturbation approach of the temperature dependence of the activation energy [13]

$$E^{(1)}(h, T) \approx [E(h, T) - E(\infty, T)] = k_B T \ln \left[\frac{\tau_\alpha(h, T)}{\tau_\alpha(\infty, T)} \right] \tag{8}$$

where h is the thickness of the film and ∞ indicates bulk values. Values of $E^{(1)}$ for all the films are given in the inset of Fig. 18 where the same symbols are used as in the main panel. $E^{(1)}$ is clearly nonzero during the latest stages of the desorption at 230 K and for the whole temperature range at 1.6 nm, where it increases upon cooling as expected [13]. These results and the concurrent absence of any transition to single molecule dynamics (scenario I) as observed for glycerol confined to silica matrices with 2.2 nm pores [58], provide compelling evidence for interfacial interactions being the dominant origin for "confinement effects" in the current situation. Differently from films probed by an electric field parallel to the direction of confinement (conventional dielectric spectroscopy), no reduction of the dielectric strength was found for the thinnest films. A likely reduction of dielectric strength due to a restriction in the molecular mobility (affecting the mean quadratic dipole moment) is possibly compensated by a more efficient coupling between the neat dipole moment of the molecule (pointing in the plane perpendicular to the alkyl chainbone) and the electric field, i.e., a macroscopic orientation effect. Such effects of surface-induced molecular order in ultrathin films of glycerol, particularly found for low temperature ($T_{dep} < T_g$) deposited films have been reported and discussed by the same authors in a recent paper [59].

In addition to a slight increase of the overall activation energy, the (local) perturbations of the molecular mobility in proximity of the substrate and the free surface will give rise to a heterogeneous mobility scenario across the film thickness, which will manifest in a broadening of the distribution of the relaxation times. As the RML and the free surface produce opposite effects on the mean molecular dynamics, we can distinguish their particular contributions by means of the "shape" parameters $m = a$ and $-n = a \cdot b$, which correspond to the limiting slope of $\log(\varepsilon'')$ versus $\log(\omega)$ at low and high frequencies respectively; see Fig. 18. The value of m is also

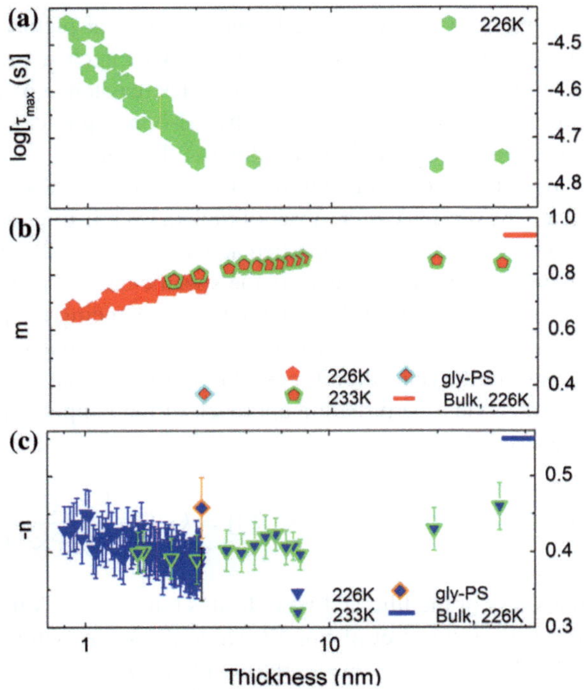

Fig. 19 *Panel a* thickness dependence of the structural relaxation time, measured during isothermal desorption at 226 K. *Panel b* parameters *m* and *n* obtained from two distinct measurements during isothermal desorption: at 233 K, in the thickness range between 2.5 and 8 nm, and at 226 K, between 3 and 0.7 nm. Reproduced from Capponi et al. [52] with permission of the American Chemical Society

related to the width of the distribution of the relaxation times, which is sharper for higher values of m and has an upper limits $m = 1$ for a system with a single relaxation time (Debye model). Values of m and n for bulk glycerol indicate a quite sharp and asymmetric peak, and show a very weak temperature dependence [54, 60]. Reductions in m or in n reflect respectively an enrichment of the distribution of relaxation times at longtime scales (slower dynamics) induced by the presence of an RML or short time scale (faster dynamics), by a free surface.

To discuss the spectral shape of the dielectric spectra, the authors used a trilayer model composed by an RML, an enhanced mobility layer (EML) in vicinity of the free surface, and a bulk-like region in between. The impact of the nonbulk component was expected to increase upon thickness reduction, resulting in a reduction of both m and n. In fact, the experimental results revealed an asymmetric broadening of structural peak toward longer relaxation times.

Figure 19 shows the thickness evolution of the values of m and n obtained from spectra acquired during two different continuous film thinning experiments (Both the values of m and n are intrinsically lower than those obtained for bulk glycerol probed with an isotropic electric field, for thicknesses at least up to 50 nm; this discrepancy can be ascribed to the different geometrical configuration adopted for the thin films), i.e., during isothermal desorption at 233 and 226 K; here, the lower temperature allowed for a higher thickness resolution due to the lower desorption rate. Good reproducibility was achieved in the overlapping thickness range covered by both experiments.

All three panels in Fig. 19 indicate the presence of two regimes. Below 3 nm the relaxation times clearly increase, while the value of m decreases already at 6–8 nm. Thus, the increase in τ is caused by a more pronounced dielectric response at low frequencies, i.e., a larger impact of the RML on the relaxation dynamics of the system upon thickness reduction. The absolute value of n, on the contrary, decreases already at few tens of nm, showing a broad minimum around 3 nm, and recovers "bulk" values below 1 nm. The different thickness dependence of the two shape parameters implies an asymmetric peak broadening upon film thinning, i.e., changes of m and n are caused by a variation in distribution of the relaxation times rather than to an intrinsic increase in the heterogeneity caused by confinement (which would induce a symmetric broadening).

The analysis of the shape parameters m and n suggests that upon film thickness reduction the free surface gradually approaches the RML and its mobility gets enslaved to the slow dynamics. This evidence is consistent with the idea that the glass transition temperature of a free surface is bound to the nature of the underlying layer [61]. In films thinner than 3 nm the more stable boundary conditions imposed by the interactions (-OH bonding) with a rigid substrate inhibit the acceleration provided by the molecules in contact with vacuum at the upper interface that alters the high frequency response of glycerol for layer thicknesses smaller than 50 nm. A more rigorous discussion of possible mobility profiles including additional experiments on "capped" glycerol layers can be found in [52].

2.5 *Correlation Between Breakdown-Scenario and Strength of the Confinement*

In the previous sections, specific glass-forming systems have been discussed that revealed "confinement effects" according to the initially suggested classification into scenarios I, II, and III. Intuitively one would expect some correlation between the strength of the confinement that increases with the number of restricted translational degrees of freedom, and the impact on the structural dynamics as classified by characteristic "deviation scenarios" ranging from a modified VFT behavior (type I) to the total breakdown of the VFT-law into Arrhenius behavior (type III).

To prove the above-made proposition, we have listed in Table 2 a couple of experimental studies reported in the literature that revealed one or more of the deviation scenarios.

Though the examples listed in Table 2 are far from being complete, they reveal a clear trend. Glass-forming liquids confined to a quasi 2D- liquid show typically

Table 2 Confined glass-forming systems classified according to the dimension of the confining host (1–3) and the deviation scenario (I–III)

Dimension confinement		Type I	Type II	Type III
	1 = layers		Smectic LC [39]	Glycerol via OMBD [52] Freely standing PS-films [62] Polymer nanocomposites [20] Supported PS-films [63]
	2 = pores	Zeolite (silicalite)/EG [25]	Salol in nanopores [14] Zeolite ($AlPO_4$-5)/*p*NA [17] PMPS/nanoporous glasses [64]	PPG/nanoporous glasses [64]
	3 = cavities	Zeolite (sodalite)/EG [25] Starch/EG [31]	Starch/EG [31]	

moderate deviations from bulk-type VFT behavior, an observation that is backed up by a substantial number of individual studies on ultrathin polymer films. On the other hand, a complete breakdown of the VFT behavior was exclusively reported for nonporous channel and cavity-type host/guest systems.

Despite the clear trend, there are a couple of systems that seemingly do not obey this correlation. The origin of such deviating behavior might be manyfold since various mechanism (others than cooperativity effects) are able to affect the structural dynamics of nanoconfined glass-forming systems:

i. Strong (specific) molecular interactions with the surface of the confining host system,
ii. Surface-induced including adsorption-driven packing anomalies [15] that might strongly depend on the surface roughness (smooth vs. corrugated surfaces),
iii. Introduction of extrinsic interfacial dynamics via "lubricated" surfaces,
iv. Confinement geometries with fractal dimensions or complex porosity,

v. Density effects due to space filling issues resulting in a "negative" pressure [16], and,
vi. Orientation effects, molecular alignment, and thus introduction of local order [59].

Nevertheless, the proposed correlation might serve as a guide to search for generic patterns in the generally very diverse wealth of experimental findings on nanoconfined glass-forming systems.

3 Summary and Conclusions

In this chapter, we have discussed the question in how far experiments on glass-forming liquids, confined to nanometer-sized geometries, can prove the cooperative nature and of the dynamic glass transition and provide quantitative information about its characteristic length scale ξ. For this purpose, we have focused on the most prominent manifestation of "confined" dynamics, viz. deviations of the temperature dependence of the structural relaxation time $\tau(T)$ from the bulk-VFT-behavior. To classify experimental findings from very diverse systems, we have introduced three major "deviation scenarios," which range from a complete breakdown of the VFT-law (type I) via a VFT-Arrhenius cross-over scenario (type II) to a scenario (III) representing a perturbed VFT behavior in terms of an overall accelerated or retarded glass transition dynamics.

Four cases were discussed in detail, all of them referring to H-bonding molecules or molecular groups. EG confined in various zeolitic hosts revealed scenario I or just bulk dynamics [25], while EG mixed with amylopectine (starch) showed an evolution through the cross-over scenario (II) toward genuine Arrhenius behavior (I) [31]. A third system, representing a H-bonding "liquid" confined to a self-assembled (smectic) layer structure, yielded a clear-cut scenario II along with a linear relation between the layer thickness and the cross-over frequency. To end with, the dynamics of ultrathin films of glycerol having one free surface was discussed as an example for a perturbated VFT-scenario (type III).

Apart from all molecular details and the diversity of confining host structures, it is evident that the strength of the confinement, expressed by the number of spatially restricted dimensions (1D = layer, 3D = cavities), should be linked to the severeness of changes in the dynamics. Inventarization of various experimental findings reported in the literature including them described above indeed confirm a correlation between the "deviation scenario" and the number of constrained axes.

Finally, the effect of mechanisms other than cooperativity replacement arguments, e.g., surface interactions, density, packing, and orientation effects, were briefly discussed to rationalize obvious exceptions from the proposed correlation.

References

1. Vogel H (1921) Z Phys 22:645
2. Fulcher GS (1925) J Am Ceram Soc 6:339
3. Tammann G, Hesse G (1926) Z Anorg Allg Chem 156:245
4. Adam G, Gibbs JH (1965) J Chem Phys 43:139
5. Lubchenko V, Wolynes PG (2007) Ann Rev Phys Chem 58:23
6. Tanaka H (2012) Eur Phys J E 35:113
7. Yamchi MZ, Ashwin SS, Bowles RK (2012) Phy Rev Lett 109:225701
8. Hedstrom J, Swenson J, Bergman R, Jansson H, Kittaka S (2007) Eur Phys J Spec Top 141:53
9. Donth E (2002) J Non-Cryst Solids 307:364
10. Hempel E, Vieweg S, Huwe A, Otto K, Schick C, Donth E (2000) J De Phys Iv 10:79
11. Cangialosi D, Alegria A, Colmenero J (2007) Phys Rev E 76:011514
12. van Turnhout J, Wübbenhorst M (2002) J. Non-Cryst. Solids 305:50
13. Napolitano S, Lupascu V, Wübbenhorst M (2008) Macromolecules 41:1061
14. Arndt M, Stannarius R, Groothues H, Hempel E, Kremer F (1997) Phys Rev Lett 79:2077
15. Napolitano S, Pilleri A, Rolla PA, Wübbenhorst M (2010) ACS Nano 4:841
16. McKenna GB (2010) Eur Phys J Spec Top 189:285
17. Wübbenhorst M, Klap GJ, Jansen JC, van Bekkum H, van Turnhout J (1999) J Chem Phys 111:5637
18. Lagrene K, Zanotti JM (2007) Eur Phys J Spec Top 141:261
19. Dobbertin J, Hensel A, Schick C (1996) J Therm Anal 47:1027
20. Anastasiadis SH, Karatasos K, Vlachos G, Manias E, Giannelis EP (2000) Phys Rev Lett 84:915
21. Wübbenhorst M, Capponi S, Napolitano S, Rozanski S, Couderc G, Behrnd NR, Hulliger J (2010) Eur Phys J Spec Top 189:181
22. Beiner M (2001) Macromol Rapid Commun 22:869
23. Starkweather HW (1991) Polymer 32:2443
24. Eyring E (1936) J Chem Phys 4:283
25. Huwe A, Kremer F, Behrens P, Schwieger W (1999) Phys Rev Lett 82:2338
26. Bibby DM, Dale MP (1985) Nature 317:157
27. Meier WM, Olson DH, Baerlocher C (1996) Atlas of Zeolite structure types. Elsevier, Amsterdam
28. Havriliak S, Negami S (1967) Polymer 8:161
29. Jordan BP, Sheppard EJ, Szwarnowski S (1978) J Phys D 11:695
30. Cusack NE (1987) The physics of structurally disordered matter. Adam Hilger, Bristol
31. Smits ALM, Wübbenhorst M, Kruiskamp PH, van Soest JJG, Vliegenthart JFG, van Turnhout J (2001) J Phys Chem B 105:5630
32. Smits ALM, Hulleman SHD, van Soest JJG, Feil H, Vliegenthart JFG (1999) Polym Adv Technol 10:570
33. Kruiskamp PH, Smits ALM, Vliegenthart JFG (2001) Ind Microbiol Biotechnol 26:90
34. Mertens IJA, Wübbenhorst M, Oosterbaan WD, Jenneskens LW, van Turnhout J (1999) Macromolecules 32:3314
35. Wübbenhorst M, van Turnhout J (2002) J Non-Cryst Solids 305:40
36. Donth E (1982) J Non-Cryst Solids 53:325
37. Jansen JC (1996) PhD thesis. Delft University of Technology, Delft
38. Wübbenhorst M, Van Koten E, Jansen J, Mijs M, van Turnhout J (1997) Macromol Rapid Commun 18:139
39. Wübbenhorst M, Van Eeten F, Van Turnhout J (2014) Manuscript in preparation
40. Yildirim Z, Wübbenhorst M, Mendes E, Picken SJ, Paraschiv I, Marcelis ATM, Zuilhof H, Sudhölter EJR (2005) J Non-Cryst Solids 351:2622
41. Jasiurkowska M, Napolitano S, Wübbenhorst M, Leys J, Juszyńska E, Massalska-Arodź M (2014) Manuscript in preparation

42. Donth E (1992) Relaxation and thermodynamics in polymers, glass transition. Akademie, Berlin
43. Hempel E, Hempel G, Hensel A, Schick C, Donth E (2000) J Phys Chem B 104:2460
44. de Gennes PG (2000) Eur Phys J E 2:201
45. Forrest JA, Dalnoki-Veress K (2001) Adv Colloid Interface Sci 94:167
46. Jones RAL (1999) Curr Opin Colloid Interface Sci 4:153
47. Keddie JL, Jones RAL, Cory RA (1994) Faraday Discuss 98:219
48. Napolitano S, Rotella C, Wübbenhorst M Acs Macro Lett 1:1189
49. Choa Y-K, Granick S (2003) J Chem Phys 119:547
50. Zheng W, Simon SL (2007) J Chem Phys 127:194501
51. Simon SL, Park JY, McKenna GB (2002) Eur Phys J E 8:209
52. Capponi S, Napolitano S, Behrnd N, Couderc G, Hulliger J, Wübbenhorst M (2010) J Phys Chem C 114:16696
53. Peters S, Napolitano S, Meyer H, Wübbenhorst M, Baschnagel J (2008) Macromolecules 41:7729
54. Lunkenheimer P, Pimenov A, Schiener B, Böhmer R, Loidl A (1996) Eur Lett 33:611
55. Mobius ME, Xia T, van Saarloos W, Orrit M, van Hecke M (2010) J Phys Chem B 114:7439
56. Zondervan R, Kulzer F, Berkhout GCG, Orrit M (2007) Proc Natl Acad Sci USA 104:12628
57. Kilburn D, Sokol PE, Sakai VG, Alam MA (2008) Appl Phys Lett 92:1
58. Levchenko AA, Jain P, Trofymluk O, Yu P, Navrotsky A, Sen S (2010) J Phys Chem B 114:3070
59. Capponi S, Napolitano S, Wübbenhorst M (2012) Nat Commun 3:1233
60. Schneider U, Lunkenheimer P, Brand R, Loidl A (1998) J Non-Cryst Solids 235:173
61. Roth CB, Torkelson JM (2007) Macromolecules 40:3328
62. Rotella C, Napolitano S, Wübbenhorst M (2009) Macromolecules 42:1415
63. Lupascu V, Picken SJ, Wübbenhorst M (2006) J Non-Cryst Solids 352:5594
64. Schonhals A, Goering H, Schick C, Frick B, Zorn R (2004) Colloid Polym Sci 282:882

Anomalous Decoupling of Translational and Rotational Motion Under 1D Confinement, Evidences from Crystallization and Diffusion Experiments

Simone Napolitano and Michael Wübbenhorst

Abstract The molecular diffusion is intrinsically bound to the viscosity of the environment via the Stokes–Einstein (SE) relation, fixing an equality between the energetic barrier for translational motion and that for rotational motion. This relation, valid for melts at equilibrium, usually breaks down upon supercooling in proximity of the glass transition. As a common feature of the glassy dynamics, diffusion is enhanced compared to the viscosity and the segmental relaxation. In this chapter, we revise recent experimental evidence on the anomalous decoupling between rotational and translation motion in thin polymer films. While the segmental relaxation is almost unperturbed down to few tens of nanometers, diffusion of small molecules and cold crystallization kinetics tremendously slow down already at thicknesses exceeding by several folds the macromolecular size. After reviewing experimental methods permitting to assess the dynamics in confinement, we propose a unifying picture on the anomaly in the SE relation based on the different impact of irreversible chain adsorption on the rotational and translation motion. In particular, we experimentally verified the validity of a relation to predict the crystallization time of thin polymer films based on finite size effects and the slowing down in the dynamics scaling according to the SE relation. Remarkably, such expression fails in correspondence of a critical size comparable to the thickness of the layer irreversibly adsorbed within the timeframe of the experiment. Similarly, we observe a severe drop in tracer diffusivity of dielectric probes into apolar matrices, not explainable in terms of perturbations in the segmental dynamics, but ascribable to the presence of adsorbed layers, limiting the Brownian movements, precursors of molecular diffusion.

S. Napolitano (✉)
Laboratory of Polymer and Soft Matter Dynamics, Faculté des Sciences,
Université Libre de Bruxelles, Boulevard du Triomphe, Bâtiment NO,
1050 Bruxelles, Belgium
e-mail: snapolit@ulb.ac.be

M. Wübbenhorst
Laboratory of Acoustics and Thermal Physics, Department of Physics and Astronomy,
KU Leuven, Celestijnenlaan 200D, 3001 Leuven, Belgium

F. Kremer (ed.), *Dynamics in Geometrical Confinement*, Advances in Dielectrics, 279
DOI: 10.1007/978-3-319-06100-9_11,

Keywords Polymer · Confinement · Crystallization · Tracer diffusivity · Dielectric spectroscopy · Finite size effect · Interfacial interactions · Physisoprtion · Irreversible chain adsorption

Abbreviations

$\alpha-$	Structural (segmental) process
$\alpha'-$	Constraint structural (segmental) process
β	Avrami exponent
Γ	Drop in $\Delta\varepsilon$ due to crystallization
δ	Drop in $\Delta\varepsilon$ due to adsorption
$\Delta\varepsilon$	Dielectric strength
ΔE	Transport energy barrier to the growth front
ΔF^*	Barrier for nucleation
ε_∞	Instantaneous dielectric constant
η	Viscosity
μ	Dipole moment
Λ	Nucleation term of the crystallization rate
ξ	Fractional coefficient
τ	Segmental relaxation time
τ_D	Diffusion time
1D	1-dimensional
a	Radius
AFM	Atomic force microscopy
Al	Aluminum
BDS	Broadband dielectric spectroscopy
C	Crystallization rate
D	Diffusion term of the crystallization rate
DSC	Differential scanning calorimetry
D_{tr}	Tracer diffusion coefficient
g	Kirkwood factor
G	Crystal growth rate
G_1	Linear crystal growth rate
h	Film thickness
h_{ads}	Thickness of the irreversibly adsorbed layer
k_B	Boltzmann constant
L	Diffusive length
l-PS	Polystyrene labeled with 4-[(4-cyanophenyl) diazenyl] phenyl}(methyl) amino
M_w	Weight average molecular weight
Ñ	Density of dipole moments
p	Pressure
PDI	Polydispersity index
PHB	Poly(hydroxy butyrate)
PET	Poly(ethylene terephthalate)

PS Polystyrene
SE Stokes–Einstein
t Time
T Temperature
T_0 Temperature where molecular motion would cease
t_{tads} Characteristic adsorption time
t_{CRY} Crystallization time
t_N Induction time
t_P Characteristic time
T_g Glass transition temperature
T_M Melting point
X_C Crystallinity
T_{CC} Cold crystallization temperature
$< \ldots >$ Statistical average

1 Introduction

The tremendous changes in the molecular dynamics occurring in confined geometry have being fascinating the scientific community for more than 20 years [1]. Pioneering work aimed at testing the validity of theories of the glass transition and predicting the presence of a characteristic length scale of a few nanometers [2–4]. However, due to the presence of interfacial interactions, masking the expression of genuine finite effects and to severe experimental difficulties connected to the preparation and the investigation of very thin free-standing films, this goal has not yet been reached. Nevertheless, the anomalous behavior displayed by polymer chains in proximity of adsorbing interfaces has recently attracted a larger and broad community of researchers [5–10]. Recent work has, in fact, demonstrated that the behavior of polymers under 1D confinement, that is, in the thin films geometry, cannot be fully explained in terms of the impact of finite size and interfacial effects on systems at thermodynamic equilibrium [11–13]. A simplistic picture based on film thickness and surface interactions cannot, for example, explain the enhancement of segmental mobility in proximity of an adsorbing interface [14, 15], or the presence of long-lasting metastable states in the liquid state [16–18].

Based on our recent relaxation, crystallization and diffusion experiments [19, 20], we discuss a new model for the dynamics of polymers confined in thin films, focusing on the description of out of equilibrium states arising from irreversibly chain adsorption. In this chapter, we focus on the impact of interfacial conformations on the fundamental correlation between translational and rotational motion, commonly known as Stokes–Einstein (SE) relation. The unique information achieved by broadband dielectric spectroscopy (BDS) evidences a anomalous decoupling between the two independent processes, not observed in bulk glassy melts.

In the following text, after reviewing current ideas on the crystallization of ultrathin polymer films, we discuss the implementation of the experimental procedure to monitor crystallization kinetics at the nanoscale level. The outcome of these measurements is discussed in terms of nucleation in finite volumes and to on the breakdown of the SE relation upon confinement. We continue describing our method to determine tracer diffusivity in thin films, which permitted to collect further evidence on the decoupling of rotational and translational motion. Finally, we propose our view on the slowing down of diffusion of small molecules into polymer matrices and on the crystallization kinetics of ultrathin polymer films, based on the impact of irreversible chain adsorption.

2 Crystallization of Ultrathin Polymer Films

Polymers do not crystallize easily [21]. The formation of ordered structures requires, in fact, stereoregular chains, and is hindered in the case of complex molecular architecture affecting the mechanisms of chain folding. Similarly as for smaller molecules [22], crystallization gets even more sluggish upon confinement at the nanoscale level [23]. In the case of 1D confinement, the crystallization rate, C, strongly decreases upon reduction of the film thickness, h, yielding a neat increase in the mean crystallization time t_{CRY} up to several orders of magnitude compared to the bulk [24]. A similar scenario is confirmed in nonisothermal experiments. The cold crystallization temperature T_{CC}, defined as the temperature where a glassy sample crystallizes upon heating, increases, in fact, upon confinement [25]. Thinner films are thus characterized by a higher stability of the glassy state [26] and in particular conditions, crystallization does not take place within the experimental timescale [27]. The overall drop in crystallization rate might yield extreme situations where the formation of crystals is not observed within experimental time scales exceeding several months, or upon annealing up to the bulk melting point. This last finding implies that the crystallization is effectively inhibited or that the crystal growth rate of interfacial chains got reduced by several orders of magnitude.

Several theoretical models predicted perturbations on the crystallization of thin films due to the reduction of both the nucleation rate and the mass transport in systems confined at the nanoscale level [28–30]. Expressing the overall crystallization rate via the classical nucleation theory [21]:

$$C(T) \sim \exp[-\Delta \mathrm{E}/k_B(T - T_0)]\exp(-\Delta \mathrm{F}^*/k_B T) \qquad (1)$$

slower crystallization kinetics can be associated to either larger work to create a critical nucleus, $\Delta \mathrm{F}^*$, or to higher T_0, that is, the temperature where molecular motion would virtually cease, or to a greater energetic barrier related to transport of segments toward the growth front ($\Delta \mathrm{E}$).

Furthermore, Schultz [31] predicted a systematic deviation from the Avrami law, the analytical expression commonly used to investigate the crystallization kinetics in bulk, due to finite size effects. By means of optical microscopy and atomic force microscopy (AFM) it was possible to directly observe a severe reduction in growth rate for crystals present on the surface of extremely thin films. However, optical methods do not permit discriminating the role of finite size effects from interfacial interactions, because they cannot access to direct information on the molecular dynamics. To overcome this problem, we developed an experiment allowing the simultaneous determination of the crystallization time and of the segmental relaxation time, τ, which shed light on the formation of ordered structures at the nanoscale level [32].

2.1 Investigation of Crystallization Kinetics by Means of Dielectric Spectroscopy: From Bulk to Thin Films

The dielectric strength of the structural process, $\Delta\varepsilon$, corresponds to the contribution of orientational polarization to the dielectric function [33]. In isothermal conditions, $\Delta\varepsilon$ is proportional to $g \cdot <\mu^2>$, where $<\mu^2>$ is the mean square dipole moment, and g is the Kirkwood correlation factor, taking into account the orientational correlation among dipole moments. g is equal to 1 in case of random orientation, and it increases with the reduction in the difference between the orientations of the single dipole moments, that is, when dipole moments are more parallel among each other. Recently, we verified that increase in $\Delta\varepsilon$ due to larger values of g indicate the presence of regions with enhanced orientational order, naturally present in supercooled isotropic liquids [34].

At constant g, we can approximate the previous expression as $\Delta\varepsilon \sim \tilde{N}\mu^2$, where $\tilde{N}$ indicates the density number of dipole moments. This relation allows assigning a perturbation in the dielectric strength to a change in the number of segments relaxing on the timescale of τ. During crystallization, incorporation of mobile segments into crystalline structures yields a reduction of the solid angle over which dipole moments can fluctuate, and thus a drop in $\Delta\varepsilon$. While in low molecular weight compounds $-\Delta\varepsilon(t)$ evolves in time as the crystalline content, X_C [35], Ezquerra and coworkers demonstrated that such a relation does not always hold in the case of polymers, because of the failure of a simple two-phase model in the presence of the complex macromolecular architecture [36–39]. During crystallization, often prior to the drop in intensity of the structural peak, a broader and more symmetric process, indicated as α'–, emerges at lower frequencies. These slower relaxation modes have often been attributed to the constrained amorphous fraction, whose presence supports the idea of the failure of simple two-phase models in semicrystalline polymers. Fukao et al. [40] verified that, for poly(ethylene terephthalate), PET, the sum of the intensities of the α- and the α'-processes (that in text, we indicate as the dielectric strength of PET) evolves in times as $X_C(t)$.

We implemented the methodology to monitor the changes in the amorphous phase during crystallization at the nanoscale, proposing a unique experiment permitting to monitor the changes in the segmental dynamics of films as thin as a few nanometers during the formation of crystalline structures [19, 32, 41–43]. In the following paragraphs, we describe the analysis of the thickness dependence of the crystallization kinetics of PHB [32] and PET [19].

Bulk samples of PHB were prepared by melting the polymer in powder (Sigma, Mw = 170 K), for 5 min 15 K above its melting point (T_m = 443 K, via Differential Scanning Calorimentry, DSC) between two brass circular electrodes. Amorphous samples were obtained by quenching the capacitor between two cold plates held at 268 K, that is, below the glass transition temperature (T_g^{DSC} = 275 K). The final thickness of the sample was fixed by the diameter of the glass fibers (50 μm) used to separate the brass electrodes to avoid shortcuts.

Ultrathin films of PHB (26–50 nm) were prepared by spincoating very dilute solutions (down to 0.2 % w/w) of the polymer dissolved in chloroform and filtered by a Teflon membrane filter (Millipore). Films of PET (1 μm to 6 nm) were prepared by spincoating solutions of the polymer (M_n = 12.20 k_g/mol, M_w/M_n = 2.7), dissolved in a trifluoroacetic acid and chloroform 5:2 mixture.

Application of an E-field to polymer films is possible via the metallization of its two surfaces. The preparation of these nanocapacitors starts via the deposition of a metallic layer (typically 50 nm of aluminum) on a solid substrate (e.g., silica glass). Subsequently, ultrathin polymer films are spincoated directly onto the thermally evaporated metal. After an annealing procedure above T_g, patterned top electrodes are evaporated in the same conditions as the lower electrode. Aluminum is preferred to other metals because of the sharpness of the interfaces between this element and several polymers [44]. Evaporated on top of a polymer film, gold, on the contrary, tends to diffuse through the organic medium. We verified that the two metal/polymer interfaces (polymer spincoated onto metal and metal evaporated onto polymer) are identical [45]. Experiments performed on multilayers containing dielectric probes showed, in fact, that the upper and the lower interfacial layers have the same dielectric strength, and the same temperature dependence of the segmental relaxation time.

In the specific case of PHB, spincoating was performed at room temperature, thus well above T_g (=275 K). After an annealing procedure of 2 h at 318 K, the samples were slowly cooled down to room temperature. After evaporation of the upper electrode, samples were amorphized by holding the nanocapacitors above T_m for 30 s, at the same annealing temperature as for bulk samples. Shorter annealing times are needed due to the extremely small mass of the samples (~10 ng).

Soon after spincoating, films of PET were annealed for 20 min at 348 K, and then placed in vacuum to favor solvent evaporation (>24 h at 7×10^{-5} bar). This procedure does alter the amorphous character of the samples, as carefully checked via AFM and X-rays diffraction.

Film thicknesses were evaluated from the electrical capacitance of the sample in the approximation of parallel electrodes with infinite lateral dimensions. Dielectric spectra were recorded in the frequency range from 10 to 10 MHz with a high resolution dielectric analyzer (ALPHA-A from Novocontrol Technologies). Measurements on PHB samples were perfomed immediately after amorphization to avoid nucleation processes due to storage of the sample above the glass transition temperature.

2.2 Analysis of the Crystallization Kinetics

The reduction in dielectric strength commonly observed during crystallization experiments can be analyzed, in bulk, using the formalism of Avrami [46]. In the case of thin films, where the volume fraction of the interfaces becomes relevant, we need to take into account another source of chain immobilization, leading to a reduction in $\Delta\varepsilon$, that is, irreversible adsorption [5, 47–49].

Both types of kinetics can be described by exponential functions of the type $\Delta\varepsilon(t) \sim 1-\exp(-t/t_P)^{\beta}$ [46, 50], where t_P is the characteristic time of the process and the exponent β provides information on the dimensionality and/or cooperative nature of the kinetics. In the case of adsorption, β assumes values between 0 and 1 (stretched exponential), while β is larger than unity (<4, compressed exponential) for crystallization kinetics following the Avrami equation. Moreover, interfacial rearrangements increasing the monomer/surface density cause a further drop in $\Delta\varepsilon$, scaling logarithmically with the annealing time.

Consequently, focusing on primary crystallization and adsorption, we proposed a relation taking into account the above-mentioned immobilization mechanisms and permitting to analyze the time evolution of the drop in dielectric strength over several decades [19]:

$$\frac{\Delta\varepsilon(t)}{\Delta\varepsilon(t_0)} = 1 - \Gamma\left[1 - \exp\left(-(t/t_{\mathrm{CRY}})^{\beta}\right)\right] - \delta\log(t/t_0) \qquad (2)$$

where Γ indicates the final drop in dielectric strength (proportionally to the crystalline content of the sample), and δ is a parameter proportional to the fraction of chains immobilized upon adsorption. An example of the thickness dependence of $\Delta\varepsilon(t)$ for PET is given in Fig. 1. Upon reduction of the thickness we observed the expected increase in t_{CRY}, accompanied by a reduction in β and in Γ. Because the condition $\beta > 1$ was always fulfilled, the observed kinetics were straightforwardly related to crystallization events. Upon reduction of the thickness, the increase in the slope at shorter times corresponds to larger values of δ, $\partial\Delta\varepsilon/\partial\log(t)$ for $t \ll t_{\mathrm{CRY}}$, a trend consistent with the larger impact of chain adsorption at smaller sample volumes. A similar scenario was found for PHB.

In the left panel of Fig. 2, we plotted the values of the Avrami exponent as a function of the crystallization time, for films PHB of different thicknesses, isothermally crystallized at different temperatures (283–293 K). A similar plot for thin films of PET annealed at 373 K is given in the right panel of Fig. 2.

Regardless the annealing temperature, for bulk samples of PHB and PET, $\beta \approx 3$ which is in agreement with heterogeneous growth in three dimensions. Smaller values of the Avrami coefficient ($\beta \to 2$ for PHB, $\beta \to 1.5$ for PET) were observed upon reduction of the thickness, as previously reported in the case of supported films. Discarding the occurrence of transitions in the heterogeneity of the crystallization process, which should manifest as an abrupt change in β upon reduction of the confinement size [51], we would be tempted to interpret the drop in β as a severe

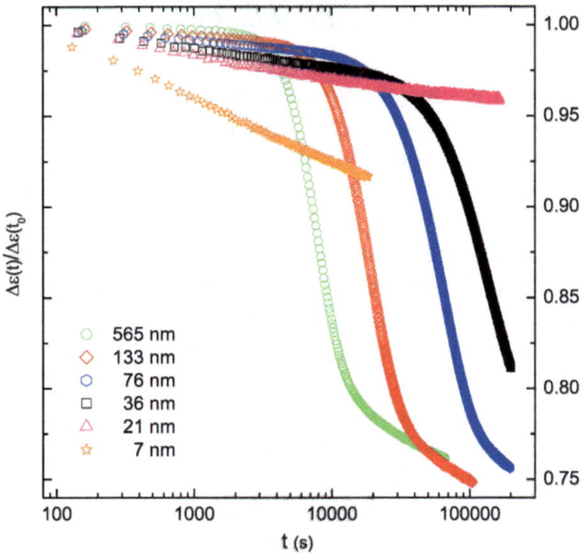

Fig. 1 Time evolution of the normalized dielectric strength of films of PET of different thickness capped between aluminum layers. Reproduced from Vanroy et al. [19] with permission of the American Chemical Society

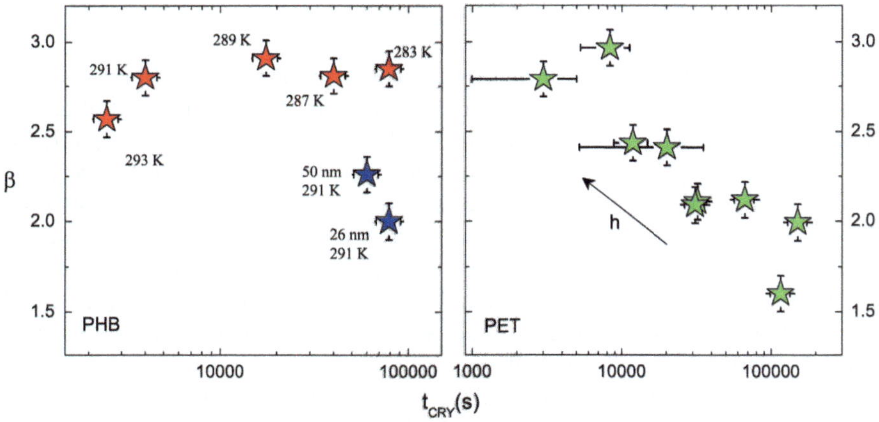

Fig. 2 *Left panel* the Avrami exponent for bulk samples (*red stars*) and ultrathin films (*blue stars*) of PHB is plotted versus the crystallization time; the temperature at which the samples were isothermally crystallized is also indicated. *Right panel* the Avrami exponent ultrathin films of PET (*green stars*, 27 nm $\leq$ h $\leq$ 1062 nm) isothermally crystallized at 373 K is plotted versus the corresponding crystallization time

decrease in the dimensionality of the crystallization process. According to this idea, upon reduction of the film thickness, the complex mechanisms of chain folding would not allow the formation of structure with three-dimensional positional and orientational order. Although intuitive, the character of this assumption is highly

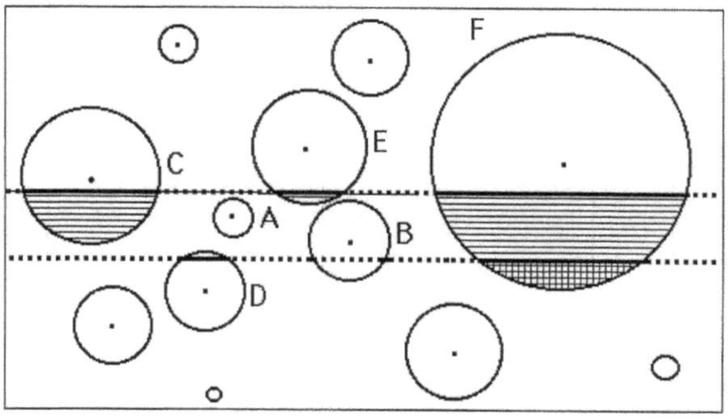

Fig. 3 Schematic illustration of homogenous nucleation in a large system. Check the original reference for more information. Reproduced from Schultz [31] with permission of the American Chemical Society

speculative, because it is restricted to the interpretation of the Avrami equation in bulk. Moreover, this hypothesis would not explain the size of the increase in t_{CRY}.

Schultz analyzed the impact of finite size effects on the formation of spherulites, confined in the thin film geometry [31], see Fig. 3. To understand his model, we considered an infinite volume with a statistical distribution of nuclei growing in three dimensions with a linear growth rate G_1, that is, the volume of isolated crystals evolves in time as $(G_1 t)^3$. In the case of sample volumes much larger than the spherulite diameter, which for melt crystallization can easily exceed 1 μm, the kinetics is well pictured by the Avrami law. The power law growth of isolated crystals smoothly slows down upon impingement, which explains the saturating exponential form of the Avrami equation.

Reduction of the distance between the two surfaces requires a correction in the total crystalline volume fraction. In particular, we need to remove the ensemble of crystals whose centers sit outside the sample volume, and the fraction of trunked crystals not grown inside the slab, because those crystals do not participate to the overall growth rate. Although the dimensionality of the crystallization kinetics is unaltered, the effect of these corrections yields a neat reduction in β. The model of Schultz predicts that the apparent drop in Avrami exponent is accompanied by an increase in t_{CRY}, in line with our findings, see Fig. 2. Deviations from bulk behavior are expected when the ratio h/G_1 deviates from infinite. This last quantity might, however, be ill-defined in the case of thin films, because the value of G_1 is thickness dependent. Moreover, the ratio h/G_1 does not provide a dimensionless parameter, but a time. Consequently, we should limit this interpretation at the quantitative level.

With these considerations in mind, in Fig. 4 we plotted the evolution of the reduced dielectric strength as a function of the annealing time for a bulk sample of PHB crystallized at 283 K and a 26 nm thick film crystallized at 291 K. Although the thin film was held at a temperature 8 K larger than for the bulk sample, the value of t_{CRY}, obtained via a fit of the experimental data to the Avrami equation is for both samples on the order of 80 ks (>22 h). The value of the Avrami exponent of the thin film is smaller than for the bulk sample, which is further in line with this model.

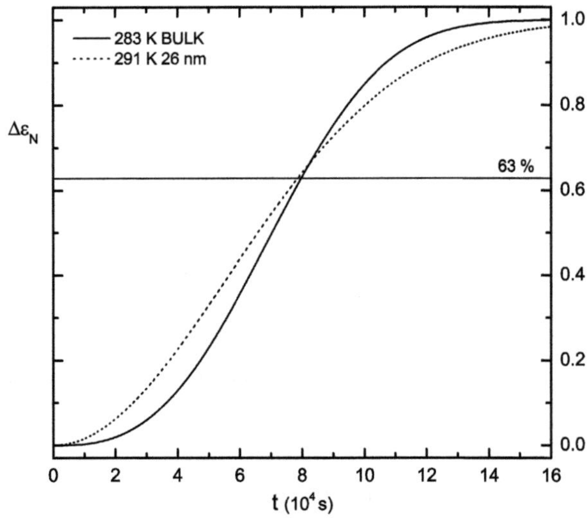

Fig. 4 Time evolution of the normalized dielectric strength, $\Delta\varepsilon_N$, for a bulk sample at 283 K and a 26 nm thick film at 291 K. To ease comparison, the contribution of adsorption was subtracted, before normalizing the data from 0 to 1. The horizontal line at $1 - e^{-1} \approx 0.63$ crosses $\Delta\varepsilon_N$ at $t = t_{CRY}$

2.3 Finite Size Effects and Interfacial Interactions are not Sufficient: The Failure of a Simplistic Model

We considered the role of finite size effects on nucleation. Assuming a constant nuclei density, that is, a constant number of nucleation events occurring per unit of time in the unit of volume, one would expect that, upon confinement, the reduction of the nucleation rate would scale with the sample volume. This hypothesis was validated in the case of homogenous nucleation in nanodroplets of poly(ethylene oxide), where the product of the nucleation rate and the droplet volume was constant [52, 53].

Extending this scenario to the case of thin films, we deduced that in case of a nucleation limited confinement effect, the crystallization time should increase with the inverse of the film thickness [19]. To reduce the impact of the role of the thickness dependence of the crystal growth rate, G, on the overall crystallization rate, we considered also the induction time t_N. We assigned this quantity to the incubation time prior to the drop in dielectric strength imputable to crystallization, a definition that permitted to discard the effect of adsorption on the drop in $\Delta\varepsilon$. In the case of significant perturbations in G, the proposed scaling is still valid for t_N, but not for t_{CRY}, as the last quantity is given by the convolution of both nucleation and growth. The data in Fig. 5 experimentally validate our picture. Reduction of the thickness from 1 μm to 20–30 nm brought to an increase in the value of t_{CRY} from 2.5×10^3 s (40 min) to 10^5 s (more than 1 day). It is noteworthy stressing that the ratio of the crystallization times of different samples simply corresponds to ratio of the film

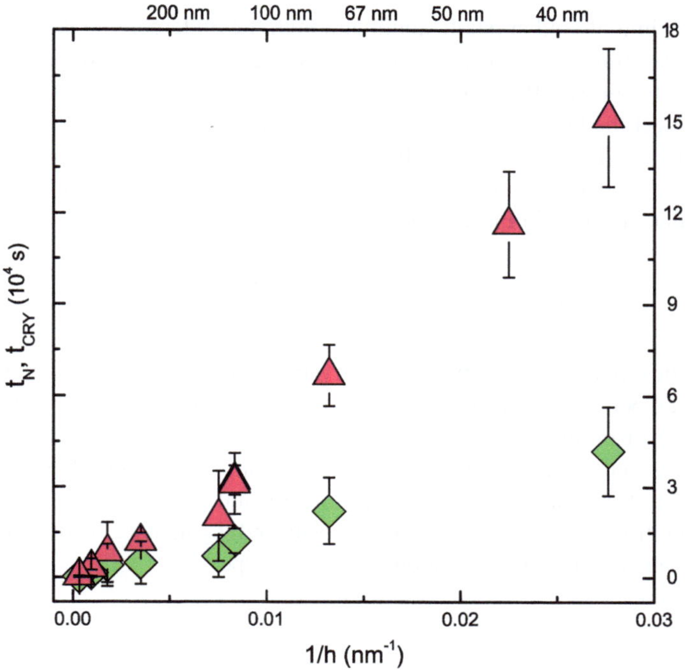

Fig. 5 Thickness dependence of the induction time. t_N (*green diamonds*), and the crystallization time, t_{CRY} (*pink triangles*), of thin films of PET capped between two aluminum layers

thicknesses, which permits using a simple scaling law to predict the crystallization rate in this regime

$$t_{CRY}(h) = \frac{h_R}{h} t_{CRY}(h_R) \tag{3}$$

where h_R is a reference thickness. Films of PET thinner than a threshold thickness of 20–30 nm did not show any significant reduction in dielectric strength within an experimental time window of 55 h. Consequently, considering the values of β in this thickness range, we estimated that the growth rate for films thinner of 20 nm should be reduced by at least three orders of magnitude compared to the bulk. To understand whether the corresponding increase in t_{CRY} could be imputable to slower interfacial dynamics, we analyzed the correlation between crystallization and segmental dynamics.

In its simplest formulation, the temperature and thickness dependence of the crystallization rate $C(h, T)$ can be written as the product of two exponential terms $\Lambda(h, T)$ and $D(h, T)$, respectively related to the barriers for nucleation and for the motion of mobile chains to the crystallization front [42], compare Eq. (1)

$$C(h, T) = D(h, T)\Lambda(h, T) \tag{4}$$

In the regime of cold crystallization, although the undercooling can easily overcome 100 K, formation of ordered structures is governed by diffusion of mobile segments to the growth front, because $|\partial D(T)/\partial T| \gg |\partial \Lambda(T)/\partial T|$. Consequently, we can neglect the temperature dependence of the nucleation barrier and replace it with a term including the effects of finite size discussed in the previous paragraphs, $\Lambda(h) = \lambda \cdot \mathrm{h}^{-1}$, where λ is a temperature dependent constant that can be determined via fits of data as those in Fig. 5. Under these conditions we can write

$$t_{\mathrm{CRY}}(h,T) \approx C^{-1}(h,T) = \frac{h}{\lambda(T)} D^{-1}(h,T) \tag{5}$$

which implies that in the regime of cold crystallization the crystallization time has the same temperature dependence as the inverse of the term taking into account the transport of mobile chains toward the growth front. We assumed that this last phenomenon could be driven by translational motion, same as for other mechanisms governing the transport of mass. Such a fundamental movement is coupled to the rotational motion, which corresponds to viscosity and segmental dynamics and thus τ.

These considerations bring to a relation of the type:

$$t_{\mathrm{CRY}}(h,T) \approx g\left[\tau(h,T)\right] \tag{6}$$

where g is a polynomial function. To test the robustness of this assumption, we considered that the translational and rotational motion of simple liquids are correlated via the SE, here written in the case of a solid sphere of radius a immersed in a fluid of viscosity η:

$$D_t = \frac{k_B T}{6\pi a \eta} \tag{7}$$

where D_t is the diffusion coefficient for translational motion. Large experimental evidence showed that such a simple relation does not hold in proximity of T_g, where it is replaced by a fractional form of the type [54]:

$$D_t = (\eta/T)^{-\xi} \tag{8}$$

where ξ is the fractional coefficient. Because $0.5 < \xi < 1.0$, upon cooling, diffusion is enhanced in comparison to the prediction of the SE relation. Consequently replacing D_t with t_{CRY}^{-1} and η with τ, we arrive at a relation between the crystallization time and the segmental time holding in bulk:

$$t_{\mathrm{CRY}} = (\tau/T)^{\xi} \tag{9}$$

We performed a series of experiments, crystallizing bulk samples of PHB at different temperatures just above T_g, see Fig. 6, and obtained $\xi = 0.748 \pm 0.044$. Such a value

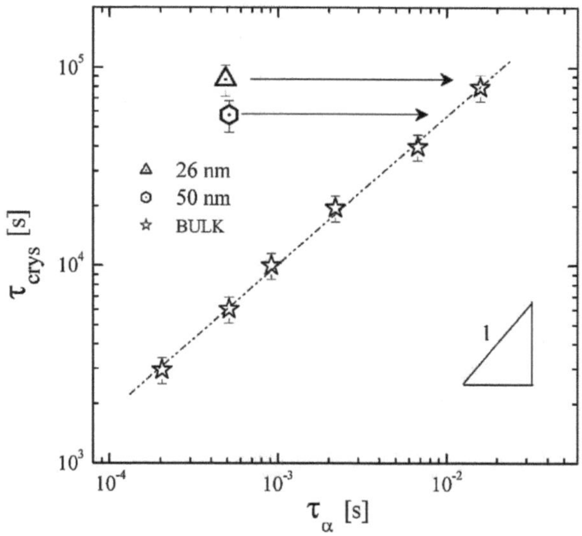

Fig. 6 Crystallization time versus the structural relaxation time for bulk and ultrathin films of poly(3-hydroxybutyrate). The *dashed-line* is the best linear fit for the data reported. The *arrows* indicate the relaxation times of bulk samples that would correspond to the crystallization times of the confined systems, according to the best linear fit. Because t varied by almost two orders of magnitude within the selected temperature window ($T_{max}/T_{min} < 4\%$), we plotted τ instead of τ/T. Reproduced from Napolitano et al. [42] with permission of the American Chemical Society

is in line with those indicated for low molecular weight and polymer glass formers in similar ranges of reduced temperatures. The value of ξ reaches 1 for $T > 1.3$–1.4 T_g, and typically it smoothly reaches ~0.7–0.8, although data in the range $0.5 \leq \xi \leq 0.95$ have been reported [55].

The physical meaning of the condition $\xi < 1$ (often referred to as fractional exponent) is still under debate, through the possible explanations it is noteworthy mentioning special thermal fluctuations, the way each observable is experimentally averaged [56], strong temperature dependence of the relaxation distributions in polymers [57]. Regardless of the different hypothesis, because the validity of Eq. (9) is independent from the numerical values of the parameters used, the same qualitative trend can be obtained from all of the series of data satisfying the requirements ensuring the physical meaning of the associated variables.

After validation of this fundamental relation in bulk samples, we introduced the thickness dependence into Eq. (9):

$$t_{cry}(h, T) = \frac{1}{\lambda(T)\, h} [\tau(h, T)/T]^{\xi} \tag{10}$$

and obtained a straightforward method to reply to the question *Is the slowing down in the crystallization kinetics related to slower interfacial dynamics?*

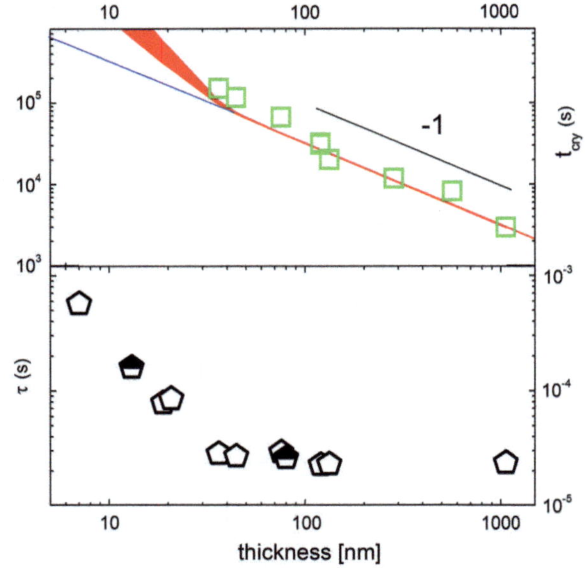

Fig. 7 Thickness dependence of the crystallization time (*upper panel*) and the segmental relaxation time (*lower panel*, semiopen symbols are data from [58]) versus thickness of thin films of PET capped between two aluminum layers. The *blue line* indicates the component of t_{CRY} due to the finite size effects on nucleation, while the *red line* is the prediction of t_{CRY} according to Eq. (10) assuming $0.5 \leq \xi \leq 1.0$. Readapted from Vanroy et al. [19] with permission of the American Chemical Society

For PHB, we verified that the crystallization time of thin films does not correspond to the predictions based on the bulk correlation between t_{CRY} and τ, see Fig. 6. However, due to the lack of the experimental data in the nucleation limited regime, we did not further proceed with the treatment of the data.

On the contrary, in the case of PET, where a much larger dataset was accessible, we could carefully check the predictions of Eq. (10). In Fig. 7, we plotted the thickness dependence of t_{CRY} and τ, for films of PET ranging from 1 μm to 6 nm. The segmental time was thickness independent for $h < 20$ nm, and then increased by almost two orders of magnitude in the thinnest films. As discussed in the previous paragraphs, the crystallization time increased with the inverse of the thickness down to 20 nm, a threshold thickness below which crystallization was not observed. We performed more experiments at higher temperatures, where the crystallization kinetics is intrinsically speed up by the increase in crystallization rate, repeatedly scanning up to $T_m + 20$ K. Scans at higher temperatures were not necessary, because regardless the entity of polymer/substrate interaction, $T_m(h)$ can only decrease in the presence of an interface.

The outcome of these experiments confirmed the previous trend. In particular, we could not detect any crystallization event for films thinner than 20 nm, and same as for the crystallization time, the cold crystallization temperature T_{CC}, defined as

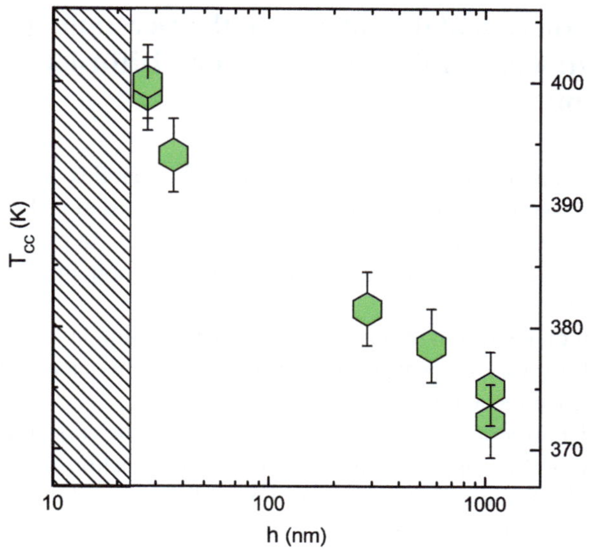

Fig. 8 Thickness dependence of the cold crystallization temperature. The *shadow area* indicates the thickness regime where crystallization was not observed

the temperature at which a frequency independent drop in the dielectric constant is measured, increased by almost 30 K, see Fig. 8. One might comment that this trend could be related to a hypothetical drop in sensitivity of our method upon reduction of the thickness. To check if this was the case, we analyzed a 15 nm thick film of PET capped between Al layers by means of X-rays reflectivity. Also this technique confirmed that the sample did not crystallize upon annealing of 24 h at 553 K, which is the temperature showing a maximum in crystal growth rate in bulk. Considering that this last quantity reaches values on the order of $\sim$75 nm/s [59], the behavior observed below 20 nm could be explained in terms of inhibition of the crystallization or a drop in crystallization by more than a factor 1,000.

In Fig. 7 we also showed the predictions of Eq. (10) based on experimentally measured values of τ; as the value of ξ was not known, we indicated a range of t_{CRY} related to $0.5 < \xi < 1.0$. The expected increase in t_{CRY} due to both finite size effects of nucleation and slower segmental dynamics is not sufficient to justify the experimental trend. In fact, in isothermal experiments, we did not observe any crystallization event at timescales exceeding by more than five times the value of t_{CRY} provided by Eq. (10). The results in Fig. 7 clearly show that the translational and orientational motion are strongly decoupled in systems confined at the nanoscale level, In order to explain the trend observed in thin PET films via Eq. (9), we should fix the value of $\xi \gg 1$, a condition opposite to the experimental observation of fraction SE relations in bulk melts, with $\xi < 1$, that is, upon confinement diffusion is suppressed, rather than enhanced, in comparison to the prediction of the SE model.

3 Further Experimental Evidence on the Decoupling of Translational and Rotational Motion: Diffusion Experiments

The previous examples provided clear evidence of a paradox in the mobility of polymer chains confined in thin films, rotational, and translational energy barriers are not strictly correlated as in bulk melts. On the contrary, in proximity of buried interfaces a severe reduction in the transport can occur without invoking a change in segmental mobility.

To shed light on the origin of this intriguing phenomenon, in the following paragraphs we quickly review other experimental evidences of this anomalous breakdown of the SE relation at the nanoscale:

Crystallization experiments. Zhang et al. investigated the crystallization kinetics of ultrathin films of PET supported on gold by means of reflection-absorption infrared spectroscopy [25]. Same as in the case of films of PET capped between aluminum layers, the crystallization time increases upon reduction of the thickness, but in the case of films supported on Au, a reduction in the glass transition temperature was reported by the same group in a previous report [60], where the same technique was used.

Confinement effect on the α-relaxation and the normal mode. Investigation by BDS of thin layers of poly(cis-1,4-isoprene) revealed that the segmental dynamics is unaffected by film thickness, while the normal mode (diffusion) is strongly reduced in the thinner films [61].

Molecular motion in irreversibly adsorbed layers. The relaxation time of layers of PS irreversibly adsorbed on aluminum oxide assume bulk values [5], while the diffusion of gold markers into layers of the same polymer adsorbed on silicon oxide is reduced by several orders of magnitude [62].

Tracer diffusivity experiments. Several experimental reports indicated a shift of the glass transition of thin supported films of PS toward lower temperatures [63]. On the contrary the diffusion of small molecules (tracer diffusion) dispersed into thin films of PS is reduced in comparison to thicker films [64]. In the case of free-standing films, reductions in T_g as large as 70 K [65] and tremendous acceleration in the segmental dynamics [66] were accompanied by the lack of thickness dependence in the tracer diffusion coefficient, at least down to 20 nm [67].

The large number of examples provided in the previous paragraphs is sufficient to discard the hypothesis that the decoupling between rotational and translation motion is due to experimental artifacts. However, the previous analysis is based on the comparison of experiments performed by means of different techniques under different investigation conditions. To overcome this problem, we designed a methodology that permitted to evaluate tracer diffusion in thin films by means of dielectric spectroscopy, via an approach that allows monitoring the changes in the segmental mobility of tracer molecules while they diffuse through matrices of different thickness, down to a couple of tens of nm [20]. Such a feature cannot be achieved by conventional methods to measure tracer diffusivity by means of neutron scattering [64] or via fluorescence nonradiative energy transfer [68], because based on the time evolution of the density profile of probe molecules diffusing inside a polymer layer. These approaches cannot, in fact, provide the exact timescale necessary to a molecule, e.g. solvent, diluent, etc. to diffuse through the interfacial layer and finally reach the substrate.

Our method is based on the direct determination of the time τ_D needed by tracer molecules to diffuse through slabs of known thickness and successively get adsorbed onto a target substrate. Dielectric spectroscopy is sensitive to adsorption because such a phenomenon occurs via immobilization, which yields a total or partial inhibition of segmental motion, resulting in drop of the dielectric strength, same as for cold crystallization. Partial or total immobilization of the probes approaching the targeted substrate limits the solid angle over which dipoles can reorient, and thus a reduction in $\Delta\varepsilon$. In annealing experiments performed on polymer melts, we carefully verified that the growth of the adsorbed layer results in correlated changes in dielectric strength [5].

Compared to neutron reflectivity measurements, our method provides multiple advantages: (a) we need smaller and more easily accessible instrumentation, such as an impedance analyzer; (b) the possibility to measure the effective time needed by tracer molecules to diffuse through the whole layer and adsorb onto the substrate, rather than the tracer diffusivity averaged over the whole slab thickness.

Our samples consist of bilayers where the polymer matrix is a thin film placed in direct contact with a layer containing a large reservoir of probe molecules, e.g., shorter chains decorated with polar moieties, so that the dielectric response of the whole sample is dominated by the α-relaxation of the probe. Depending on the selectivity of the solvents used for spin coating, we prepared samples in two configurations (A and B), see Fig. 9.

Polystyrene (weight average molecular mass M_w = 97 and 640 kg/mol; PDI = 1.01 and 1.11, T_g via differential scanning calorimetry = 372 ± 2 K and 374 ± 2 K) was purchased from Scientific Polymer Products and used as matrix without any further treatment. The low molecular weight (M_w 20 kg/mol PDI = 2.5, T_g = 371±3 K), chromophore-functionalized polymer (l-PS), used as tracer in this study, is a random copolymer of styrene and methyl methacrylate whose longer side chain is decorated with the polar group {4-[(4-cyanophenyl)diazenyl]phenyl}(methyl) amino. The final content of chromophore was smaller than 2 % mol, which corresponded to dielectric strength 25 folds larger than that neat PS.

Samples containing matrices thicker than 19 nm were prepared as trilayers (configuration A), placing the films of l-PS in between an impenetrable adsorbed layer (Guiselin brush [69]) and a matrix of PS640. This configuration permitted to avoid diffusion toward the lower electrode, protected by the 8 nm thick adsorbed layers of PS97 (where 97 indicates the M_w in kg/mol) with high surface coverage, prepared by spincoating filtered solution of the polymer in chloroform onto thermally evaporated aluminum substrates (~50 nm of Al 99.5 % Goodfellow, $p \leq 10^{-6}$ mbar, evaporation rate ~10 nm/s), successively annealed at 423 K for 20 h to favor adsorption, and finally washed in chloroform to remove the unbound chains. Washed films were dried overnight in high vacuum ($p \leq 10^{-6}$ mbar). Films of l-PS (~50 nm) were spincoated (chloroform) directly onto the impenetrable adsorbed layer, which is not soluble in chloroform. Matrices of PS640 of desired thickness (>19 nm) previously spincoated

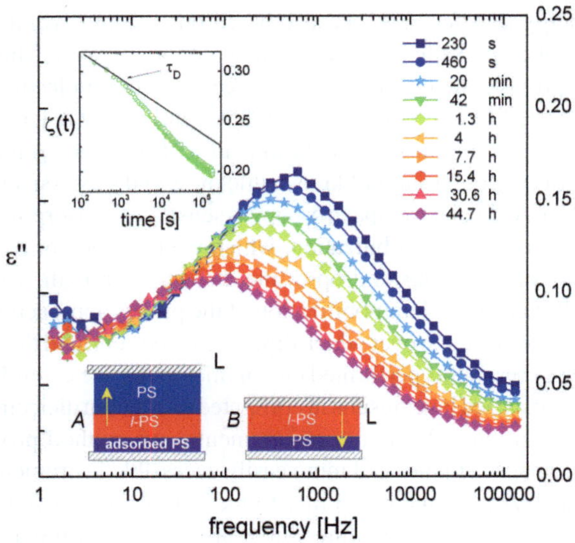

Fig. 9 Frequency dependence of the dielectric loss during diffusion of labeled PS into a 7.5 nm thick matrix of neat PS640 at 398 K, a scheme of the assembled multilayers in the two configurations used is provided, *arrows* indicate diffusion of labeled PS. In the inset, time evolution of the ratio between the dielectric strength and the dielectric constant. Reproduced from Napolitano et al. [20] with permission from Wiley-VCH

(chloroform) on mica and floated onto a reservoir of ultrapure water, were finally transferred on top of the bilayer (brushes/l-PS). Water occasionally trapped between consecutive layers was allowed to evaporate at ambient conditions for at least a couple of days before the deposition of the next layer. To permit the direct application of an electric field, the upper surface was finally metalized following the same procedure used for the lower electrode. To avoid perturbations in the chain conformations due to film formation at thicknesses much smaller than the gyration radius, the thinnest film was prepared in bilayer (configuration B) following Guiselin's experiment [69]. Thick films (~200 nm) of PS640, were spincoated (chloroform) and annealed at $T_g + 50$ K for 6 h. The unbound fraction was removed as described above. Films of l-PS (20 nm) were directly spincast on the residual film (7.5 nm) and metalized via the same procedure described above. The thickness of the different layers was evaluated via their geometrical capacitance, using nanocapacitors of single layers obtained by the same solution used in the multilayers under identical spincoating conditions.

During prolonged annealing in the liquid state, the dielectric strength decreases due to both rearrangements of chains adsorbed already during spincoating and metal evaporation (short times, ~segmental time), and adsorption of new chains (long times, $\sim\tau_D$). The two contributions provide different slopes in a plot of $\Delta\varepsilon$ versus log(t), see inset Fig. 9, and can be thus easily disentangled. This procedure is licit because of their different physical origin of the two contributions and their very large separation in timescale. The first logarithmic decay of the dielectric strength reflects increase of surface coverage, that is, the number of monomers irreversibly adsorbed per unit of substrate area. This process, mainly related to those chains already in direct contact with the aluminum oxide (at the nondiffusive interface), starts already before reaching the temperature at which experiments are performed and appears as a linear background in plots of $\Delta\varepsilon$ versus $\log(t)$. It is noteworthy that this time dependence resembles that of crystal perfectioning during secondary crystallization. Chains of PS already adsorbed on the metallic surface form an incomplete layer hindering the insertion of the labeled chains. The corresponding energetic barrier retards the diffusion of probe molecules in the last few nanometers of the matrix. When segments of l-PS finally reach the substrate, the contact with the metallic layer limits the solid angle over which dipole moments can reorient, which causes a drop in the mean square dipole moment and thus a lower $\Delta\varepsilon$.

Consequently, we could attribute the diffusion time, τ_D, to the onset of the larger reduction rate in the dielectric strength. To avoid systematic errors arising from variation of the sample size upon adsorption, we preferred analyzing the time evolution of $\zeta = \Delta\varepsilon/\varepsilon_\infty(\zeta_{\text{l-PS}} = 0.4;\ \zeta_{\text{PS}} = 0.016)$, where ε_∞ is the high frequency limit of the dielectric constant, a quantity free from uncertainties in the film thickness and the electrodes surface. Assuming a behavior following the laws of Fick, the tracer diffusion coefficient D_{tr} was estimated as $L^2 \times (4\tau_D)^{-1}$, where L is the thickness of the matrix. The approximation is based on the solution of Fick's equations in one dimension, considering L proportional to the diffusive length $(=2\sqrt{D_{\text{tr}}\tau_D})$. Tracer diffusivity experiments were performed at $T_g + 20\,\text{K}$, where the adsorption kinetics of PS is not relevant within the measurement time. Reduction of the diffusive length does not affect the sensitivity of the measurement, because of the inverse proportionality between the thickness of the matrix and its electric capacitance. However, a lower contrast between the value of the dielectric strength before and after chain adsorption is expected upon increase in the thickness of the matrix; regardless this issue, due to the large value of $\Delta\varepsilon_{\text{l-PS}}/\Delta\varepsilon_{\text{PS}}$ we could promptly assign the value of τ_D for films as thick as 230 nm.

In Fig. 10 we plotted the thickness dependence of τ_D and the corresponding variation in tracer diffusivity. Constant values of D_{tr} are observed for thicknesses larger than 100–120 nm (4–5 R_g, $D_{\text{tr}}^{\text{BULK}} = 5\ 10^{-15}\text{cm}^2/\text{s}$), while further reduction of the matrix size corresponded to a smooth drop in tracer diffusivity, which reached 5 % of the bulk value for films of the order of one macromolecular size. This tremendous reduction implies that in this thickness range the diffusion time was ~100 times longer than expected in the case of tracer diffusivity independent on the distance from the metallic oxide surface, that is, in the last 7 nm probe molecules diffuse at a rate two orders of magnitude lower than in bulk.

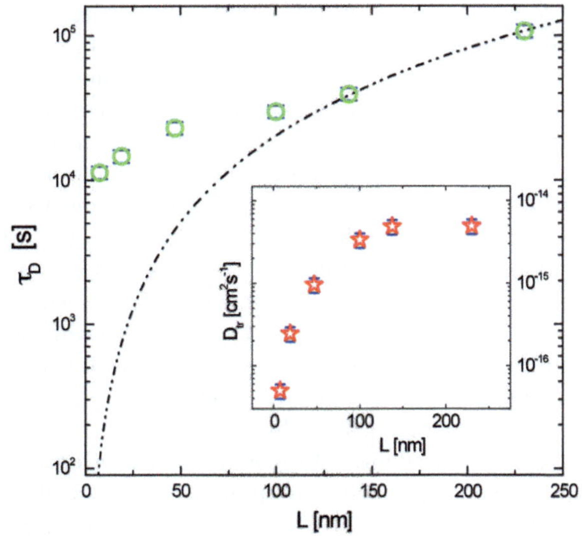

Fig. 10 Thickness dependence of the diffusion time of l-PS in matrices of PS640. The resulting tracer diffusivity is plotted in the inset. The dash-dotted line provides the value of the diffusion time in the case of position independent tracer diffusivity. Reproduced from Napolitano et al. [20] with permission from Wiley-VCH

3.1 On the Molecular Mechanisms Responsible of the Extreme Decoupling

What is the molecular origin of this phenomenon? The segmental time of the probe and the relaxation rate of the matrix did not show any significant thickness dependence (<2 K), in line with other investigations of thin films of PS via BDS and calorimetry [70] and with the minor shifts in T_g detected by Tress et al. in isolated coils of poly(2-vynil piridine) [71]. To get a quantitative estimation of the shift in T_g corresponding to the reduction in D_{tr} of Fig. 10, we considered that the temperature dependence of the tracer diffusivity of polystyrene can be described in terms of a Vogel Fulcher Tammann equation of the form [72]:

$$\text{In}\left(\frac{D_{tr}^{BULK}}{T}\right) = A - \frac{B}{T - T_0^{BULK}} \tag{11}$$

where A and B are constants and T_0 is a reference temperature. Assuming that the reduction in D_{tr} is a result in a perturbation of T_0 (constant fragility), we can express the thickness dependence of the tracer diffusivity as

$$\text{In}\left[\frac{D_{tr}(h)}{T} = A - \frac{B}{T - T_0(h)}\right] \tag{12}$$

An expression for $T_0(h)$ can be obtained subtracting Eq. (11) from Eq. (12) :

$$T_0(h) = T - \frac{B}{\text{In}\left[\frac{D_{\text{tr}}^{\text{BULK}}}{D_{\text{tr}}(L)}\right] + \frac{B}{T - T_0^{\text{BULK}}}} \tag{13}$$

We considered $B = 710$ K and $T_0 = 332$ K as determined by Green and Kramer [72]. At constant fragility (A and B constant) the ratio between T_g and T_0 is thickness independent, and the effective T_g finally reads:

$$T_g(h) = T_g^{\text{BULK}} \frac{T_0(h)}{T_0^{\text{BULK}}} \tag{14}$$

We obtained an increase in effective T_g of 8 K for 7 nm thin films, a result in contrast with the typical depressions observed for example by ellipsometry and fluorescent spectroscopy. This comparison brings us to a similar paradox as that remarked during crystallization experiments: in interfacial layers, the rotational barriers (τ) are almost unperturbed while the translational mobility of probes (D_{tr}) is highly suppressed.

The diffusion experiments provided further information on this intriguing behavior. We noticed a strong correlation between the thickness dependence of the T_g determined via tracer diffusivity and that of T_g probed by capacitive dilatometry [73], see Fig. 11. The last technique senses the temperature dependence of the dielectric constant in absence of molecular mobility and is thus couple to density fluctuations providing variations in the electric polarizability. Moreover, we observed a shift of the structural peak toward lower frequencies during diffusion experiments, corresponding to an increase by more than 10 folds of the relaxation time of the probes as approach the metallic interface. In the next paragraph, we discussed these experimental evidences and the results obtained in crystallization experiments in relation to the impact of chain adsorption on the SE relation.

3.2 The Different Impact of Irreversible Adsorption on Rotational and Translational Motion

We propose that the origin of the paradox in mobility related to the severe breakdown of the SE upon confinement is due to the impact of interfacial adsorption on the behavior of the whole film. The different conformations assumed by chains in the adsorbed layer might not particularly affect molecular relaxations on the length scale of the cooperative motion (2–4 nm for the dynamic glass transition), but lowers the probability to find available space where diffusing in proximity of the interface. This phenomenon reduces the efficiency of the random walk, that is, it limits the mean square displacement per unit of time, and it consequently leads to a reduction of tracer diffusivity.

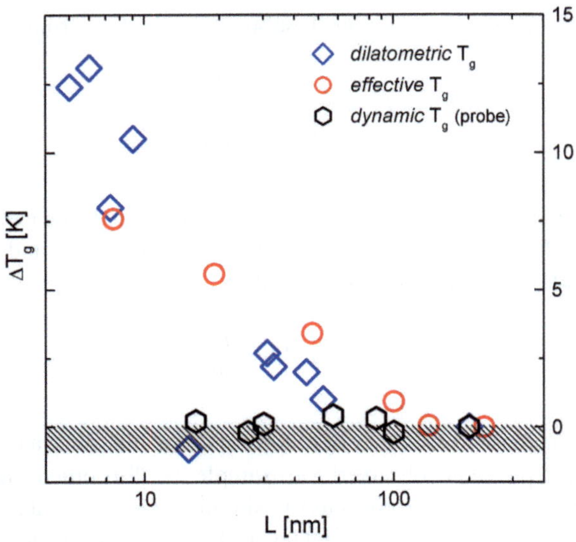

Fig. 11 Comparison between the increase in "effective" T_g of the PS (matrix) related to the measured reduction in tracer diffusivity, T_g obtained by capacitive dilatometry for single layers of PS160 capped between aluminum electrodes [73], and the "dynamic" T_g of single layers of 1-PS (probe) as deduced by the temperature dependence of the segmental mobility. The shadow area indicates the variation in the "dynamic" T_g of ultrathin film of PS of different molecular weight [70]. Errors are smaller than the symbol. Reproduced from Napolitano et al. [20] with permission from Wiley-VCH

A similar argument can help understanding an increase in t_{CRY} larger than the correlated shift in τ. In the case of thin films of PET, we performed further annealing experiments at 363 K where crystallization does not occur during the whole measurement time (~96 h) and the adsorption kinetics is slower, see Fig. 12. After 4 days of prolonged annealing the thickness h_{ads} of the adsorbed layer, obtained via Guiselin's experiment [69], reached values comparable to those found at 373 K after 10 min. We could fit the data at 363 K with an exponential curve of the type $h_{ads}(t) = h(0) + h_f[1 - \exp(-t/t_{ads})]$, where $h_f = 10.0 \pm 1.2$ nm, $t_{ads} \sim 28$ h and built up the dimensionless parameter $t^* = t/t_{ads}$, a useful probe of confinement effects. Regardless the annealing temperature, the onset of crystallization, for films thinner than 100 nm, corresponds to $t^* \gg 1$. In this regime, due to a potential barrier exerted by chains already immobilized onto the substrate, adsorption of new chains is accompanied by a severe entropic penalty [74, 75]. In fact, due to the lesser available space the number of possible conformations permitting adsorption at this stage is strongly limited.

Consequently, at $t^* \gg 1$, the crystallization of chains in the irreversibly adsorbed layer is highly hindered. Formation of crystals provides a remarkable enthalpy gain, but ordering via chain folding requires an initial entropic loss that cannot be compensated by the adsorbed chains. The entropic barrier of stem formation increases due to the restriction in conformational degrees of freedom upon chain immobilization.

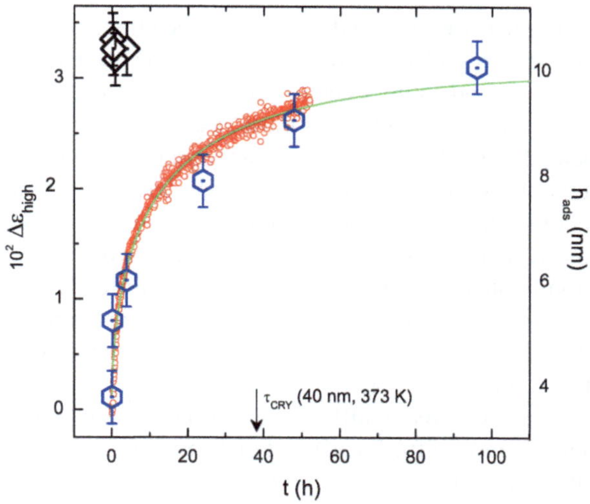

Fig. 12 Time evolution of the component of the dielectric strength with higher decaying rate, $\Delta\varepsilon_{high}$ (*red circles, left axis*), and the thickness of the irreversibly adsorbed layer, hads, obtained in the same annealing conditions (*blue hexagons, right axis*) at 363 K. Value of h_{ads} collected under the isothermal crystallization conditions, 373 K, are shown for comparison (*black diamonds, right axis*). Reproduced from Vanroy et al. [19] with permission of the American Chemical Society

Thus for those systems where the onset of crystallization occurs in the regime $t^* \gg 1$ (e.g., for low M_w polymers), we expect a severe reduction in the overall crystallization rate at thicknesses comparable to that of the irreversibly adsorbed layer grown during the experiment, $\sim h_f$. Considering that in our capped samples we have two adsorbing interfaces, we expect the occurrence of the huge drop crystallization rate at $2h_f$ Our hypothesis is validated considering that $2h_f \approx h^*$, that is, the thickness below which we did not observe crystallization during our experimental timescale.

4 Summary, Conclusions, Outlook

In this chapter, we revised recent work on the tremendous slowing down in crystallization kinetics and the severe drop in tracer diffusivity observed in ultrathin polymer films. This experimental evidence correspond to an anomalous decoupling of rotational and translation motion in polymer chains under 1D confinement. While diffusion and viscosity are intrinsically correlated via the SE relation in bulk melts at equilibirium, $D\eta = \text{const}/T$, the anomalous decoupling in thin films follows a trend opposite the fractional SE model, characteristic of the glassy dynamics, where diffusion is enhanced in comparison to viscosity, that is $D\eta^{\xi} = \text{const}/T$ with $\xi < 1$.

We discussed on the impact of finite size and interfacial interactions on the crystallization rate, considering that this quantity is related to the convolution of nucleation and growth. In particular, we showed that for Al capped films of PET thicker than a threshold value h^*, the crystallization time scales with the inverse of the sample volume, that is $t_{CRY} \sim h^{-1}$, indicating a nucleation limited regime. This trend can be, in fact, rationalized considering the effect of finite size on a phase transition occurring at a constant nuclei density. Below h^* we did not observe any crystallization event, which implies inhibition of the crystallization or a drop in crystallization rate by more than three orders of magnitude. We estimated the increase in t_{CRY}, correlated to the slower segmental dynamics observed at $h < h^*$, via the SE model and combined it to the nucleation limited trend. Consequently, we obtained Eq. (10) permitting to estimate the thickness dependence of t_{CRY}. At constant temperature, this expression can be approximated as $t_{CRY}(h)\alpha h[\tau(h)]^{\xi}$. Under this framework, a large increase in t_{CRY} at constant τ implies $\xi \gg 1$.

A similar trend was observed in diffusion experiments, that permitted us to measure the coefficient of tracer diffusion, D_{tr}, of dielectric probes in polymer matrices, by means of BDS. Our method is based on the direct determination of the time necessary to probe molecules to diffuse via the matrix, before being finally adsorbed on a target substrate. Under these conditions, at a reduction of the matrix thickness, L, should correspond short diffusion times $\tau_D \sim L^{-2}$. We observed values of τ_D exceeding those expected at constant tracer diffusivity, that is, a reduction in D_{tr}. Same as for crystallization experiments, the increase in barrier for translatinal motion was accompained by an almost constant τ, and thus $\xi \gg 1$.

We rationalized this peculiar behavior considering the different impact of irreversible chain adsorption on the molecular mechanisms responsible for segmental mobility and diffusion. We propose that this anomalous decoupling could be related to the content and distribution of free volume in proximity of free surfaces [76] and adsorbing interfaces [48]. Recent experimental evidence and theoretical models support, in fact, the idea that the deviation from bulk behavior are strictly connected to the local content in free volume. Under this framework, a reduction in T_g should be related to an increase in packing frustration [1], e. g. as in the first stages of adsorption, and the consequent larger value of free volume compared to the bulk [15]. This hypothesis is in line with recent simulations by Tito et al. showing that the glassy dynamics is affected by the propagation of colonies of free volume [77, 78]. Free volume can, in fact, be transported inside a film, via a mechanism of propagation of "kinks" along loops formed on the chains, as proposed by de Gennes [79]. The size of the reduction in T_g observed in thin films is quantitatively reproduced by this mechanism [80]. Inhibition of the kink-propagation-mechanism yields an increase of T_g, as shown by Glynos et al. in high-functionality and low arms-molecular-weight star-shaped polymers [81]. Finally Cangialosi and coworkers [13, 82–86] demonstrated, by means of the free volume holes diffusion model, that the shift in T_g of thin films and polymer-based nanocomposites is univocally correlated to the effective surface available for diffusion of free volume holes.

While τ and T_g should be related to the free volume content, we might imagine that the barrier for translational motions should be related to the spatial distribution of free volume, that is, to the effective tortuosity for the diffusion of free volume holes. The experimental evidence described in this chapter could, for example, be explained considering interfacial layers with free volume content as in bulk (constant τ), but a distribution of free volume yielding larger tortuosity, and thus bigger values of τ_D. Future work addressing this hypothesis could solve the molecular origin of the anomalous decoupling of the fundamental relation between translation and rotational motion in polymers confined at the nanoscale level.

Acknowledgments S.N. acknowledges financial support from the funds FER of the Université Libre de Bruxelles. M.W. acknowledge financial support from the Research Council of the KU Leuven, Project No. OT/11/065, and financial support from FWO (Fonds Wetenschappelijk Onderzoeks-Vlaanderen), Project G.0642.08.

References

1. Napolitano S, Capponi S, Vanroy B (2013) Glassy dynamics of soft matter under 1D confinement: how irreversible adsorption affects molecular packing, mobility gradients and orientational polarization in thin films. Eur Phys J E 36:61
2. Adam G, Gibbs JH (1965) On temperature dependence of cooperative relaxation properties in glass-forming liquids. J Chem Phys 43:139
3. Watanabe K, Kawasaki T, Tanaka H (2011) Structural origin of enhanced slow dynamics near a wall in glass-forming systems. Nat Mater 10:512–520
4. Arndt M, Stannarius R, Groothues H, Hempel E, Kremer F (1997) Length scale of cooperativity in the dynamic glass transition. Phys Rev Lett 79:2077–2080
5. Napolitano S, Wubbenhorst M (2011) The lifetime of the deviations from bulk behaviour in polymers confined at the nanoscale. Nat Commun 2:260
6. Koga T, Jiang N, Gin P, Endoh MK, Narayanan S, Lurio LB, Sinha SK (2011) Impact of an irreversibly adsorbed layer on local viscosity of nanoconfined polymer melts. Phys Rev Lett 107:225901
7. Gin P, Jiang N, Liang C, Taniguchi T, Akgun B, Satija SK, Endoh MK, Koga T (2012) Revealed architectures of adsorbed polymer chains at solid-polymer melt interfaces. Phys Rev Lett 109:265501. doi:10.1103/PhysRevLett.109.265501
8. Nguyen HK, Labardi M, Lucchesi M, Rolla P, Prevosto D (2013) Plasticization in ultrathin polymer films: the role of supporting substrate and annealing. Macromolecules 46:555–561
9. Nguyen HK, Labardi M, Capaccioli S, Lucchesi M, Rolla P, Prevosto D (2012) Prevosto Interfacial and annealing effects on primary α-relaxation of ultrathin polymer films investigated at nanoscale. Macromolecules 45:2138–2144
10. Krutyeva M, Wischnewski A, Monkenbusch M, Willner L, Maiz J, Mijangos C, Arbe A, Colmenero J, Radulescu A, Holderer O, Ohl M, Richter D (2013) Effect of nanoconfinement on polymer dynamics: surface layers and interphases. Phys Rev Lett 110:108303
11. Napolitano S, Rotella C, Wubbenhorst M (2012) Can thickness and interfacial interactions univocally determine the behavior of polymers confined at the nanoscale? Acs Macro Lett 1:1189–1193
12. Reiter G, Napolitano S (2010) Possible origin of thickness-dependent deviations from bulk properties of thin polymer films. J Polym Sci Part B-Polym Phys 48:2544–2547
13. Napolitano S, Cangialosi D (2013) Interfacial free volume and vitrification: reduction in tg in proximity of an adsorbing interface explained by the free volume holes diffusion model. Macromolecules 46:8051–8053
14. Fukao K, Miyamoto Y (2000) Glass transitions and dynamics in thin polymer films: dielectric relaxation of thin films of polystyrene. Phys Rev E 61:1743–1754
15. Napolitano S, Pilleri A, Rolla P, Wubbenhorst M (2010) Unusual deviations from bulk behavior in ultrathin films of poly(tert-butylstyrene): can dead layers induce a reduction of t-g? ACS Nano 4:841–848

16. Lu HY, Chen W, Russell TP (2009) Relaxation of thin films of polystyrene floating on ionic liquid surface. Macromolecules 42:9111–9117
17. Thomas KR, Chenneviere A, Reiter G, Steiner U (2011) Nonequilibrium behavior of thin polymer films. Phys Rev E 83:021804
18. Barbero DR, Steiner U (2009) Nonequilibrium polymer rheology in spin-cast films. Phys Rev Lett 102:248303
19. Vanroy B, Wubbenhorst M, Napolitano S (2013) Crystallization of thin polymer layers confined between two adsorbing walls. Acs Macro Lett 2:168–172
20. Napolitano S, Rotella C, Wuebbenhorst M (2011) Is the reduction in tracer diffusivity under nanoscopic confinement related to a frustrated segmental mobility? Macromol Rapid Commun 32:844–848
21. Hu W (2013) Polymer physics. Springer, Vienna
22. Grigoriadis C, Duran H, Steinhart M, Kappl M, Butt HJ, Floudas G (2011) Suppression of phase transitions in a confined rodlike liquid crystal. ACS Nano 5:9208–9215
23. Liu YX, Chen EQ (2010) Polymer crystallization of ultrathin films on solid substrates. Coord Chem Rev 254:1011–1037
24. Frank CW, Rao V, Despotopoulou MM, Pease RFW, Hinsberg WD, Miller RD, Rabolt JF (1996) Structure in thin and ultrathin spin-cast polymer films. Science 273:912–915
25. Zhang Y, Lu YL, Duan YX, Zhang JM, Yan SK, Shen DY (2004) Reflection-absorption infrared spectroscopy investigation of the crystallization kinetics of poly(ethylene terephthalate) ultrathin films. J PolymSci Part B Polym Phys 42:4440–4447
26. Nascimento MLF, Souza LA, Ferreira EB, Zanotto ED (2005) Can glass stability parameters infer glass forming ability? J Non-Cryst Solids 351:3296–3308
27. Capitan MJ, Rueda DR, Ezquerra TA (2004) Inhibition of the crystallization in nanofilms of poly(3-hydroxybutyrate). Macromolecules 37:5653–5659
28. Hu WB, Cai T, Ma Y, Hobbs JK, Farrance O, Reiter G (2009) Polymer crystallization under nano-confinement of droplets studied by molecular simulations. Faraday Discuss 143:129–141
29. Sommer JU, Reiter G (2000) Polymer crystallization in quasi-two dimensions. Ii. Kinetic models and computer simulations. J Chem Phys 112:4384–4393
30. Sommer JU, Reiter G (2005) Crystallization in ultra-thin polymer films-morphogenesis and thermodynamical aspects. Thermochim Acta 432:135–147
31. Schultz JM (1996) Effect of specimen thickness on crystallization rate. Macromolecules 29:3022–3024
32. Napolitano S, Wubbenhorst M (2006) Slowing down of the crystallization kinetics in ultrathin polymer films: a size or an interface effect? Macromolecules 39:5967–5970
33. Bottcher C (1973) Theory of dielectric polarization. Elsevier, Amsterdam
34. Capponi S, Napolitano S, Wubbenhorst M (2012) Supercooled liquids with enhanced orientational order. Nat Commun 3:1233
35. Massalska-Arodz M, Williams G, Thomas DK, Jones WJ, Dabrowski R (1999) Molecular dynamics and crystallization behavior of chiral isooctyloxycyanobiphenyl as studied by dielectric relaxation spectroscopy. J Phys Chem B 103:4197–4205
36. Nogales A, Denchev Z, Sics I, Ezquerra TA (2000) Influence of the crystalline structure in the segmental mobility of semicrystalline polymers: poly(ethylene, naphthalene-2,6-dicarboxylate). Macromolecules 33:9367–9375
37. Ezquerra TA, Majszczyk J, Baltacalleja FJ, Lopezcabarcos E, Gardner KH, Hsiao BS (1994) Molecular-dynamics of polymers during crystallization as revealed by dielectric-spectroscopy. Phys Scripta 55:212–215
38. Ezquerra TA, Liu F, Boyd RH, Hsiao BS (1997) Crystallization of poly(aryl ether ketone) polymers as revealed by time domain dielectric spectroscopy. Polymer 38:5793–5800
39. Alvarez C, Sics I, Nogales A, Denchev Z, Funari SS, Ezquerra TA (2004) Structure-dynamics relationship in crystallizing poly(ethylene terephthalate) as revealed by time-resolved x-ray and dielectric methods. Polymer 45:3953–3959
40. Fukao K, Miyamoto M (1997) Dynamical transition and crystallization of polymers. Phys Rev Lett 79:4613–4616

41. Napolitano S, Wubbenhorst M (2007) Deviation from bulk behaviour in the cold crystallization kinetics of ultrathin films of poly(3-hydroxybutyrate). J Phys Condens Matter 19:205121
42. Napolitano S, Wubbenhorst M (2007) Effect of a reduced mobility layer on the interplay between molecular relaxations and diffusion-limited crystallization rate in ultrathin polymer films. J Phys Chem B 111:5775–5780
43. Napolitano S, Wubbenhorst M (2007) Monitoring the cold crystallization of poly(3-hydroxy butyrate) via dielectric spectroscopy. J Non-Cryst Solids 353:4357–4361
44. Bebin P, Prud'homme RE (2003) Comparative xps study of copper, nickel, and aluminum coatings on polymer surfaces. Chem Mater 15:965–973
45. Rotella C, Napolitano S, De Cremer L, Koeckelberghs G, Wubbenhorst M (2010) Distribution of segmental mobility in ultrathin polymer films. Macromolecules 43:8686–8691
46. Avrami M (1940) Kinetics of phase change. Ii transformation-time relations for random distribution of nuclei. J Chem Phys 8:212
47. Rotella C, Napolitano S, Vandendriessche S, Valev VK, Verbiest T, Larkowska M, Kucharski S, Wubbenhorst M (2011) Adsorption kinetics of ultrathin polymer films in the melt probed by dielectric spectroscopy and second-harmonic generation. Langmuir 27:13533–13538
48. Rotella C, Wubbenhorst M, Napolitano S (2011) Probing interfacial mobility profiles via the impact of nanoscopic confinement on the strength of the dynamic glass transition. Soft Matter 7:5260–5266
49. Yin H, Napolitano S, Schoenhals A (2012) Molecular mobility and glass transition of thin films of poly(bisphenol a carbonate). Macromolecules 45:1652–1662
50. Granick S (2002) Perspective: kinetic and mechanical properties of adsorbed polymer layers. Eur Phys J E 9:421–424
51. Duran H, Steinhart M, Butt HJ, Floudas G (2011) From heterogeneous to homogeneous nucleation of isotactic poly(propylene) confined to nanoporous alumina. Nano Lett 11:1671–1675
52. Massa MV, Carvalho JL, Dalnoki-Veress K (2006) Confinement effects in polymer crystal nucleation from bulk to few-chain systems. Phys Rev Lett 97:247802
53. Massa MV, Dalnoki-Veress K (2004) Homogeneous crystallization of poly(ethylene oxide) confined to droplets: the dependence of the crystal nucleation rate on length scale and temperature. Phys Rev Lett 92:255509
54. Mapes MK, Swallen SF, Ediger MD (2006) Ediger Self-diffusion of supercooled o-terphenyl near the glass transition temperature. J Phys Chem B 110:507–511
55. Ngai KL, Magill JH, Plazek DJ (2000) Flow, diffusion and crystallization of supercooled liquids: revisited. J Chem Phys 112:1887–1892
56. Ngai KL (1999) Alternative explanation of the difference between translational diffusion and rotational diffusion in supercooled liquids. J Phys Chem B 103:10684–10694
57. Hall DB, Dhinojwala A, Torkelson JM (1997) Translation-rotation paradox for diffusion in glass-forming polymers: the role of the temperature dependence of the relaxation time distribution. Phys Rev Lett 79:103–106
58. Napolitano S, Prevosto D, Lucchesi M, Pingue P, D'Acunto M, Rolla P (2007) Influence of a reduced mobility layer on the structural relaxation dynamics of aluminum capped ultrathin films of poly(ethylene terephthalate). Langmuir 23:2103–2109
59. Palys LH, Phillips PJ (1980) Microkinetics of crystallization of poly(ethylene terephthalate). J Polym Sci Part B Polym Phys 18:829–852
60. Zhang Y, Zhang JM, Lu YL, Duan YX, Yan SK, Shen DY (2004) Glass transition temperature determination of poly(ethylene terephthalate) thin films using reflection-absorption FTIR. Macromolecules. 37:2532–2537
61. Mapesa EU, Tress M, Schulz G, Huth H, Schick C, Reiche M, Kremer F (2013) Segmental and chain dynamics in nanometric layers of poly(cis-1,4-isoprene) as studied by broadband dielectric spectroscopy and temperature-modulated calorimetry. Soft Matter 9:10592–10598
62. Koga T, Li C, Endoh MK, Koo J, Rafailovich M, Narayanan S, Lee DR, Lurio LB, Sinha SK (2010) Reduced viscosity of the free surface in entangled polymer melt films. Phys Rev Lett 104:066101

63. Forrest JA (2002) A decade of dynamics in thin films of polystyrene: where are we now? Eur Phys J E 8:261–266
64. Zheng X, Rafailovich MH, Sokolov J, Strzhemechny Y, Schwarz SA, Sauer BB, Rubinstein M (1997) Long-range effects on polymer diffusion induced by a bounding interface. Phys Rev Lett 79:241–244
65. Forrest JA, Dalnoki-Veress K, Stevens JR, Dutcher JR (1996) Effect of free surfaces on the glass transition temperature of thin polymer films. Phys Rev Lett 77:2002–2005
66. Rotella C, Napolitano S, Wuebbenhorst M (2009) Segmental mobility and glass transition temperature of freely suspended ultrathin polymer membranes. Macromolecules 42: 1415–1417
67. Pu Y, White H, Rafailovich MH, Sokolov J, Patel A, White C, Wu WL, Zaitsev V, Schwarz SA (2001) Probe diffusion in thin ps free-standing films. Macromolecules 34:8518–8522
68. Hall DB, Torkelson JM (1998) Small molecule probe diffusion in thin and ultrathin supported polymer films. Macromolecules 31:8817–8825
69. Guiselin O (1991) Irreversible adsorption of a concentrated polymer solution. Europh Lett 17:225–230
70. Tress M, Erber M, Mapesa EU, Huth H, Muller J, Serghei A, Schick C, Eichhorn K-J, Voit B, Kremer F (2010) Glassy dynamics and glass transition in nanometric thin layers of polystyrene. Macromolecules 43:9937–9944
71. Tress M, Mapesa EU, Kossack W, Kipnusu WK, Reiche M, Kremer F (2013) Glassy dynamics in condensed isolated polymer chains. Science 341:1371–1374
72. Green P, Kramer E (1986) Temperature dependence of tracer diffusion coefficients in polystyrene. J Mater Res 202–204
73. Napolitano S, Wubbenhorst M (2007) Dielectric signature of a dead layer in ultrathin films of a nonpolar polymer. J Phys Chem B 111:9197–9199
74. Ligoure C, Leibler L (1990) Thermodynamics and kinetics of grafting end-functionalized polymers to an interface. J De Phys 51:1313–1328
75. Zajac R, Chakrabarti A (1995) Irreversible polymer adsorption from semidilute and moderately dense solutions. Phys Rev E 52:6536–6549
76. Napolitano S, Wubbenhorst M (2010) Structural relaxation and dynamic fragility of freely standing polymer films. Polymer. 51:5309–5312
77. Tito NB, Lipson JEG, Milner ST (2013) Lattice model of dynamic heterogeneity and kinetic arrest in glass-forming liquids. Soft Matter. 9:3173–3180
78. Tito NB, Lipson JEG, Milner ST (2013) Lattice model of mobility at interfaces: free surfaces, substrates, and bilayers. Soft Matter 9:9403–9413
79. de Gennes PG (2000) Glass transitions in thin polymer films. Eur Phys J E 2:201–203
80. Lipson JEG, Milner ST (2009) Percolation model of interfacial effects in polymeric glasses. Eur Phys J B 72:133–137
81. Glynos E, Frieberg B, Oh H, Liu M, Gidley DW, Green PF (2011) Role of molecular architecture on the vitrification of polymer thin films. Phys Rev Lett 106:128301
82. Boucher VM, Cangialosi D, Alegria A, Colmenero J, Pastoriza-Santos I, Liz-Marzan LM (2011) Physical aging of polystyrene/gold nanocomposites and its relation to the calorimetric t-g depression. Soft Matter 7:3607–3620
83. Cangialosi D, Boucher VM, Alegria A, Colmenero J (2011) Free volume holes diffusion to describe physical aging in poly(mehtyl methacrylate)/silica nanocomposites. J Chem Phys 135:014901
84. Boucher VM, Cangialosi D, Alegria A, Colmenero J (2012) Enthalpy recovery in nanometer to micrometer thick polystyrene films. Macromolecules 45:5296–5306
85. Boucher VM, Cangialosi D, Yin HJ, Schonhals A, Alegria A, Colmenero J (2012) T-g depression and invariant segmental dynamics in polystyrene thin films. Soft Matter 8:5119–5122
86. Cangialosi D, Boucher VM, Alegria A, Colmenero J (2012) Enhanced physical aging of polymer nanocomposites: the key role of the area to volume ratio. Polymer 53:1362–1372

Dynamic Calorimetric Glass Transition in Thin Polymer Films

Christoph Schick

Abstract Calorimetry is an effective analytical tool to characterize the glass transition under confinement. The step in heat capacity at the dynamic glass transition in nanometer thin films can effectively be measured with low addenda heat capacity chip calorimeters. Since AC-calorimetry measures a compliance (C_p) in the linear response regime it probes molecular dynamics in the super-cooled liquid state. Down to a few nanometers of film thickness, AC-calorimetry always reveals bulk dynamics in the vicinity of the dynamic glass transition. This is different from experiments probing the transition from the glassy to the liquid state where liquid-like layers at the surface of the glass or free volume diffusion in the glass may influence the observed behavior. Temperature modulated or AC-calorimetry offers a broad dynamic range. Today 11 orders of magnitude in frequency are covered for bulk samples by combining various kinds of calorimeters. This broad dynamic range, akin to broadband dielectric spectroscopy, is significantly important for the study of confinement effects. Varying the frequency allows to change the corresponding length scale of the probed segmental dynamics. In a relaxation map, again, bulk behavior is observed for all investigated films down to a few nanometer film thickness.

Keywords Calorimetry · Temperature modulation (AC) · Glass transition · Confinement · Thin film

Abbreviations

ACDSC	Alternating Current Differential Scanning Calorimetry
CRR	Cooperative Rearranging Region
DSC	Differential Scanning Calorimetry

C. Schick (✉)
Institute of Physics, University of Rostock, Wismarsche Str. 43-45, 18051 Rostock, Germany
e-mail: christoph.schick@uni-rostock.de

F. Kremer (ed.), *Dynamics in Geometrical Confinement*, Advances in Dielectrics, 307
DOI: 10.1007/978-3-319-06100-9_12,

i-PMMA	Isotactic Poly(methyl methacrylate)
PPO	Poly(propylene oxide)
PS	Polystyrene
PS/PPO	*Polystyrene/ poly(propylene oxide)
PVAc	Poly(vinyl acetate)
PVME/PS	Poly(vinyl methyl ether)/ polystyrene
P2VP	Poly(2-vinylpyridine)
TMDSC	Temperature Modulated Differential Scanning Calorimetry
VFTH	Vogel-Fulcher-Tammann-Hesse

1 Introduction

It is now proven that calorimetry can provide useful information about glass transition and its related phenomena, like enthalpy relaxation. Three kinds of experiments are usually employed in calorimetry: (i) stepwise increasing temperature, e.g., adiabatic calorimetry, (ii) scanning at constant or variable rate, e.g., differential scanning calorimetry (DSC), and (iii) periodic temperature oscillations, e.g., alternating current (AC) calorimetry or temperature modulated DSC (TMDSC). The relation between the scanning rate and the timescale of the molecular processes under study at the glass transition is not easily established and controversially discussed [1–3]. For AC-calorimetric measurements, the temperature oscillation's frequency can be linked in a straight-forward manner to the dynamics of the molecular processes being investigated. At the dynamic glass transition the relation $\omega\tau = 1$ holds for the maximum of the frequency dependent imaginary part of complex heat capacity (where $\omega = 2\pi f$ is the angular frequency of the perturbation and τ is the mean relaxation time). Heat capacity in the glass transition regime is related to entropy fluctuations by $c_p = k\overline{\Delta S^2}$ [1]. Therefore, only relaxation processes which are linked to entropy fluctuations appear in calorimetric curves. This, generally speaking, is true for the segmental relaxation (dynamic glass transition) but commonly not for secondary relaxation processes. For exceptions from this rule see, e.g., [4] and references therein. Given that the glass transition has a more pronounced calorimetric signature compared to secondary relaxation processes, calorimetry—unlike dielectric spectroscopy—makes it easy to distinguish these processes [5–7]. Therefore, calorimetric studies in a broad frequency range are poised to avail invaluable knowledge to the on-going discussion about the glass transition in confinement.

Using microemulsions, Angell et al. [8, 9] performed one of the first investigations on nanoconfined glass forming systems. Amazingly, the authors did not find significant differences between bulk and droplet T_g, even for droplet sizes in the range of only 2–6 nm. They could only observe a broadening of the glass transition. Confinement effects on the glass transition were identified to cause a similar broadening of the calorimetric glass transition in amorphous layers, which were confined between neighboring crystal lamellae in [10]. Jackson and McKenna [11] first reported from calorimetric studies how the glass transition temperature T_g was influenced through nano-confinement by a porous matrix. The seminal heat capacity measurements of polymeric thin films were reported by Efremov [12] employing an ultra-fast scanning calorimeter developed by Allen et al. [13]. Even down to 2 nm of film thickness, no shift of the calorimetric glass transition temperature could be observed in this pioneering work. The calorimeter used is operated under basically adiabatic conditions. Therefore, controlled cooling at constant rate is not possible as would be preferred in glass transition studies in order to obtain glasses with well-defined history, see [3] for a detailed exposition. To overcome this limitation and to carry out experiments at longer times in order to follow the relaxation processes below the glass transition after cooling, Efremov performed annealing experiments [14]. Recently, these results were compared with variable cooling rate ellipsometric measurements [15]. It was claimed that a change from Vogel-Fulcher-Tammann-Hesse (VFTH) to Arrhenius-like behavior takes place in dependence on the cooling rate applied in ellipsometric measurements; however, no shifts— with decreasing film thickness— in the calorimetric glass transition were observed after long annealing. In [15], it is held that the calorimetric measurements are performed at moderate to high rates. This view was supported by Simon et al. with fast scanning calorimetric measurements on thin polystyrene films [16]. Further on Cangialosi et al. [17–19] applied the free volumes holes diffusion model [20–22] to successfully describe these data. Details of the model are available in this book from the contribution by Cangialosi et al. [23]. The main assumption there is a film thickness independent segmental mobility as observed in many dielectric and AC-calorimetric studies on thin polymer films. Such dielectric data are reviewed in this book in the chapters by Kremer et al. [24] and Schönhals et al. [25]. Ediger and Forrest [26] review a large number of different findings in support of a mobile surface layer of a few nanometers in glassy polymers and also assume bulk segmental mobility in the core of the film. Since the mobile layer disappears for the supercooled liquid state bulk behavior is expected above the conventional T_g. In these both attempts describing the glass transition behavior of thin films bulk-like segmental mobility is assumed even for nanometer thin films. Evidence for such film thickness independent segmental mobility was obtained from AC-calorimetric data too and will be summarized in this contribution.

Here, we describe how to obtain information about glass transition dynamics by AC-calorimetry in a wide frequency range. The chapter is structured as follows: First, the AC-calorimetric technique for investigation of thin films is described. Then results on polymeric and nonpolymeric thin films in the glass transition interval are summarized. Finally, an outlook is provided.

2 Calorimetry

2.1 Basic Considerations

In many studies probing the effects of confinement, most attention is paid to effects on the glass transition temperature T_g. Nonetheless, using calorimetry, one is able to determine other useful quantities in addition to the T_g. From the absolute data of c_p, which is no-doubt the most valuable outcome of calorimetric experiments at the glass transition, one can obtain (i) the step height Δc_p (relaxation strength), (ii) the width of the transition interval δT, and (iii) the T_g. For the case of complex heat capacity, the determination of these quantities is illustrated in Fig. 1. There the situation is quite simple. The relaxation process is described, like other compliances, by the maximum (T_g) and the width (δT) of the peak in the imaginary part, as well as the step height of the step of the real part (Δc_p), which is related to the area under the peak in the imaginary part. The width of the transition interval $2\delta T$ is usually found as the width of a Gaussian function, which fits the imaginary part or the phase angle or the temperature derivative of the real part versus temperature. The case where the transition from the supercooled liquid to the nonequilibrium glassy state is measured (e.g., total heat capacity by DSC or adiabatic calorimetry)—as opposed to studying the relaxation process in the supercooled liquid via measurement of complex heat capacity—is more complex and the glass transition is described using different quantities, see e.g., [3]. The concept of limiting fictive temperature, T_f, as introduced by Tool [27] and developed further by Moynihan [28] is oft preferred and applied because it has a sound basis.

Figure 1 displays a typical example, for polystyrene, of specific complex heat capacity ($c_p^*(T, w) = c_p'(T, w) - ic_p''(T, w)$) at the dynamic glass transition and the total specific heat capacity ($c_{p\,\mathrm{total}}(T, \beta_0) = 1/m \, {}^{dH(T,\beta_0)}/_{dT}$, with β_0 the underlying scanning rate and H the enthalpy) describing the transformation from the supercooled liquid to the glassy state.

The half step temperature of c'_p coincides, as usual with the maximum of $c''_{p.}$ The position of both quantities depends on the measurement frequency, like other compliances. The Vogel-Fulcher-Tammann-Hesse (VFTH) function can well describe it, see Fig. 15. The step in $c_{p\,\mathrm{total}}$, related to the change from the supercooled liquid to the nonequilibrium glassy state on cooling, is rate and thermal history-dependent [30]. Compared to the relaxation process in the supercooled liquid, as measured by complex heat capacity, the step in $c_{p\,\mathrm{total}}$ is due to the increasing deviation from equilibrium (as reviewed in [3]).

From both the complex and total heat capacities, the strength of the relaxation process is obtained from the height of the step in heat capacity Δc_p. The importance of this quantity is underscored by the fact that it counts the degrees of freedom as the system goes from a relaxed to a frozen state upon cooling though the glass transition. Since liquid and the glassy heat capacity curves have different slopes Δc_p is temperature-(and frequency, scanning rate) dependent. Assuming the concept of

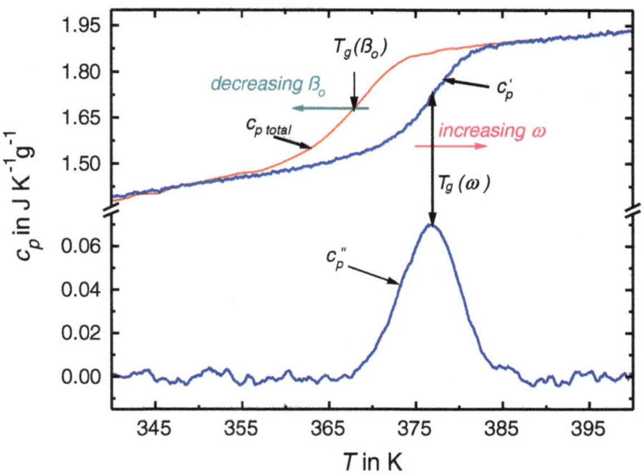

Fig. 1 Real and imaginary part of complex specific heat capacity for polystyrene measured by TMDSC ($c'_p(\omega)$, $c''_p(\omega)$) and total heat capacity from the underlying linear scan at cooling rate β_0 ($c_{p\ \mathrm{total}}(T, \beta_0)$). Measuring conditions: Temperature modulation amplitude $A_T = 0.5$ K, angular frequency $\omega = 2\pi/P$, period $P = 60$ s, underlying scanning rate $\beta_0 = -0.25$ K min^{-1}. All data are taken from Ref. [29]

cooperative rearranging regions (CRR), as presented by Adam and Gibbs [31], it is possible to estimate the fraction of the confined material participating in the glass transition on the level of the CRR size. Here, it is taken that a CRR contributes either to the glassy or the liquid heat capacity. This permits the identification of immobilized or liquid-like layers in nanoconfined systems at the common glass transition, see e.g., [32]. In addition, the size of a CRR can be estimated from the parameters of the glass transition by an equation derived by Donth [33]. For such estimates, which require more information about confinement beyond just the T_g value, appropriate determination of heat capacity is inevitable.

Adiabatic or alternating current (AC) calorimeters perform precise measurements of heat capacity at low temperatures. At temperatures below room temperature, uncertainties less than 1 % are reached. At higher temperature or for low thermal conducting materials or if heat capacity becomes time dependent, some extra problems emerge, increasing significantly the experimental errors in heat capacity measurements.

Differential scanning calorimetry (DSC) at linear scanning has proven as a versatile technique for obtaining, in a reasonably short time, heat capacities at elevated temperatures. DSC also avails the possibility to study—in a wide dynamic range—the kinetics of transitions. Because of its simplicity and ease of use, DSC finds wide application in materials science [34–39]. As already briefly discussed in the Introduction, applying a periodic power to generate a periodic temperature oscillation has some advantages, not just because of increased accuracy in the heat capacity determination, but also due to the chance to measure the frequency dependence of the complex heat capacity. The three techniques already mentioned are briefly discussed below.

A majority of studies concerning the effects of confinement on the glass transition are collected on ultra-thin films (<100 nm). However, because of the diminished mass of a single film, DSC experiments are not easy to perform on ultra-thin films. A few successful attempts have been achieved by Wang and Zhou [40], Simon et al. [41, 42], Fukao et al. [43, 44], and Cangialosi et al. [18]. A stack of many of such films allows to increase sample mass to an appropriate level. TMDSC (2.2.1) or conventional DSC was applied to investigate such stacked ultra-thin films. A T_g depression with decreasing film thickness was recorded. Surprizingly, recovery of the bulk behavior was not realized from several heating-cooling runs, but only by applying some pressure [41, 42] or annealing at elevated temperatures [44].

Allen et al. described a calorimetric scanning technique for the measurement of the glass transition in single ultra-thin supported films [12, 14]. In spite of its superior sensitivity for glass transition measurements, the technique has a serious drawback: the calorimeter is operated under quasi-adiabatic conditions. Cooling can hence not be well controlled. However, as earlier mentioned, knowledge of the sample thermal history is paramount for glass transition studies. Since this is also the case for crystallization studies, and not only for the investigation of the glass transition, a versatile non-adiabatic thin film calorimeter was developed [45–48]. Here, the sensor is operated in gas, instead of vacuum, as surrounding medium, an innovation which yields sufficient heat losses for samples in the sub-nanogram range for cooling rates up to 1 $MK\,s^{-1}$. Only a small area of the heated membrane (ca. 10 μm in diameter) can ensure an effective enough fast cooling. Because of the small heated area, there exist non-negligible temperature gradients outside the heater area [49, 50]. Therefore, reliable measurements cannot be carried out when the whole membrane area is covered by the spin-coated sample. This is different from the large area sensors employed by Allen et al. [12, 14, 51] and Rodríguez-Viejo et al. [52–54] where the edge effects can either be neglected or avoided by a deposition mask. Recently, Simon et al. [16] published fast scanning calorimetric data on supported thin polystyrene films using a Mettler Toledo Flash DSC 1 utilizing a medium-sized chip sensor (500 μm diameter of the active area) [55, 56].

Sample preparation is always a critical aspect in calorimetry, like in all other techniques applied to the study of glass transition under nanometric confinement. Because of the large surface area of nanosized objects, interaction of the sample with nanoparticles, a porous host matrix, substrates for thin films, etc. may result in substantial alterations of the behavior at glass transition. Often a surface layer with increased mobility [26, 57] or a dead layer [58] arising from strong interfacial interactions are discussed. Furthermore, these interfaces may be attractive to solvents or other substances which would influence the relaxation behavior. Residual stresses in the sample need to be carefully eliminated through drying and annealing above T_g [59, 60]. The scenario becomes more compounded if chemical reactions occur in the course sample preparation on the nanoscale. The complexity of the glass transition behavior of polycyanorates cured under nanoconfinement is discussed in [61, 62]. Large deviations in the range of ~60 K in 10 nm pores from the bulk calorimetric T_g value were observed for these particular systems. In summary, slight variations in sample preparation may result in significant changes in the glass transition behavior of the nanoconfined system, and demand careful consideration [44, 60, 63].

2.2 Temperature Modulation (TMDSC, AC-calorimetry)

For a long time now, oscillating power (heating rate, temperature) is used in calorimetry, see the review by Kraftmakher [64]. Periodic perturbations have found application in calorimetry since 1910 when Corbino [65, 66] made use of the so-called 3ω-method [67, 68] to deduce the heat capacity of electrically conducting wires. In the 1960s, Kraftmakher [69] and Sullivan and Seidel [70] put forward AC-calorimetry. Although it was generally known from ultrasound propagation in gases that heat capacity should be considered as a frequency-dependent complex quantity [71], they considered it as a real-valued quantity. If heat is not instantaneously dispersed to all degrees of freedom in the material under study, the ratio between heat and temperature change becomes time-dependent and hence heat capacity too. The transition from time to frequency domain delivers a complex heat capacity comprising real and imaginary parts, which are both frequency-dependent. This response is analogous to other dynamic perturbations such as in dielectric spectroscopy and can be treated in an similar manner, see e.g., Birge et al. [67, 72], Christensen [73], Jeong et al. [74], and Richert et al. [75, 76].

In 1971, using an inorganic polymer, Gobrecht et al. [77] carried out the first direct measurement of complex heat capacity in the glass transition range. Theirs was not just the first direct measurement of complex heat capacity, but also the pioneering application of a temperature modulated DSC (TMDSC). This idea to combine DSC with periodic temperature perturbations was reconsidered by Reading et al. [78–80] and Salvetti et al. [81] in 1992. Due to developments in computer technology, it was possible then to overcome the limitations of the setup put forward by Gobrecht et al. [77]. Therefore, TMDSC became available as a standard technique in thermal analysis and has found wide application in glass transition studies [80, 82, 83].

2.2.1 Temperature Modulated Differential Scanning Calorimetry

Anny temperature change $\mathrm{d}T/\mathrm{d}t$ in a sample with sample mass m and specific heat capacity c_p yields a heat flow rate $\Phi(t)$

$$\Phi(t) = m \cdot c_p \cdot \frac{\mathrm{d}T}{\mathrm{d}t} \tag{1}$$

For any temperature perturbation, $\mathrm{d}T/\mathrm{d}t \neq 0$, a nonzero heat flow rate $\Phi(t)$ takes place. For any heating rate profile (see Eq. (1)), the measured heat flow rate contains information about heat capacity $C_p = m\,c_p$. Here, a discussion of only sinusoidal heating rate profiles is given, although the formalism is applicable to other signal shapes [84–86] and nonperiodic perturbations [83, 87–90]. This sought of DSC measurements are termed as modulated temperature (MTDSC) or temperature modulated (TMDSC). However, it must be underscored that the perturbation in scanning calorimetry is not temperature, but rather (heating; cooling) scanning rate. A calorimetric signal can only be obtained if the temperature changes (i.e., scan rate $\neq 0$). Any heating rate perturbation (periodic, stochastic, harmonic, anharmonic) may be added to any temperature-time profile, like scan or isothermal. In the simplest case, a harmonic perturbation superimposes to the underlying linear heating (or cooling):

$$T(t) = T_0 + \beta_0 t + A_T \sin(\omega t) \tag{2}$$

where ß$_0$ is the underlying heating (cooling) rate, T_0 the start temperature, A_T the amplitude of temperature perturbation and $\omega = 2\pi/t_p$ is the angular frequency with t_p being the modulation period. The heat flow rate thus measured allows calculating the (effective reversing or) complex heat capacity given by:

$$C_{p\,effectiv}(\omega) = \frac{A_\Phi(\omega)}{A_\beta(\omega)} = \frac{A_\Phi(\omega)}{\omega A_T} \tag{3}$$

where A_Φ is complex heat flow rate amplitude and $A_ß = \omega A_T$ is heating rate amplitude. Under conditions of linearity and stationarity [91, 92], this quantity is equal to the so-called reversing or complex heat capacity [83]. The complex amplitudes, required for heat capacity determination, can be found with high sensitivity and accuracy using frequency selective techniques like lock-in amplifiers or Fourier analysis. Compared to conventional DSC measurements, the influence of instrumental drift (low frequency noise) together with electrical noise at high frequencies is dramatically reduced. Absolute values of the heat capacity can therefore be measured with an accuracy superior to 1 % in a wide temperature range using TMDSC [89]. Consequently, slight changes in heat capacity, for instance due to confinement, can be detected. For ß$_0 = 0$, the mean temperature is constant and only the oscillating component contributes to the heat flow rate. This condition is referred to as "quasi isotherm" and allows a time-dependent determination of heat capacity.

2.2.2 AC Chip Calorimetry of Thin Films

For conventional differential scanning calorimeters (DSC) in TMDSC mode a sample mass of about 1 mg is needed. Otherwise the signal-to- noise ratio becomes very bad. For experiments with films below 100 nm many of these films must be stacked in the DSC pan (2.1). To overcome this limitation and to measure single films highly sensitive calorimetric techniques are needed. Chip-based thin film sensors offer an attractive alternative to reach the goal (see 2.1 for some references regarding high sensitive fast scanning calorimetry). The same high sensitivity was realized by AC-calorimetry utilizing the same thin film sensors, again in non-adiabatic conditions. Large scanning rate amplitudes are required even for sensitive AC-calorimetric measurements, because the scanning rate is the perturbation. Simultaneously, small temperature amplitudes must be applied to ensure that the system is not enormously disturbed. This was realized by using small temperature amplitudes (<1K) and high modulation frequencies, see [93] for a more detailed discussion. In the following, description of such a thin film AC-calorimetric technique is given. It realizes the needed sensitivity and because of the applied frequency a well-defined timescale is straight forward assigned to the measurements. The developed AC calorimetric technique for glass transition studies covers 11 orders of magnitude in frequency for bulk samples [93] and three orders of magnitude for nanometer thin films [94–96].

Hellmann et al. [97, 98] pioneered the use of free standing silicon nitride membranes for AC-calorimetry and employed them as AC-calorimeters in relaxation mode. Recently, experiments on a few 100 nm [97] and 30 nm [99–101] films making use of these calorimeters were reported.

In order to achieve fast and controlled cooling, our approach to nanocalorimetry [47, 48] is based on non-adiabatic conditions too. The thin film calorimeter operated non-adiabatically permits heating (and cooling) at rates up to $\mathrm{MK\,s^{-1}}$ [47]. Using such high rates often yields nonequilibrium states of the probed sample. For instance, it is possible to prevent crystallization and reach the amorphous glassy state for several fast crystallizing polymers on cooling [48]. It is also possible to superheat polymer crystals by fast heating [102]. Often, however, one would rather measure thermal properties of minute samples at (or close to) thermodynamic equilibrium. This is achievable by combining chip calorimetry with AC calorimetry. As usual in AC calorimetry, a small periodic heat flow is delivered to the sample and the resulting complex temperature amplitude measured at slow scanning or at constant bath temperature. The timescale of the experiment is well-defined by the chosen frequency. In several instances, e.g., at glass transition, it is possible to make a direct comparison with results from other dynamic methods like dielectric spectroscopy [6, 7, 103–112]. A detailed description of such an AC-chip calorimeter for small samples using a single sensor under non-adiabatic conditions is available in [113]. The sensitivity of the setup is about 10 $\mathrm{nJ\,K^{-1}}$ at room temperature and hence the measurement of the dynamic glass transition for films down to 500 nm of thickness is possible; for much thinner polymer films, the sensitivity of the calorimeter had to be enhanced.

A differential AC-chip-calorimeter which is based on a commercially available chip sensor is discussed in [114]. Because of the differential setup, sensitivity in the pico-Joule per Kelvin range is achieved, allowing measurement of samples less than one nanogram. Therefore, the film thickness limit is in the range of a few nanometers [94–96, 110–112, 115–122]. Because of the small total heat capacity (addenda + sample) and the chance to cool fast, two main gains are made: high sensitivity, and AC measurements at relatively high frequencies [123]. This calorimeter can measure sample heat capacity ($\mathrm{J\,K^{-1}}$) in the frequency range 1–50,000 Hz, and with laser heating up to 1 MHz [93].

The sample heat capacity C_S is much smaller than addenda heat capacity C_0 for ultra-thin films. Using a differential setup can be one way of minimizing the influence of the addenda heat capacity. The basic equations for this differential setup based on

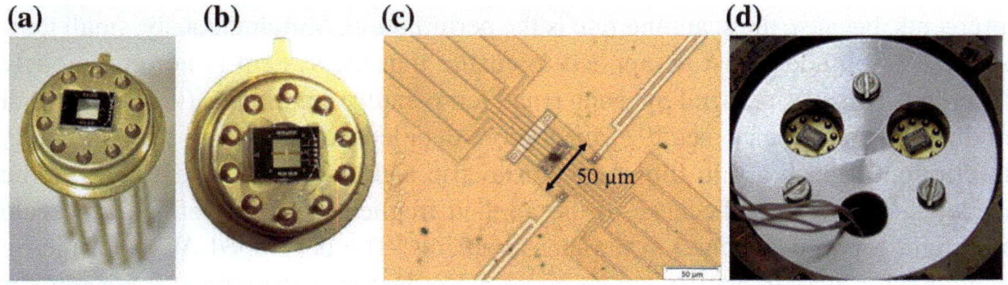

Fig. 2 Chip sensor for the AC calorimeter. **a** and **b** The calorimetric sensor from Xensor Integrations, NL, on a TO5 housing. **c** Magnified center area of the membrane with the two heater stripes at the end of the double arrow scale bar and the six hot junctions of the thermopile (*squares* in the *center*). **d** Chip sensors in a differential setup placed in an oven. The aluminum block, which is mounted in a cryostat, has a diameter of 36 mm. Photographs are taken from Ref. [95]

thin film sensors from Xensor Integrations, NL [114], see Fig. 2, are detailed in [94, 95, 122].

The above considerations made use of a differential AC chip calorimeter. A small heater is placed in the center of a free standing 500 nm thin silicon nitride membrane; the heater comprises two 50 μm long parallel strips. It is shown in [49, 50] that the temperature distribution in the inter-stripes area is nearly uniform. Six hot junctions of a thermopile are arranged in the heater area. A 700 nm silicon oxide layer protects the thermopile, the heater, and the conducting strips. The sample is either placed in the heated area (for small samples) or, for the case of thin films, spread over the whole sensor. The temperature amplitude due to the AC power is of the order of 0.1 K. Therefore, the temperature gradient toward the frame from the heater [49] does not contribute to the uncertainty of the temperature reading, very much unlike the use of the same device for fast scanning where the temperature of the heater can be several 100 K above the constant frame (oven) temperature.

The only region of interest for AC-calorimetric experiments is the small heated area, which can be approximated as a point heat source. The two sensors are positioned adjacent to each other in a thermostat at temperature T_B. Either, measurements are performed at different frequencies keeping the temperature constant (frequency scans), or at a single frequency while continuously changing the temperature (temperature scans). In the first case, a stepwise change of temperature can be applied. The heating rate of the thermostat is limited to 5 K/min in order to realize (i) an almost uniform temperature distribution and (ii) small errors in temperature. This calorimeter can be operated at ambient or slightly reduced pressure in air or an inert gas (e.g., nitrogen or helium) in a gas with controlled humidity [124, 125], or even in high vacuum [119–122].

Figure 3 provides the electrical scheme of the device. The heaters on the chips are driven by an AC-voltage having frequency $\omega/2$ from the oscillator of the lock-in amplifier, resulting in an oscillating power at frequency ω. The phase and amplitude of the resulting oscillating temperature (at frequency ω) is measured by the thin film thermopile employing a digital lock-in amplifier [94, 95].

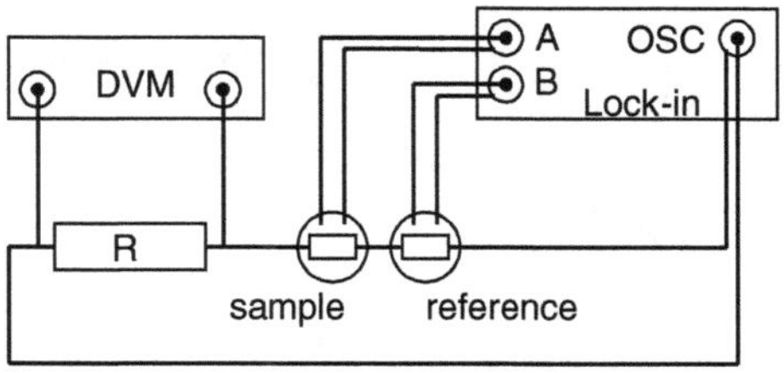

Fig. 3 Schematic picture of the electric setup. The voltage of the internal oscillator (OSC) drives the current through the heaters. The differential signal A–B of the thermopiles is analyzed and further processed. The voltage over the known resistor R is measured with a digital multimeter (DVM) to calculate the heater power. The scheme is taken from Ref. [95]

Depending on the surrounding gas, sample thickness, and sensor type (area heated region), the available frequency range is between 1 Hz and 50 kHz. The heater power per sensor is about 10–50 μW, which results, in a temperature oscillation amplitude and a temperature bias less than 1 K, depending on the frequency. The heat capacity and thermal conductivity of the surrounding gas determine the frequency of maximal sensitivity. A maximum sensitivity is observed between 10 and 100 Hz for N_2 at ambient pressure. For lower frequencies, losses through the gas become dominant while the distance between the thermopile and the heater causes a decay of the signal at higher frequencies. The frequency range can be stretched to lower values through minimizing the losses by lowering pressure [122].

It is possible to increase the sensitivity of this differential setup by a factor 100 or more (compared to a single sensor) by use of sensors which have minimal asymmetry [113]. Furthermore, ambient disturbances are minimized by the differential setup. This delivers a highly sensitive calorimetric tool able to measure in the pJ K^{-1} range and sub-nanograms of matter [94–96, 110–112, 115–117, 119, 125, 126]. Consequently, the step in heat capacity at the dynamic glass transition—for films with thickness in the order of 1 nm—can be measured [120].

As already described above, there are two possible modes of operating the AC chip calorimeter: either scanning the frequency at constant temperature and increasing (or decreasing) the temperature stepwise, or temperature scans at constant frequency. Both modes were applied to measure frequency and thickness-dependence of the dynamic glass transition temperature for a series of thin films.

Figure 4 shows data for a 15 nm thin polystyrene film. Frequency sweeps from 0.6 to 1,280 Hz were performed and the temperature changed in steps of 5 K from 323 to 423 K. The measured signal is presented here, which is, for thin films, proportional to sample heat capacity [95, 118, 122]. The glass transition temperature and changes of the signal can easily be detected from the data. The additive term from an empty measurement of the system is usually an order of magnitude or more lower than the signal with sample. As a result, contributions from asymmetry change only slightly the overall picture in such a small temperature range, and bear no effect on the determination of the glass transition temperature. For a 15 nm (Fig. 4) and a 4 nm

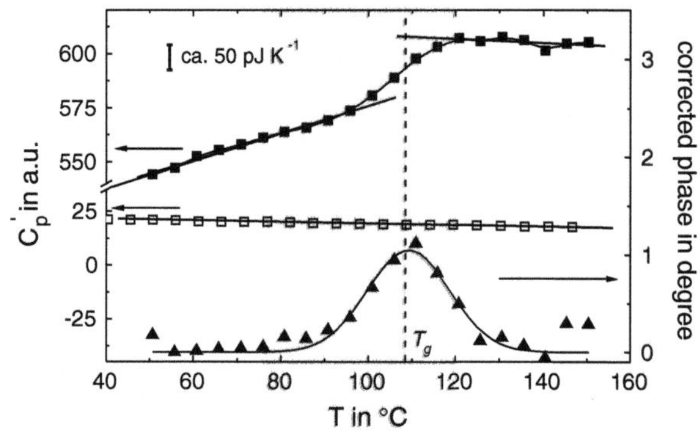

Fig. 4 Real part of complex heat capacity, $C_p{}'$, for a polystyrene film of 15 nm thickness (*solid squares*) and for an empty sensor (*open squares*) and the corrected phase angle (*solid triangles*) at a frequency of 80 Hz. The measurements are isothermally performed in steps of 5 K. The tangent construction for determining the dynamic glass transition temperature from the real part as half step temperature is shown. A peak in the phase appears in the glass transition range as expected. The *vertical dashed line* indicates T_g. All data are taken from Ref. [95]

film, the signal from the empty sensor is, respectively, about 20 and 5 times smaller than that for the sensor with film.

The probed sample mass of the 15 nm film, which is frequency-dependent only at frequencies above 100 Hz, can be estimated to be about 1 nanogram, given an assumed effective heated area of $0.2 \times 0.2\ mm^2$. With the heat capacity of polymers being in the order of 1 $Jg^{-1}K^{-1}$, a sample heat capacity of 1 $nJ\,K^{-1}$ is measured. From this estimation, and the scatter in Fig. 4, the sensitivity of the setup at the chosen frequency (80 Hz) can be approximated to about 50 $pJ\ K^{-1}$.

The frequency-dependent glass transition temperature is determined as the half step temperature in the real part of the thermopile signal thus measured. The tangent construction for its determination is displayed in Fig. 4. As expected in the phase of the complex heat capacity, a peak is found at the dynamic glass transition. The maximum in phase and the half step of the real part occur at the same temperature. There is an underlying step in the phase which is proportional to the real part. By subtracting this contribution, the phase signal can be corrected, as described in [127]. For the 15 nm sample the peak in the phase is detectable. It can be employed to determine the glass transition temperature in the frequency range of 20–320 Hz, while the step at the glass transition can be evaluated in the range 0.6–1,280 Hz. Within the error limit of ± 3 K, both values—the temperature at half step height and the peak temperature—are in good agreement.

3 Results

3.1 *Sample Preparation*

Thin films were prepared by spin-coating from solutions with different concentrations in order to vary their thicknesses. For calorimetric measurements and thickness determination, respectively, the filtered (minipore, 0.2 μm) solution was spin-coated directly onto the center position of a cleaned and annealed sensor and a cleaned silica wafer. A same spinning rate of 3,000 rpm and spinning time 60 s were applied for all samples. The thickness of the film on the silica wafer was measured by X-ray reflectivity [96, 128], by ellipsometry and AFM [117, 129] or determined from the measured capacitance of a capacitor [115] and was found to be in agreement with the calculated value that is based on viscosity, spinning rate and solvent volatility [130]. Due to the similar surface (SiO_2) between the sensor and the wafer, we assumed that thicknesses were the same given the identical sample preparation procedure.

The sample preparation details can strongly affect the measured glass transition [110, 131–133]. In order to avoid oxidation during high temperature annealing, both the measurements and annealing were carried out in a flow of dry nitrogen [110]. The multi-exponential like decay of the signal amplitude during annealing of the thin films can be followed by quasi-isothermal AC-calorimetric measurements. Such decay speeds up with increasing temperature, but is always found if the sample is not annealed long enough above the bulk glass transition temperature. At the moment, we cannot establish whether the decay originates from the relaxation of the metastable state of the spin-coated film or from residual solvent [60, 134, 135]. Unless otherwise stated, the heating/cooling traces in this work are obtained after they become stabilized. Prior to and after the calorimetric measurements, the sensor surface was checked by AFM and no dewetting was detected.

3.2 *Film Thickness Independent Dynamic Glass Transition in Nanometer Polymer Films from AC-Calorimetry*

The AC-chip calorimeter makes a certain frequency range accessible for glass transition studies in thin polymer films. This is applied to study the frequency dependence of the dynamic glass transition in order to probe the influence of film thickness on the dynamic behavior in a wider frequency range. Figure 5 confirms the linearity between the measured thermopile voltage amplitude (proportional to heat capacity for films below 500 nm [120]).

In Fig. 6, the real part of complex heat capacity for a 40 nm thin polystyrene film and the corrected phase of the thermopile signal (at three frequencies) are shown.

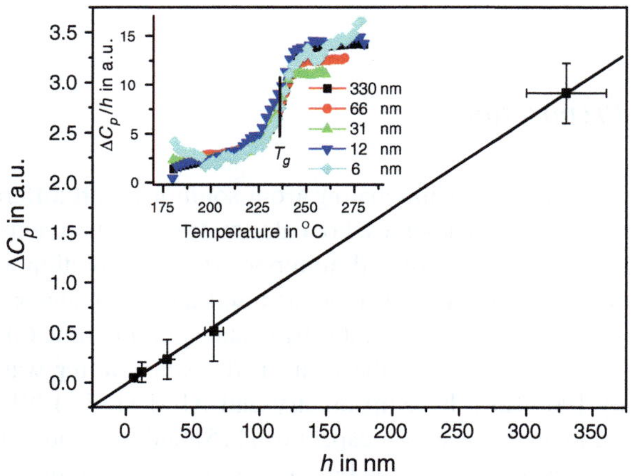

Fig. 5 The step in heat capacity at T_g as a function of film thickness as determined from X-ray diffraction of similar samples on silicon wafers. The inset displays the heat capacity step normalized by film thickness. Only 50 points for each *curve* are shown. The accuracy in the determination of ΔC_p is 25 %; the resulting uncertainty for the measurement of T_g is ±3 K. All data are taken from Ref. [96]

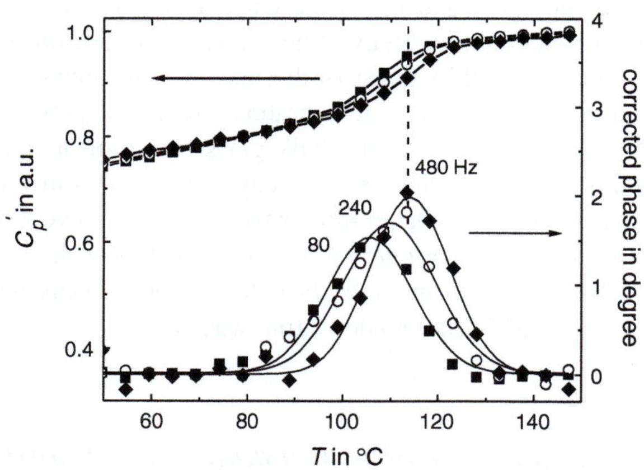

Fig. 6 Real part of complex heat capacity and corrected phase for a 40 nm thin polystyrene film for frequencies 80 Hz ■, 240 ○ and 480 Hz ◇. The scatter of the points around 60 °C (outside the transition region) reflects the uncertainty of the measurement. All data are taken from Ref. [95]

The frequency dependence is further measured for polystyrene films with thicknesses of 40, 15, and 8 nm, respectively. The frequency range is from 0.6 to 1,280 Hz for the thicker films and reduced to 6–1,280 Hz for the thinnest film due to the smaller signal. Figure 7 shows a relaxation map where the temperature position at half step height is plotted. The maximum position of the peak, ω_p, respectively the half step of real part shifts as described by the VFTH equation.

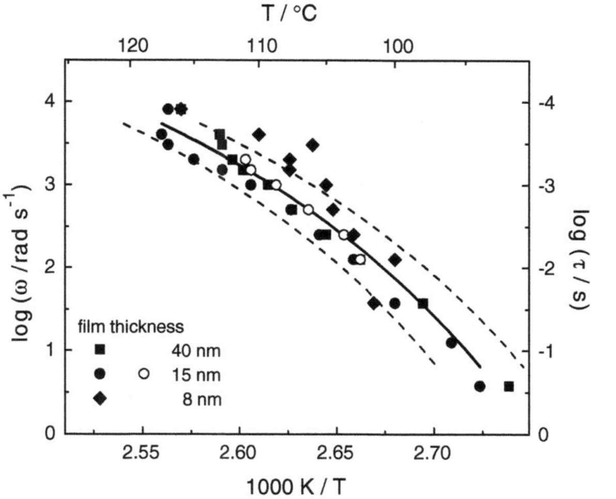

Fig. 7 Relaxation rate versus inverse temperature for PS films with thicknesses as indicated. The *solid symbols* show the frequency (ω) dependence of the half step temperature of the real part and the *open circles* from the peak in the phase for a film thickness of 15 nm. The right axis shows the corresponding relaxation time with $\tau = 1/\omega$. The *curve* is a fit to the data of the 15 nm film by VFTH function. The two *dashed lines* correspond to the uncertainty of the T_g determination of ±3 K. The increasing scatter of the points with decreasing film thickness corresponds to the increasing uncertainty of T_g determination. All data are taken from Ref. [95]

$$\log \omega_p = \omega_\infty - \frac{A}{T - T_0} \tag{4}$$

ω_∞ and A are constants and T_0 is the so-called Vogel temperature. The data can be described by a single set of parameters within the error limits. For the high as well as for the low frequency range, the limitations of the method become visible from the stronger scattering of the data. These limits are also visible from the scatter in the data for the thinnest sample (8 nm).

Similar data were obtained for PPO films (Fig. 8). Again, in the whole accessible frequency range the data for the films of different thicknesses between 330 and 6 nm are well described by one single VFTH function.

To determine the thickness dependence of the glass transition temperature for thin films in more detail measurements at the frequency of optimum sensitivity were performed for PS [18, 19, 115, 136], PPO [96], PPO/PS 50/50 blend, PMMA [94, 111], toluene [120], poly(cis-1,4-isoprene) [137], poly(vinyl methyl ether) [138], and poly(bisphenol A carbonate) [139]. The measurements were performed in a temperature range from $T_g - 50$ K to $T_g + 50$ K at a scanning rate of 2 K/min and fixed frequency. Evaluation is based on the second heating-cooling cycle to avoid influences from sample preparation (like residual solvent evaporation). The glass transition temperature is determined as half step temperature using the tangent construction shown above. Some results are shown in Figs. 9, 10, and 11. For all

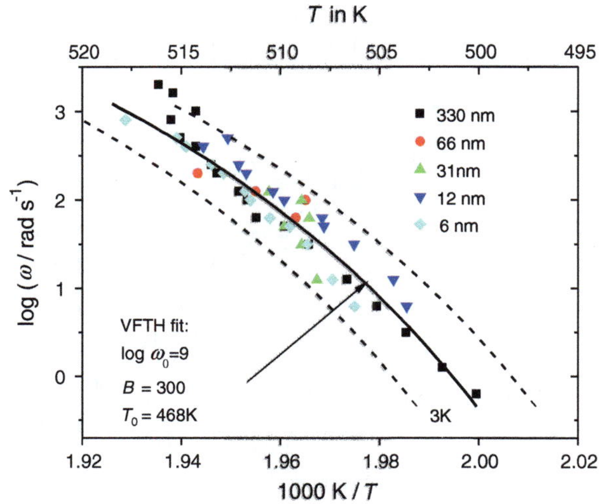

Fig. 8 Relaxation rate versus inverse temperature for PPO films with thicknesses as indicated. The *solid curve* is the VFTH fit whose parameters are indicated in the graph. The two *dashed lines* correspond to the uncertainty of the T_g determination of ±3 K. All data are taken from Ref. [96]

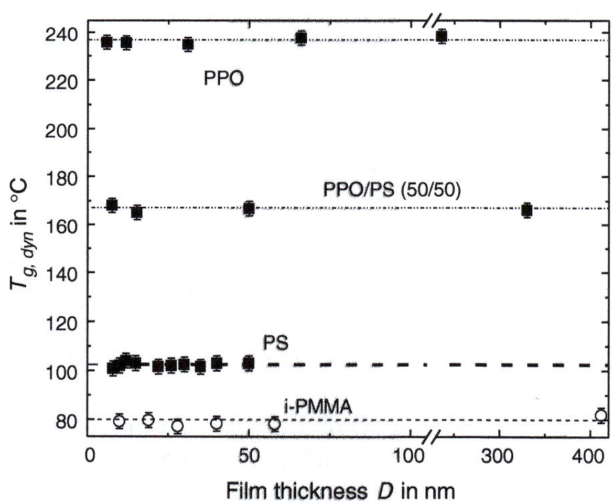

Fig. 9 The dynamic glass transition temperature plotted as a function of film thickness at a frequency of 80 Hz for polymer films ranging in thickness from 400 to 4 nm. All data are taken from Ref. [140]

studied polymers and film thicknesses, no change in the dynamic glass transition temperatures is observed.

The glass transition temperature remains within the error limits of ±3 K equal to the bulk value, as shown in Figs. 9, 10, and 11. This is similar to the fast scanning

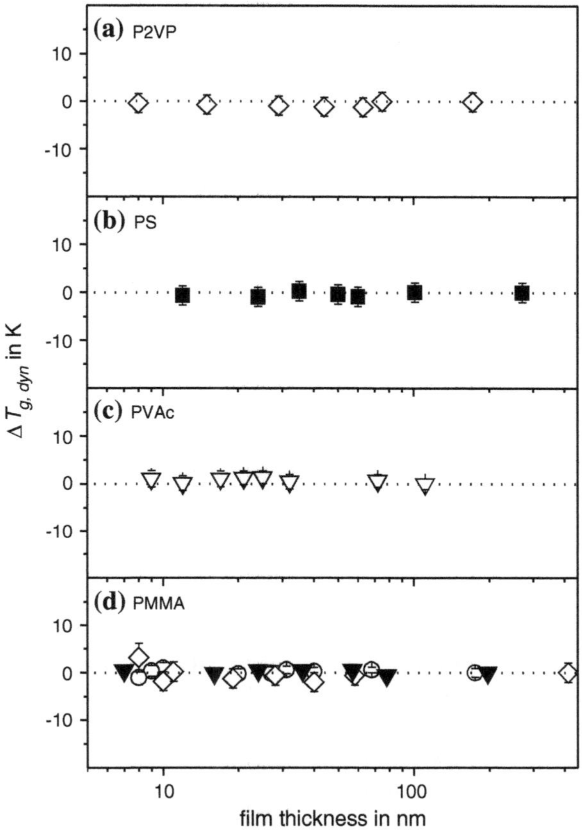

Fig. 10 Shift of the temperature position of the dielectric α relaxation peak in respect to the value corresponding to the maximum film thickness in dependence on the film thickness: **a** P2VP (0.75 Hz); **b** PS (380 Hz); **c** PVAC (0.75 Hz); **d** (O) PMMA-144K (0.12 Hz), (♦) PMMA-300K (0.12 Hz), and (◇)) calorimetric measurements on PMMA-144K (80 Hz). All data are taken from Ref. [111]

results at a few thousand kelvin per second presented by Efremov and Allen [12, 14] and for molecular glass formers by Rodriguez-Viejo et al. [52, 141]. For a film thickness range from 100 nm down to 4 nm, reductions of more than 25 K in the glass transition temperature were reported for polystyrene of similar molecular weight in other studies [142]. This reduction is significantly larger than the uncertainty for the glass transition temperature determination for the AC-calorimetric measurements. Here, we have to consider that AC-calorimetry probes molecular dynamics in the linear response regime in the supercooled liquid state. Similar to dielectric spectroscopy molecular motions are probed in thermodynamic (quasi) equilibrium. As detailed by Cangialosi et al. [17–20, 143–152] and Chap. 14 in this book, segmental relaxation in the supercooled liquid state, as probed by dielectric spectroscopy and AC-calorimetry, is not influenced by film thickness. On the other hand measurements

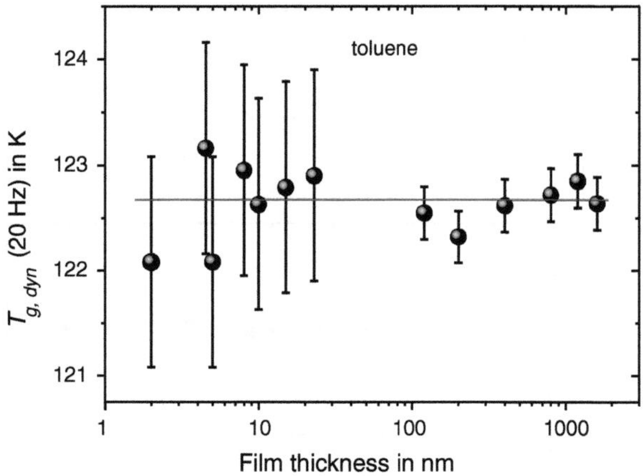

Fig. 11 Film thickness dependence of the dynamic glass transition temperature of vapor-deposited toluene. The as-deposited samples were heated to 125 K to ensure that they were in the supercooled liquid state. Next, the samples were cooled at 0.67 K min^{-1} to 105 K to form a slowly cooled ordinary glass. The data presented were obtained from the second heating to 125 K at a frequency of 20 Hz. All data are taken from Ref. [120]

probing volume (e.g., ellipsometry, X-ray reflectivity) or enthalpy of the glassy state (e.g., DSC) show a thickness dependency which is explained by the free volume holes diffusion model [20–22].

The results from Figs. 7, 8, 9, 10, and 11 confirm the thickness independent cooperative dynamics measured by specific heat spectroscopy. As highlighted by Kremer et al. [63, 111, 136, 137, 153–155] also dielectric spectroscopy in most cases does not show any thickness-dependent shift in T_g. A comparison can also be made based on the relaxation times of the α-process obtained from dielectric and specific heat spectroscopy. Figures 12 and 15 show such a comparison; evidently, there's an excellent agreement between the relaxation time data $\tau(T)$ from both techniques for films of 15–20 nm in thickness and bulk samples, respectively.

This result makes us confident that also under geometrical confinement both the dielectric α-process and the specific heat relaxation data probe the same cooperative fluctuations associated to the dynamic glass transition in the bulk, see also [5, 6, 104, 136, 137, 156].

3.3 Polymer Blends in Thin Films

Polymer blend thin films are interesting both for their technological applications (e.g., coatings, ...), and from an academic point of view. Over the last two decades, new materials generated through the blending of polymers found widespread, both scientifically as well as for commercial utilization. The performance of these mate-

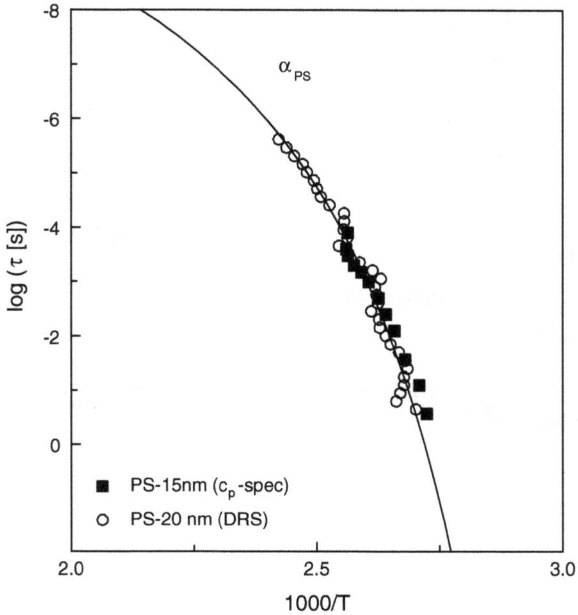

Fig. 12 Logarithm of relaxation time τ for PS obtained from the peak maxima in the dielectric loss ε''(T) (*open circles*) and temperature of the half step of the real part of complex heat capacity (C_p'') from AC-calorimetry (*filled squares*) versus inverse temperature. Reprinted from [115], Copyright (2005), with permission from Elsevier

rials is strongly influenced through the miscibility of the polymers. This is also true for ultra-thin films.

Investigation of the composition dependence of the glass transition is the generally accepted method to study the miscibility of two polymers in bulk. A blend is said to be miscible when there is only one glass transition. The reverse, however, is not necessarily true because of the so-called "self-concentration effect", which is inherent to chain-like connectivity of the polymers [157]. Using calorimetric methods to study glass transition and miscibility in polymer blends is attractive. In principle, it can allow direct measurement of the heat capacity. In some practical cases, the thermal property is of the utmost importance for film application.

We used AC chip calorimetry to study the miscibility of PS and PPO confined into ultra-thin films [128]. We found that within a margin of 3 K, there is no obvious shift of T_g even for the thinnest film of the blend (Fig. 9). This is the same as for pure polymers. For all the films, only one glass transition is observed. The Couchman or Gordon-Taylor mixing laws, that are commonly used for miscible bulk blends, can well describe the composition dependence of T_g. Another interesting finding is that of the fact that PS/PPO miscibility is not affected even for the thinnest film with 6 nm thickness.

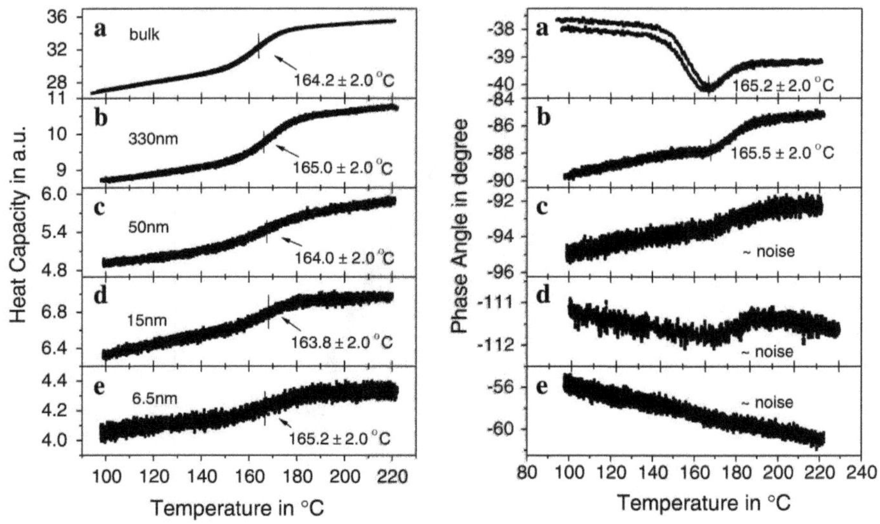

Fig. 13 Amplitude (*left*) and phase angle (*right*) of the complex heat capacity of the $PS_{50}PPO_{50}$ blend. For bulk and thick film samples, the T_g are estimated by the half step construction from the amplitude or from the peak in the phase angle, both give quite similar values. For the thin films, the T_g is estimated from the half step construction or the peak value of the first derivative of the amplitude. The thickness of the *curves* is dependent on the sampling frequency. For the *curves* of thin films the sampling frequency is fixed to about 30 points K^{-1}, so the *curve* thickness indicates the decreasing signal-to-noise ratio with decreasing film thickness. All data are taken from Ref. [128]

Figure 13 shows the dynamic glass transition of the blend $PS_{50}PPO_{50}$ in a bulk sample (a) and in thin film samples with thicknesses 300 nm (b), 50 nm (c), 15nm (d), and 6.5 nm (e), respectively. Both PS and PPO contribute 50 % in mass, as indicated by the formula $PS_{50}PPO_{50}$. The step-like change at T_g is clearly seen in the left panels, which display the heat capacity. A peak is shown superposed onto an underlying step for the bulk and thick film samples in the right panels (showing the phase angle between temperature and power oscillation) [127]. It represents the contribution of the imaginary part of complex heat capacity at the dynamic glass transition [64, 127].

The intensity of the relaxation peak reduces with decreasing film thickness so much so that for the thinnest film of 6.5 nm, the relaxation peak is at the noise level. Nevertheless, even in this case, the step in the amplitude is clearly observable. To extract T_g, either the half step construction for the real part of heat capacity or the peak value in the phase angle can be used. Similar T_g values with an uncertainty of ±3 K are obtained from both evaluations.

From Fig. 13, it is seen that for $PS_{50}PPO_{50}$ blends, whether in the bulk or thin films, only one glass transition at ~165°C is observed. Because we are measuring at a higher frequency (20 Hz), this temperature is about 20 K higher than obtained from DSC. The variation of T_g between different thicknesses is smaller than 3 K, which is within the uncertainty of the measurements and similar to the results obtained

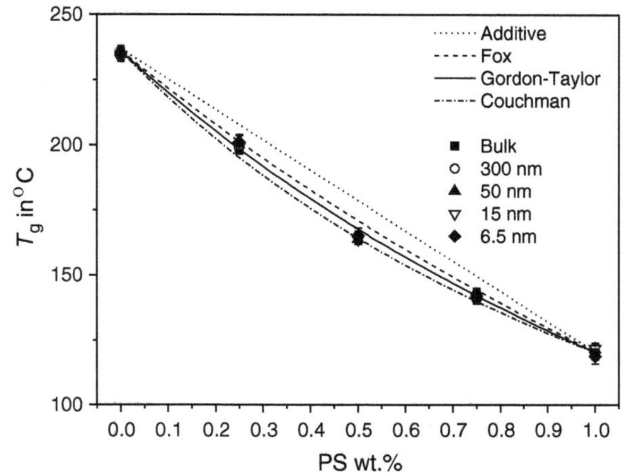

Fig. 14 Composition dependence of T_g of PS/PPO blends with different thickness. The uncertainty of ±3 K for the T_g determination is slightly larger than the symbol size. Calculated *curves* based on the Fox, Gordon-Taylor Couchman, and the additive mixing law are plotted to guide the eyes [159]. All data are taken from Ref. [128]

for pure polymers in thin film geometry (3.2). Nevertheless, here our focus is on the concentration dependence of T_g of thin polymer blend films, because it avails sufficient information about miscibility under spatial confinement.

In Fig. 14 the glass transition temperatures, T_g, of films with different composition and thickness are plotted together. The Fox, Couchman, or Gordon-Taylor mixing laws are able to describe the composition dependence of T_g for the miscible blends. From this plot, we can reliably state that the good miscibility between PS and PPO is maintained in the ultra-thin films. Kim and Zin [158], by use of ellipsometry, also observed that PS and PPO were miscible in thin films, although they found a dramatic depression of T_g for film thickness below 50 nm for the same blend spin-coated on a Si wafer.

The initial purpose of the work presented in [128] was to study the self-concentration effect, which is expected to be more significant in ultra-thin films. The PS/PPO blend was chosen because it has only one calorimetric glass transition, but two relaxation glass transitions (by DMTA and NMR) in bulk. We hoped, therefore, to establish whether the self-concentration effect is enhanced enough to produce two calorimetric glass transitions in the ultra-thin films. However, such a result was not observed. It is known that miscibility in a polymer system is strongly related to the length scale of inspection; any polymer pair is immiscible on the length scale of the Kuhn length. On the other hand, it is generally believed that the characteristic length scale of the calorimetric method is in the order of a few nanometers. Calorimetric miscibility is thus not sensitive to investigate the local concentration or dynamics heterogeneity below this length scale. Alternatively, we state, without risk of contradiction, that PS and PPO are still calorimetrically miscible even in the thinnest film of only a few nanometers.

While our results and those of Kim and Zin (using ellipsometry) on the miscibility between PS and PPO in ultra-thin films are in agreement, we have different results on the thickness-dependence of T_g. Since the two studies employed the same blend system, substrate and sampling procedure, a systematic difference between the two techniques as explained by the free volume holes diffusion model must be the origin of the contradiction between the calorimetric and ellipsometric (dilatometric) glass transition temperatures [20–22].

4 Summary

A large number of studies investigated the glass transition under conditions of geometrical confinement by calorimetry. Two categories of experiments can be identified: (i) Confinement of the glass forming regions in a stiff (or sometimes soft) matrix. In this manner, collecting of large amounts (milligram) of confined material allows application of conventional calorimetric techniques, e.g., adiabatic calorimetry or DSC. Typical examples for this group of systems are phase separated copolymers and blends, or confinement by porous structures or in nanocomposites, or self-assembled confined amorphous regions inside a semicrystalline structure. (ii) Confinement in thin film geometries. Samples can be considered as essentially free standing films, when large amounts of thin films are collected (typically milligrams). Calorimetric measurements can now be performed on single a few nanometer thick supported films through the recent developments in silicon nitride membrane-based nanocalorimetry.

These devices allow for scanning as well as AC-calorimetry on nanometer thick films. The effectively heated mass of such films can be in the ng-range. By this a sensitivity of a few 10 pJ/K can be achieved with differential setups, which is sufficient to measure the step at the dynamic glass transition of polymeric samples of about 1 nm thickness.

The AC-chip calorimetry allows investigations in the frequency range 1–50,000 Hz. Different groups investigated glass transition temperatures of thin polymer films from 400 to 2 nm. The dynamic glass transition temperature of supported thin films is found to be thickness independent, within the experimental uncertainty of a few K. The frequency dependence was also measured for selected film thicknesses using an AC-chip calorimeter. In the accessible frequency range (starting from 0.6 Hz) no significant deviations from the VFTH behavior are observed.

These results are contrary to many other studies on thin polymer films, e.g. [142]. A possible explanation was given by Cangialosi et al. [17–19]. They applied the free volumes holes diffusion model [20–22] to successfully describe previously conflicting data.

Further, we used the differential AC chip calorimeter to study the miscibility of the PS/PPO blend confined in thin films. The good miscibility between PS and PPO is preserved even for only a few nanometers of film thickness. Finally, but not described here, we show that the sensitivity of the chip calorimeter can also be used to study the inter-layer diffusion in ultra-thin films. Starting from a double layer system of about 55 nm PPO and 100 nm PS, we end up with a mixed layer of $PS_{43}PPO_{57}$ blend thin film which contains all PPO, and a basically pure PS layer of about 30 nm, which does not diffuse into the $PS_{43}PPO_{57}$ blend thin film even staying in its liquid state for more than 10 h [128].

AC chip calorimetry in combination with AFM was also demonstrated to be a powerful tool for studying phase separation in thin polymer blend films too. A state diagram was constructed for PVME/PS in 100 nm thin films [129]. For that purpose a new methodology was developed. Very short (10 ms to s) temperature jumps to temperatures above the cloud point temperature could be applied using the on-chip heater. This made the simulation of laser patterning possible [117]. 10 ms jumps into the two phase area of the state diagram showed evidence for phase separation. The resulting morphology was very fine and remixed very rapidly. Using short temperature pulses, AC chip calorimetry—as compared to AFM—becomes a more sensitive characterization tool for the detection of the early stages of phase separation in thin polymer films.

A particular promising application of AC chip calorimetry is in the field of extra-ordinary stable glasses as pioneered by Ediger [160]. We described an AC chip calorimeter for in situ heat capacity measurements of as-deposited nanometer thin films of organic glass formers [93]. The calorimetric system is based on a differential AC chip calorimeter which is placed in the vacuum chamber for physical vapor deposition. The sample is directly deposited onto one calorimetric chip sensor while the other sensor is protected against deposition.

Phase transformations in biopolymers are another subject of AC calorimetry. Often the water content determines the properties of the biopolymers. AC-calorimetry allows operation under controlled environmental conditions like humidity or all kind of gasses or gas mixtures. An AC calorimeter with humidity controlled environment is described and applied for the study of protein crystals [124, 125] and a study on thin polymer films under different humidity is under way.

5 Future Directions and Outlook

For the thin film measurements the frequency range is limited to the values shown in Figs. 7 and 8. To obtain more complete information about the segmental relaxation a wider frequency range is preferable. So far a significant extension to higher and lower frequencies is only possible for bulk samples. The general concepts for heat capacity spectroscopy covering 11 orders of magnitude in frequency are discussed next.

The complex heat capacity can be measured in a wide range of frequencies through the combination of different calorimeters. A common problem for all these calorimeters is that heat transfer from a heated surface into the sample requires a temperature gradient and consequently time. It can therefore not be avoided that the measured phase angle contains a contribution from the real part of heat capacity. This can on the other hand be utilized to measure heat capacity if it is real-valued [64]. A careful correction of the measured phase angle for this contribution is required for determining the imaginary part of complex heat capacity [127, 161]. In classical calorimetry

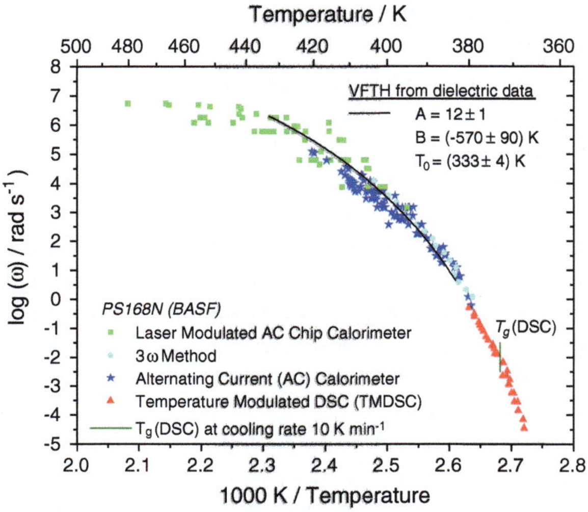

Fig. 15 Relaxation rate versus inverse temperature for PS from TMDSC (*triangles*), AC-calorimetry (*squares*) and 3ω – *method* (HCS) (*circles*). The *lines* correspond to a VFTH fit to data from dielectric spectroscopy (ε"). The *vertical bar* shows the conventional T_g from DSC at 10 K min^{-1}

heat is first transferred to the fast degrees of freedom (phonons) before it also distributes over the slow modes. Richert et al. [162, 163] introduced a so-called "reverse" calorimetry: here energy is transferred first to the "slow" configurational modes in a glass forming liquid by a nonlinear electrical perturbation at the chosen frequency. Through a relaxation process it then distributes over the "fast" degrees [162, 163]. In both cases, the distinctive representation of a relaxation process in the supercooled liquid—a step in real and a peak in the imaginary part of the susceptibility—is observed, see Fig. 1 above. Measurements of the complex heat capacity are nowadays possible over the wide frequency of range of more than 11 orders of magnitude through the combination of different calorimeters. The calorimetric dynamic glass transition temperature for bulk PS is shown in Fig. 15 in comparison to dielectric data in such an extended frequency range.

Application of heat capacity spectroscopy in such a wide frequency range to thin glass forming films seems to be very promising. Particularly the high frequency range requires thin samples and is therefore attractive for the study of thin films [93].

Another field of on-going activity is the study of phase transitions and glass transitions in polymers and biopolymers in dependence on the water, more general the solvent or plasticizer, content in thin films. The water uptake of polyvinyl acetate (PVAc) at different relative humidity for bulk-like and ultra-thin films (>20 nm) is under studied by AC-chip calorimetry. For a micrometer-sized sample of about 200 ng on the chip the plasticizing effect of water vapor is the same as for bulk samples. The water dissolved (absorbed) in the film results in a shift of the glass transition temperature (plasticizing effect) and an increase in heat capacity. With decreasing film thickness the water uptake becomes more complex. An increasing amount of adsorbed water is observed. The adsorbed water does not plasticize the film but contributes to the measured heat capacity (JK^{-1}). AC calorimetry may help to gain a better understanding of the distribution of the absorbed water.

Going away from the nanometer thin films, other applications are envisaged too. As an example, the drying or curing of micrometer thick coatings or joints can be studied by AC-calorimetry [164].

Acknowledgments The valuable contributions of all my co-workers, especially A. Minakov, M. Merzlyakov, H. Huth, M. Ahrenberg, Y.-Z. Chua, D. Zhou, S. Adamovsky, E. Zhuravlev, G. Schulz, and A. Wurm as well as financial support from German Science Foundation (DFG) and European Union made this work possible. The author acknowledges extensive language polishing by Emmanuel Urandu Mapesa.

References

1. Donth E (2001) Glass transition. Springer, Berlin
2. Hensel A, Schick C (1998) Relation between freezing-in due to linear cooling and the dynamic glass transition temperature by temperature-modulated dsc. J Non-Crystal Solids 235-237:510–516
3. Hodge IM (1994) Enthalpy relaxation and recovery in amorphous materials [review]. J Non-Crystal Solids 169:211–266
4. Suga H (2001) Adiabatic calorimeter as an ultra-low frequency spectrometer: interplay between phase and glass transitions in solids. Thermochim Acta 377:35–49
5. Huth H, Wang LM, Schick C, Richert R (2007) Comparing calorimetric and dielectric polarization modes in viscous 2-ethyl-1-hexanol. J Chem Phys 126:104503
6. Brás AR, Dionísio M, Huth H, Schick C, Schönhals A (2007) Origin of glassy dynamics in a liquid crystal studied by broadband dielectric and specific heat spectroscopy. Phys Rev E 75:061708
7. Schick C, Sukhorukov D, Schönhals A (2001) Comparison of the molecular dynamics of a liquid crystalline side group polymer revealed from temperature modulated dsc and dielectric experiments in the glass transition region. Macromol Chem Phys 202:1398–1404
8. Dubochet J, Adrian M, Teixeira J, Alba CM, Kadiyala RK, Macfarlane DR, Angell CA (1984) Glass-forming microemulsions - vitrification of simple liquids and electron-microscope probing of droplet-packing modes. J Phys Chem 88:6727–6732
9. Angell CA, Kadiyala RK, Macfarlane DR (1984) Glass-forming microemulsions. J Phys Chem 88:4593–4596
10. Schick C, Donth E (1991) Characteristic length of glass transition: experimental evidence. Physica Scripta 43:423–429
11. Jackson CL, McKenna GB (1991) The glass transition of organic liquids confined to small pores. J Non-Crystal Solids 131:221–224
12. Efremov MY, Olson EA, Zhang M, Zhang Z, Allen LH (2003) Glass transition in ultrathin polymer films: calorimetric study. Phys Rev Lett 91:085703
13. Allen LH, Ramanath G, Lai SL, Ma Z, Lee S, Allman DDJ, Fuchs KP (1994) 1000 000 "c/s thin film electrical heater: in situ resistivity measurements of al and ti/si thin films during ultra rapid thermal annealing. Appl Phys Lett 64:417–419
14. Efremov MY, Olson EA, Zhang M, Zhang Z, Allen LH (2004) Probing glass transition of ultrathin polymer films at a time scale of seconds using fast differential scanning calorimetry. Macromolecules 37:4607–4616

15. Fakhraai Z, Forrest JA (2005) Probing slow dynamics in supported thin polymer films. Phys Rev Lett 95:025701
16. Gao S, Koh YP, Simon SL (2013) Calorimetric glass transition of single polystyrene ultrathin films. Macromolecules 46:562–570
17. Boucher VM, Cangialosi D, Alegría A, Colmenero J (2014) Accounting for the thickness dependence of the tg in supported ps films via the volume holes diffusion model. Thermochim Acta 575:233–237
18. Boucher VM, Cangialosi D, Yin H, Schonhals A, Alegria A, Colmenero J (2012) T g depression and invariant segmental dynamics in polystyrene thin films. Soft Matter 8:5119–5122
19. Yin H, Cangialosi D, Schönhals A (2013) Glass transition and segmental dynamics in thin supported polystyrene films: the role of molecular weight and annealing. Thermochim Acta 566:186–192
20. Cangialosi D, Wübbenhorst M, Groenewold J, Mendes E, Schut H, van Veen A, Picken SJ (2004) Physical aging of polycarbonate far below the glass transition temperature: evidence for the diffusion mechanism. Phys Rev B 70:224213
21. McCaig MS, Paul DR, Barlow JW (2000) Effect of film thickness on the changes in gas permeability of a glassy polyarylate due to physical agingpart ii. Math Model Polym 41:639–648
22. McCaig MS, Paul DR (2000) Effect of film thickness on the changes in gas permeability of a glassy polyarylate due to physical agingpart i. Exp Observ Polymer 41:629–637
23. Cangialosi D (2014) Chapter 14: equilibrium and out-of-equilibrium dynamics in confined polymers by dielectric spectroscopy and calorimetric techniques. In: Kremer F (ed) Dynamics in confinement. Springer, Berlin
24. Kremer F (2014) Chapter 1: molecular dynamics in confinement - a silver lining at the horizon? In: Kremer F (ed) Dynamics in confinement, Springer, Berlin
25. Schönhals A, Yin H (2014) Chapter 3: glass transition of ultra-thin polymeric films - a combination of relaxation spectroscopy with surface analytic. In: Kremer F (ed) Dynamics in confinement. Springer, Berlin
26. Ediger MD, Forrest JA (2013) Dynamics near free surfaces and the glass transition in thin polymer films: a view to the future. Macromolecules
27. Tool AQ (1946) Relation between inelastic deformability and thermal expansion of glass in its annealing range. J Am Ceram Soc 29:240
28. Moynihan CT, Easteal AJ, De Bolt MA, Tucker J (1976) Dependence of the fictive temperature of glass on cooling rate. Am Ceram Soc 59:12–16
29. Schick G, Lexa D, Leibowitz L (2012) Differential scanning calorimetry and differential thermal analysis. In: Kaufmann EN (ed) Characterization of materials. John Wiley & Sons Inc, New York, pp 483–495
30. Gutzow IS, Schmelzer JWP (2013) The vitreous state thermodynamics, structure, rheology, and crystallization. Springer, Heidelberg
31. Adam G, Gibbs JH (1965) On the temperature dependence of cooperative relaxation properties in glass-forming liquids. J Chem Phys 43:139–146
32. Sargsyan A, Tonoyan A, Davtyan S, Schick C (2007) The amount of immobilized polymer in pmma sio2 nanocomposites determined from calorimetric data. Eur Polym J 43:3113–3127
33. Donth E (1982) The size of cooperatively rearranging regions at the glass transition. J Non-Crystal Solids 53:325–330
34. Hemminger W, Hohne GWH (1984) Calorimetry - fundamentals and practice. Vch, Weinheim
35. Mathot VBF (1994) Calorimetry and thermal analysis of polymers. Hanser Publishers, München
36. Hohne GWH, Hemminger W, Flammersheim HJ (1996) Differential scanning calorimetry - an introduction for practitioners. Springer, Berlin
37. Wunderlich B (2005) Thermal analysis of polymeric materials. Springer, Berlin
38. Brown ME, Gallagher PK (2008) Recent advances, techniques and applications. Elsevier B. V, New York

39. Schick C (2012) Calorimetry. In: Matyjaszewski K, Möller M (eds) Polymer science: a comprehensive reference. Elsevier B.V., Amsterdam, pp 793–823
40. Wang X, Zhou W (2002) Glass transition of microtome-sliced thin films. Macromolecules 35:6747–6750
41. Koh YP, McKenna GB, Simon SL (2006) Calorimetric glass transition temperature and absolute heat capacity of polystyrene ultrathin films. J Polym Sci B: Polym Phys
42. Koh YP, Simon SL (2008) Structural relaxation of stacked ultrathin polystyrene films. J Polym Sci Part B: Polym Phys 46:2741–2753
43. Fukao K, Oda Y, Nakamura K, Tahara D (2010) Glass transition and dynamics of single and stacked thin films of poly(2-chlorostyrene). Eur Phys J Special Topics 189:165–171
44. Fukao K, Terasawa T, Nakamura K, Tahara D (2013) Heterogeneous and aging dynamics in single and stacked thin polymer films. In: Kanaya T (ed) Glass transition, dynamics and heterogeneity of polymer thin films. Springer, Berlin, pp 65–106
45. Zhuravlev E, Schick C (2010) Fast scanning power compensated differential scanning nanocalorimeter: 1. The device. Thermochim Acta. doi:10.1016/j.tca.2010.03.019
46. Zhuravlev E, Schick C (2010) Fast scanning power compensated differential scanning nanocalorimeter: 2. Heat capacity analysis. Thermochim Acta. doi:10.1016/j.tca.2010.03.020
47. Minakov AA, Schick C (2007) Ultrafast thermal processing and nanocalorimetry at heating and cooling rates up to 1 mk/s. Rev Sci Instrum 78:073902–073910
48. Adamovsky SA, Minakov AA, Schick C (2003) Scanning microcalorimetry at high cooling rate. Thermochim Acta 403:55–63
49. Minakov A, Morikawa J, Hashimoto T, Huth H, Schick C (2006) Temperature distribution in a thin-film chip utilized for advanced nanocalorimetry. Meas Sci Technol 17:199–207
50. Minakov AA, Adamovsky SA, Schick C (2005) Non adiabatic thin-film (chip) nanocalorimetry. Thermochim Acta 432:177–185
51. Efremov MY, Olson EA, Zhang M, Allen LH (2003) Glass transition of thin films of poly (2-vinyl pyridine) and poly(methyl methacrylate): nanocalorimetry measurements. Thermochim Acta 403:37–41
52. Leon-Gutierrez E, Garcia G, Lopeandia AF, Clavaguera-Mora MT, Rodriguez-Viejo J (2010) Size effects and extraordinary stability of ultrathin vapor deposited glassy films of toluene. J Phys Chem Lett 1:341–345
53. León-Gutierrez E, Garcia G, Clavaguera-Mora MT, Rodríguez-Viejo J (2009) Glass transition in vapor deposited thin films of toluene. Thermochim Acta 492:51–54
54. Leon-Gutierrez E, Garcia G, Lopeandia AF, Fraxedas J, Clavaguera-Mora MT, Rodriguez-Viejo J (2008) In situ nanocalorimetry of thin glassy organic films. J Chem Phys 129:181101–181104
55. van Herwaarden S, Iervolino E, van Herwaarden F, Wijffels T, Leenaers A, Mathot V (2011) Design, performance and analysis of thermal lag of the ufs1 twin-calorimeter chip for fast scanning calorimetry using the mettler-toledo flash dsc 1. Thermochim Acta 522:46–52
56. Mathot V, Pyda M, Pijpers T, Vanden Poel G, van de Kerkhof E, van Herwaarden S, van Herwaarden F, Leenaers A (2011) The Flash DSC 1, a power compensation twin-type, chip-based fast scanning calorimeter (FSC): first findings on polymers. Thermochim Acta 522:36–45
57. Yang Z, Fujii Y, Lee FK, Lam C-H, Tsui OKC (2010) Glass transition dynamics and surface layer mobility in unentangled polystyrene films. Science 328:1676–1679
58. Napolitano S, Wubbenhorst M (2007) Dielectric signature of a dead layer in ultrathin films of a nonpolar polymer. J Phys Chem B 111:9197–9199
59. Anatoli Serghei FK (2008) Metastable states of glassy dynamics, possibly mimicking confinement-effects in thin polymer films. Macromole Chem Phys 209:810–817
60. Reiter G (2013) Probing properties of polymers in thin films via dewetting. In: Kanaya T (ed) Glass transition, dynamics and heterogeneity of polymer thin films. Springer, Berlin, pp 29–63
61. Li Q, Simon SL (2009) Surface chemistry effects on the reactivity and properties of nanoconfined bisphenol m dicyanate ester in controlled pore glass. Macromolecules 42:3573–3579

62. Koh YP, Li Q, Simon SL (2009) Tg and reactivity at the nanoscale. Thermochim Acta 492:45–50
63. Serghei A (2008) Challenges in glassy dynamics of polymers. Macromole Chem Phys 209:1415–1423
64. Kraftmakher Y (2002) Modulation calorimetry and related techniques. Phys Reports 356:1–117
65. Corbino OM (1910) Thermische Oszillationen wechselstromdurchflossener Lampen mit dünnem Faden und daraus sich ergebende Anwesenheit geradzahliger Oberschwingungen. Physik Zeitschr XI(1910):413–417
66. Corbino OM (1911) Periodische Widerstandsänderungen feiner Metallfäden, die durch Wechselströme zum Glühen gebracht werden, sowie Ableitung ihrer thermischen Eigenschaften bei hoher Temperatur. Physik Zeitschr XII(1911):292–295
67. Birge NO, Nagel SR (1985) Specific-heat spectroscopy of the glass transition. Phys Rev Lett 54:2674–2677
68. Christensen T (1985) The frequency dependence of the specific heat at the glass transition. J Phys (Paris) 46(1985):C8–635-C638-637
69. Kraftmakher YA (1962) Modulation method of heat capacity measurement. Zhurnal prikladnoj mechaniki i techniyeceskoj fiziki 5:176–180
70. Sullivan P, Seidel G (1966) An ac temperature technique for measuring heat capacities. Ann Acad Sci Fennicae A VI(1966):58–62
71. Schäfer K (1940) Die stoßanregung intramolekularer schwingungen in gasen und gasmischungen vii. theorie der schalldispersion bei vorhandensein mehrerer normalschwingungen. Z Phys Chem 46:212–228
72. Birge NO (1986) Specific-heat spectroscopy of glycerol and propylene glycol near the glass transition. Phys Rev B 34:1631–1642
73. Christensen T, Olsen NB (1994) Determination of the frequency-dependent bulk modulus of glycerol using a piezoelectric spherical-shell. Phys Rev B 49:15396–15399
74. Jeong YH (1997) Progress in experimental techniques for dynamic calorimetry. Thermochim Acta 305:67–98
75. Richert R (2011) Calorimetry based on energy absorbed from time-dependent fields. J Non-Crystal Solids 357:726–730
76. Huang W, Richert R (2009) Dynamics of glass-forming liquids. xiii. microwave heating in slow motion. J Chem Phys 130:194509–194514
77. Gobrecht H, Hamann K, Willers G (1971) Complex plane analysis of heat capacity of polymers in the glass transition region. J Phys E: Sci Instrum 4:21–23
78. Sauerbrunn S, Crowe B, Reading M (1992) Modulated differential scanning calorimetry. Am Labor 24:44–47
79. Reading M (1993) Modulated differential scanning calorimetry - a new way forward in materials characterization. Trends Polym Sci 8:248–253
80. Reading M, Hourston DJ (2006) Modulated temperature differential scanning calorimetry: theoretical and practical applications in polymer characterisation. Springer, Berlin
81. Cassettari M, Papucci F, Salvetti G, Tombari E, Veronesi S, Johari GP (1993) Simultaneous measurements of enthalpy and heat-capacity of a thermosetting polymer during the curing process. Rev Sci Instrum 64:1076–1080
82. Wunderlich B (2003) Reversible crystallization and the rigid-amorphous phase in semicrystalline macromolecules. Prog Polym Sci 28:383–450
83. Schick C (2002) Temperature modulated differential scanning calorimetry (TMDSC) - Basics and applications to polymers. In: Brown ME (ed) Handbook of thermal analysis and calorimetry, series editor: Gallagher PK, Elsevier, Amsterdam, pp 713–810
84. Merzlyakov M, Schick C (2001) Simultaneous multi-frequency tmdsc measurements. Thermochim Acta 377:193–204
85. Androsch R, Wunderlich B (1999) Temperature-modulated dsc using higher harmonics of the fourier transform. Thermochim Acta 333:27–32

86. Schawe JEK (1996) Modulated temperature dsc measurements: the influence of the experimental conditions. Thermochim Acta 271:127–140
87. Cassel B (2000) A stepwise specific heat technique for dynamic dsc. Am Labor 32:23–26
88. Merzlyakov M, Schick C (2001) Step response analysis in dsc - a fast way to generate heat capacity spectra. Thermochim Acta 380:5–12
89. Kamasa P, Merzlyakov M, Pyda M, Pak J, Schick C, Wunderlich B (2002) Multi-frequency heat capacity measured with different types of tmdsc. Thermochim Acta 392-393:195–207
90. Schawe J, Hütter T, Heit C, Alig I, Lellinger D (2006) Stochastic temperature modulation: a new technique in temperature modulated dsc. Thermochim Acta 446:147–155
91. Merzlyakov M, Schick C (1999) Complex heat capacity measurements by tmdsc part 1. influence of non-linear thermal response. Thermochim Acta 330:55–64
92. Schick C, Merzlyakov M, Hensel A (1999) Nonlinear thermal response at the glass transition. J Chem Phys 111:2695–2700
93. Shoifet E, Chua YZ, Huth H, Schick C (2013) High frequency alternating current chip nano calorimeter with laser heating. Rev Sci Instrum 84:073903–073912
94. Huth H, Minakov AA, Serghei A, Kremer F, Schick C (2007) Differential ac-chip calorimeter for glass transition measurements in ultra thin polymeric films. Eur Phys J Special Topics 141:153–160
95. Huth H, Minakov AA, Schick C (2006) Differential ac-chip calorimeter for glass transition measurements in ultrathin films. J Polym Sci B Polym Phys 44:2996–3005
96. Zhou D, Huth H, Gao Y, Xue G, Schick C (2008) Calorimetric glass transition of poly(2,6-dimethyl-1,5-phenylene oxide) thin films. Macromolecules 41:7662–7666
97. Revaz B, Zink BL, Hellman F (2005) Si-n membrane-based microcalorimetry: heat capacity and thermal conductivity of thin films. Thermochim Acta 432:158–168
98. Denlinger DW, Abarra EN, Allen K, Rooney PW, Messer MT, Watson SK, Hellman F (1994) Thin-film microcalorimeter for heat-capacity measurements from 1.5-k to 800-k. Rev Sci Instrum 65:946–958
99. Queen DR, Hellman F (2009) Thin film nanocalorimeter for heat capacity measurements of 30 nm films. Rev Sci Instrum 80:063901–063907
100. Queen DR, Liu X, Karel J, Metcalf TH, Hellman F (2013) Excess specific heat in evaporated amorphous silicon. Phys Rev Lett 110:135901
101. Cooke DW, Hellman F, Groves JR, Clemens BM, Moyerman S, Fullerton EE (2011) Calorimetry of epitaxial thin films. Rev Sci Instrum 82:023904–023908
102. Minakov A, Wurm A, Schick C (2007) Superheating in linear polymers studied by ultrafast nanocalorimetry. Eur Phys J E Soft Matter 23:43–53
103. Huth H, Beiner M, Weyer S, Merzlyakov M, Schick C, Donth E (2001) Glass transition cooperativity from heat capacity spectroscopy - temperature dependence and experimental uncertainties. Thermochim Acta 377:113–124
104. Schonhals A, Schick C, Huth H, Frick B, Mayorova M, Zorn R (2007) Molecular dynamics in glass-forming poly(phenyl methyl siloxane) as investigated by broadband thermal, dielectric and neutron spectroscopy. J Non-Crystal Solids 353:3853–3861
105. Schönhals A, Goering H, Schick C, Frick B, Zorn R (2005) Polymers in nanoconfinement: what can be learned from 3 relaxation and scattering experiments? J Non-Crystal Solids 351:2668–2677
106. Schönhals A, Goering H, Schick C, Frick B, Zorn R (2004) Glass transition of polymers confined to nanoporous glasses. Colloid Polym Sci 282:882–891
107. Schönhals A, Goering H, Schick C, Frick B, Zorn R (2003) Glassy dynamics of polymers confined to nanoporous glasses revealed by relaxational and scattering experiments. Eur Phys J E 12:173–178
108. Schönhals A, Goering H, Brzezinka KW, Schick C (2003) The molecular dynamics of polymers in random nanometer confinements investigated by relaxational and scattering techniques. J Phys Cond Mater 15:S1139–S1148
109. Schönhals A, Goering H, Schick C (2002) Segmental and chain dynamics of polymers: from the bulk to the confined state. J Non-Crystal Solids 305:140–149

110. Serghei A, Huth H, Schellenberger M, Schick C, Kremer F (2005) Pattern formation in thin polystyrene films induced by an enhanced mobility in ambient air. Phys Rev E 71:061801-061801-061801-061804
111. Serghei A, Huth H, Schick C, Kremer F (2008) Glassy dynamics in thin polymer layers having a free upper interface. Macromolecules 41:3636–3639
112. Serghei A, Mikhailova Y, Huth H, Schick C, Eichhorn KJ, Voit B, Krerner F (2005) Molecular dynamics of hyperbranched polyesters in the confinement of thin films. Eur Phys J E 17:199–202
113. Minakov AA, Roy SB, Bugoslavsky YV, Cohen LF (2005) Thin-film alternating current nanocalorimeter for low temperatures and high magnetic fields. Rev Sci Instrum 76:043906
114. van Herwaarden AW (2005) Overview of calorimeter chips for various applications. Thermochim Acta 432:192–201
115. Lupascu V, Huth H, Schick C, Wübbenhorst M (2005) Specific heat and dielectric relaxations in ultra-thin polystyrene layers. Thermochim Acta 432:222–228
116. Kearns KL, Ediger MD, Huth H, Schick C (2010) One micrometer length scale controls kinetic stability of low-energy glasses. J Phys Chem Lett 1:388–392
117. Gotzen N-A, Huth H, Schick C, van Assche G, Neus C, van Mele B (2010) Phase separation in polymer blend thin films studied by differential ac chip calorimetry. Polymer 51:647–654
118. Kearns KL, Whitaker KR, Ediger MD, Huth H, Schick C (2010) Observation of low heat capacities for vapor-deposited glasses of indomethacin as determined by AC nanocalorimetry. J Chem Phys doi:10.1063/1.3442416
119. Whitaker K, Scifo D, Ediger MD, Ahrenberg M, Schick C (2013) Highly stable glasses of cis-decalin and cis/trans-decalin mixtures. J Phys Chem B 117:12724–12733
120. Ahrenberg M, Chua YZ, Whitaker KR, Huth H, Ediger MD, Schick C (2013) In situ investigation of vapor-deposited glasses of toluene and ethylbenzene via alternating current chip-nanocalorimetry. J Chem Phys 138:024501–024511
121. Whitaker KR, Ahrenberg M, Schick C, Ediger MD (2012) Vapor-deposited alpha, alpha, beta-tris-naphthylbenzene glasses with low heat capacity and high kinetic stability. J Chem Phys 137:154502–154507
122. Ahrenberg M, Shoifet E, Whitaker KR, Huth H, Ediger MD, Schick C (2012) Differential alternating current chip calorimeter for in situ investigation of vapor-deposited thin films. Rev Sci Instrum 83:033902–033912
123. Merzlyakov M (2003) Integrated circuit thermopile as a new type of temperature modulated calorimeter. Thermochim Acta 403:65–81
124. Svanidze AV, Huth H, Lushnikov SG, Schick C (2012) Study of phase transition in tetragonal lysozyme crystals by ac-nanocalorimetry. Thermochim Acta 544:33–37
125. Svanidze AV, Huth H, Lushnikov SG, Seiji K, Schick C (2009) Phase transition in tetragonal hen egg-white lysozyme crystals. Appl Phys Lett 95:263702
126. Ahrenberg M, Brinckmann M, Schmelzer JWP, Beck M, Schmidt C, Kragl KU, Verevkin SP, Schick C (2014) Determination of volatility of ionic liquids at the nanoscale by means of ultra-fast scanning calorimetry. Phys. Chem. Chem. Phys.
127. Weyer S, Hensel A, Schick C (1997) Phase angle correction for tmdsc in the glass-transition region. Thermochim Acta 305:267–275
128. Jiang W, Du M, Gu Q, Jiang J, Huth H, Zhou D, Xue G, Schick C (2010) Calorimetric study of blend miscibility of polymers confined in ultra-thin films. Eur Phys J Special Topics 189:187–195
129. Gotzen N-A, Van Assche G, Van Mele B (2011) Construction of the state diagram of polymer blend thin films using differential ac chip calorimetry. Polymer 52:4277–4283
130. Hall DB, Underhill P, Torkelson JM (1998) Spin coating of thin and ultrathin polymer films. Polym Eng Sci 38:2039–2045
131. Serghei A, Kremer F (2008) Metastable states of arrested glassy dynamics, possibly mimicking confinement effects in thin polymer films. Macromol Chem Phys 209:810–817
132. Serghei A, Kremer F (2008) Challenges in glassy dynamics of polymers. Macromol Chem Phys 209:1415–1423

133. Serghei A, Kremer F (2006) Unexpected preparative effects on the properties of thin polymer films. Prog Colloid Polym Sci 132:33–40
134. Reiter G, de Gennes PG (2001) Spin-cast, thin, glassy polymer films: highly metastable forms of matter. Eur Phys J E 6:25–28
135. Thomas KR, Chenneviere A, Reiter G, uuml, nter, Steiner U (2011) Nonequilibrium behavior of thin polymer films. Phys Rev E 83:021804
136. Tress M, Erber M, Mapesa EU, Huth H, Müller J, Serghei A, Schick C, Eichhorn K-J, Voit B, Kremer F (2010) Glassy dynamics and glass transition in nanometric thin layers of polystyrene. Macromolecules 43:9937–9944
137. Mapesa EU, Tress M, Schulz G, Huth H, Schick C, Reiche M, Kremer F (2013) Segmental and chain dynamics in nanometric layers of poly (cis-1,4-isoprene) as studied by Broadband Dielectric Spectroscopy and temperature-modulated Calorimetry. Soft Matter
138. Yin H, Schönhals A (2013) Calorimetric glass transition of ultrathin poly(vinyl methyl ether) films. Polymer 54:2067–2070
139. Yin H, Schonhals A (2012) Calorimetric glass transition of ultrathin poly(bisphenol a carbonate) films. Soft Matter 8:9132–9139
140. Schick C (2010) Glass transition under confinement-what can be learned from calorimetry. Eur Phys J Special Topics 189:3–36
141. Sepúlveda A, Leon-Gutierrez E, Gonzalez-Silveira M, Rodríguez-Tinoco C, Clavaguera-Mora MT, Rodríguez-Viejo J (2011) Accelerated aging in ultrathin films of a molecular glass former. Phys Rev Lett 107:025901
142. Keddie JL, Jones RAL, Cory RA (1994) Size-dependent depression of the glass transition temperature in polymer films. Europhys Lett 27:59–64
143. Zhang C, Boucher VM, Cangialosi D, Priestley RD (2013) Mobility and glass transition temperature of polymer nanospheres. Polymer 54:230–235
144. Napolitano S, Cangialosi D (2013) Interfacial free volume and vitrification: reduction in tg in proximity of an adsorbing interface explained by the free volume holes diffusion model. Macromolecules 46:8051–8053
145. Cangialosi D, Boucher VM, Alegría A, Colmenero J (2013) Direct evidence of two equilibration mechanisms in glassy polymers. Phys Rev Lett 111:095701
146. Cangialosi D, Boucher VM, Alegría A, Colmenero J (2013) Volume recovery of polystyrene/silica nancomposites. J Polym Sci Part B Polym Phys 51:847–853
147. Cangialosi D, Boucher VM, Alegria A, Colmenero J (2013) Physical aging in polymers and polymer nanocomposites: recent results and open questions. Soft Matter
148. Cangialosi D, Boucher VM, Alegría A, Colmenero J (2012) Enhanced physical aging of polymer nanocomposites: the key role of the area to volume ratio. Polymer 53:1362–1372
149. Boucher VM, Cangialosi D, Alegría A, Colmenero J (2012) Enthalpy recovery in nanometer to micrometer thick polystyrene films. Macromolecules 45:5296–5306
150. Cangialosi D, Boucher VM, Alegría A, Colmenero J (2011) Free volume holes diffusion to describe physical aging in poly(mehtyl methacrylate)/silica nanocomposites. J Chem Phys 135:014901
151. Boucher VM, Cangialosi D, Alegria A, Colmenero J, Pastoriza-Santos I, Liz-Marzan LM (2011) Physical aging of polystyrene/gold nanocomposites and its relation to the calorimetric tg depression. Soft Matter 7:3607–3620
152. Boucher VM, Cangialosi D, Alegrìa A, Colmenero J, González-Irun J, Liz-Marzan LM (2011) Physical aging in pmma/silica nanocomposites: enthalpy and dielectric relaxation. J Non-Crystal Solids 357:605–609
153. Tress M, Mapesa EU, Kossack W, Kipnusu WK, Reiche M, Kremer F (2013) Glassy dynamics in condensed isolated polymer chains. Science 341:1371–1374
154. Kremer F, Mapesa EU, Tress M, Reiche M (2013) Molecular dynamics of polymers at nanometric length scales: from thin layers to isolated coils. In: Kalmykov YP (ed) Recent advances in broadband dielectric spectroscopy. Springer, Netherlands, pp 163–178
155. Serghei A, Hartmann L, Kremer F (2007) Molecular dynamics in thin films of isotactic poly(methylmethacrylate) - revisited. J Non-Crystal Solids 353:4330–4333

156. Schönhals A, Schick C, Huth H, Frick B, Mayorova M, Zorn R (2007) Molecular mobility of poly(phenyl methyl siloxane) investigated by thermal, dielectric and neutron spectroscopy. Polym Mater Sci Eng 97:948–949
157. Lodge TP, McLeish TCB (2000) Self-concentrations and effective glass transition temperatures in polymer blends. Macromolecules 33:5278–5284
158. Kim JH, Jang J, Lee DY, Zin WC (2002) Thickness and composition dependence of the glass transition temperature in thin homogeneous polymer blend films. Macromolecules 35:311–313
159. An LJ, He DY, Jing JK, Wang ZG, Yu DH, Jiang BZ, Jiang ZH, Ma RT (1997) Effects of molecular weight and interaction parameter on the glass transition temperature of polystyrene mixtures and its blends with polystyrene/poly(2,6-dimethyl-p-phenylene oxide). Eur Polym J 33:1523–1528
160. Swallen SF, Kearns KL, Mapes MK, Kim YS, McMahon RJ, Ediger MD, Wu T, Yu L, Satija S (2007) Organic glasses with exceptional thermodynamic and kinetic stability. Science 315:353–356
161. Jiang Z, Imrie CT, Hutchinson JM (1998) Temperature modulated differential scanning calorimetry - part-i - effects of heat-transfer on the phase-angle in dynamic adsc in the glass-transition region. Thermochim Acta 315:1–9
162. Huang W, Richert R (2009) Reverse dynamic calorimetry of a viscous ionic liquid. J Chem Phys 131:184501–184507
163. Wang L-M, Richert R (2007) Measuring the configurational heat capacity of liquids. Phys Rev Lett 99:185701
164. Alig I, Lellinger D, Oehler H, Adamovsky SA, Schick C (2008) Microcalorimetry for characterization of film formation and cure of coatings and adhesives. Prog Org Coatings 61:166–175

Equilibrium and Out-of-Equilibrium Dynamics in Confined Polymers and Other Glass Forming Systems by Dielectric Spectroscopy and Calorimetric Techniques

Daniele Cangialosi

Abstract Glassy dynamics under nanoscale confinement is currently a topic under intense debate in soft matter physics. The reason is that this kind of studies may deliver important insight on the glassy dynamics in general. Furthermore, from a technological point of view, there exists a rising interest in the understanding of how properties are modified at the nanoscale in comparison to the corresponding bulk system. Within this context, this chapter critically discusses the experimental findings in the field. The vast majority of results concerns thin polymer films. However, other geometries of confinement, such as polymer nanocomposites and nanospheres, are considered as well. Special attention is devoted to the kind of information achieved by a specific technique. Within this context, the ability of dielectric and calorimetric techniques is highlighted. Particular attention is devoted to the determination of the different aspects of glassy dynamics in confinement, that is, the equilibrium dynamics in terms of the rate of spontaneous fluctuations as probed by experiments where a perturbation in the linear regime is applied, on the one hand, and the out-of-equilibrium dynamics in terms of thermal glass transition temperature (T_g) and the physical aging on the other. In the latter case, the application of a temperature ramp for T_g measurements and the recovery of equilibrium in physical aging imply the application of large perturbations, in particular with amplitude well beyond that of spontaneous fluctuations. It is demonstrated how, in view of numerous experimental results, the two aspects are not one-to-one related in confinement. Specifically, the reduction in T_g and the acceleration of equilibrium recovery in the aging regime does not imply a concomitant speed-up of the rate of spontaneous fluctuations, which is in several cases found to be unaltered in comparison to the bulk. Finally, a description of suitable frameworks to describe such phenomenology is presented with special attention to the free volume hole diffusion (FVHD) model. This is shown to quantitatively

D. Cangialosi (✉)
Centro de Física de Materiales (CSIC-UPV/EHU), Paseo Manuel de Lardizabal 5,
20018 San Sebastián, Spain
e-mail: swxcacad@sw.ehu.es

F. Kremer (ed.), *Dynamics in Geometrical Confinement*, Advances in Dielectrics, 339
DOI: 10.1007/978-3-319-06100-9_13, © Springer International Publishing Switzerland 2014

catch the acceleration of physical aging and the T_g depression with no need to assume any acceleration on the intrinsic molecular mobility of the glass former.

Keywords Glass transition · Molecular mobility · Linear response · Out -of-Equilibrium dynamics

Abbreviations

AG theory	Adam-Gibbs Theory
CD	Capacitive Dilatometry
CRR	Cooperative Rearranging Region
FVHD	Free Volume Hole Diffusion Model
NEXAFS	Near-Edge X-Ray Absorption Fine Structure
PALS	Positron Annihilation Lifetime Spectroscopy
PMMA	Poly(methyl methacrylate)
RFOT	Random First-order Theory
SMFM	Shear Modulation Force Microscopy
VFT	Vogel-Fulcher-Tammann

1 Introduction

Modern technology often requires the employment of materials confined at the nanoscale, that is, with typical dimensions in the submicron range. In such configuration, properties can be dramatically affected due to the dominant role of those portions of the material close to the interface. Therefore, the knowledge of how the interface interferes with the overall material performance is crucial. Among those properties that can be deeply affected under nanoscale confinement, those related to glass transition phenomena have been intensively investigated in the last years. The reason is that glass forming materials are a widely employed class of materials since the beginning of civilization (their use among phoenicians dating back to the third millennium a.C. is reported). Among them a subclass is represented by glass forming polymers and most of the studies of the glass transition in confinement involve these systems. The one-dimensional confinement, namely that based on polymer thin films, is by far the most investigated, although in recent years much effort has been devoted to the investigation of confined glass formers in other geometries. In particular polymer nanocomposites and nanospheres, and glass formers in nanopores have recently received considerable attention.

In this chapter, we first briefly introduce the main aspects of the phenomenology of the glass transition. In particular, the existence of a typical length scale associated with density fluctuations relevant for glassy dynamics will be introduced. This, among the different facets of the phenomenology of the glass transition, has historically represented one of the main rationale to study glass forming systems confined at length scales approaching those relevant for glassy dynamics. After introducing these aspects, we provide a review on the debate regarding the effect of confinement on such phenomena. This discussion involves the apparently controversial views based on different experimental observations, where the glassy dynamics is seen to either speed-up, slow-down, or remain equal to that of the bulk system. In doing so, the following aspects will be emphasized: (i) the conceptual difference among the different kind of determinations delivering information on nonequivalent aspects of the glass transition phenomenon; (ii) how—given a certain confinement length scale, e.g., the thickness in thin polymer films—the applied cooling rate, the kind of substrate, and the preparation conditions affect the deviation of glass transition phenomena from the bulk behavior. In this context, the numerous experimental contributions will be reviewed. Particular emphasis will be provided to the study of glassy dynamics by dielectric relaxation and calorimetric techniques. These allow the determination of the different aspects of glassy dynamics, in some cases in a single experiment. The critical analysis emphasizing the different factors affecting the glass transition in confinement of such contributions can in principle provide a suitable framework to understand what kind of deviations should be expected in given confining conditions and experimental protocols. Finally, some possible theoretical frameworks, recently introduced into the scientific debate, to account for the different aspects of glassy dynamics in confinement will be reported.

2 Glassy Dynamics: Established Facts

From a thermodynamic point of view, the fate of a liquid cooled down below the melting temperature is that of transforming to the most stable crystalline phase. However, for kinetic reasons, a considerable number of liquids can be supercooled, that is, they remain amorphous over large time scales [29, 110]. Among glass forming systems, polymers represent an important subclass, due the fact that crystallization is often hindered by chain connectivity and, in some cases, conformational irregularities.

Further temperature reduction of a supercooled liquid at a given rate leads to the formation of a glass, that is, a disordered system with the mechanical properties of a solid. The temperature marking the liquid to glass transformation is commonly addressed as the glass transition temperature (T_g). At such temperature, a jump in thermodynamic coefficients (the specific heat, the compressibility, the coefficient of thermal expansion, etc.) occurs. This is reminiscent of a second-order thermodynamic transition in the Ehrenfest classification [110]. However, several experimental aspects of the glass transition indicates that this is actually not the case. Among them, probably the most evident is the fact that the T_g depends on the cooling rate, that is, the timescale of the experiment. Nowadays, it is well accepted that the glass transition is intimately linked to the dramatic slowing down of density fluctuations associated to the glassy dynamics, the so-called α relaxation, with decreasing temperature occurring in the supercooled state. This is often described by the Vogel-Fulcher-Tammann (VFT) equation: $\tau = \tau_0 \exp(B/(T - T_0))$. Here, τ is the relaxation time relevant for spontaneous fluctuation, τ_0 a pre-exponential factor, and B and T_0 the Vogel activation energy and temperature, respectively.

For bulk glass formers, it has been shown that, not only there exists a relation between the time scale of spontaneous fluctuations τ and the T_g, but that the two magnitudes are unequivocally related to each other [34, 125]. This is reflected in the dependence of the T_g with the cooling rate, exhibiting the same VFT behavior as that of τ. Hence, cooling down a supercooled melt below its T_g implies that, in the timescale of the experiment, dictated by the applied cooling rate, the rate of spontaneous fluctuations is too small to maintain equilibrium. However, once a glass is brought into the glassy state, it spontaneously evolves toward equilibrium, an aspect of glassy dynamics commonly known as physical aging. As for the T_g, in bulk glass formers, it has been shown that the rate of approach to equilibrium is exclusively determined by that of spontaneous fluctuations [55, 66]. Nevertheless, it has to be remarked that, from a conceptual point of view, measuring τ, on the one hand, and the T_g or the time to equilibrate in the physical aging regime, on the other, are two separate aspects of the dynamics of the glass transition. In particular, measuring τ requires the application of a perturbation in the linear regime, that is, with amplitude smaller than those of spontaneous fluctuations, in order to fulfill the fluctuation-dissipation theorem [17, 91]. Conversely, measuring the T_g at a given rate or the recovery of equilibrium in the physical aging regime entails the application of perturbations beyond the linear regime (e.g., a change of temperature with a certain cooling ramp in T_g measurements). As will be seen in the next sections of the chapter, this conceptual difference constitutes an important ingredient in the understanding of experimental results on glassy dynamics in confinement.

A long-debated aspect of the drastic slowing down of the dynamics of the glass transition when decreasing the temperature is the possible existence of a concomitant growing length scale, that is, the spatial extent of the relaxation. This has been long ago put forward by Adam-Gibbs (AG) [1], who theorized that the magnitude controlling the slowing down of the α relaxation with decreasing temperature is the configurational entropy (S_c). According to the AG theory, the α relaxation occurs via cooperative rearrangement of several basic structural units. The AG theory suggests that the number of units (z^*) involved in cooperatively rearranging regions (CRR) increases with decreasing temperature and such temperature variation is also related to the configurational entropy: $z^* \sim S_c^{-1}$.

Since the AG theory, numerous approaches have been presented in the search for the relevant length scale of the α relaxation. As the AG theory, some of them, such as that proposed by Donth [33, 56] and the random first-order theory (RFOT) [76], rely on thermodynamics. Conversely others, such as that based on string-like motion [32] and the four point dynamic susceptibility [8], are based on the estimation of dynamically correlated structural units. Among the different approaches seeking for the relevant length scale of the α process, within the context of this chapter, it is worth considering that based on the self-concentration [73]. Such concept was introduced to explain the presence of two glass transitions in miscible polymer blends. It relies on the fact, that due to the limited size of CRR(s), the effective concentration within the cooperative volume differs from the macroscopic one [18, 19, 25]. In particular, it is richer in the component of the target unit at the center of the CRR. As such the self-concentration approach provides information on how far the dynamics of a given structural unit is affected by the surrounding. This is especially relevant once the effect of an interface in nanoscale confinement is considered. In all approaches providing an estimation of the relevant length of the α relaxation, this is generally found to be in the order of several nanometers or, for some polymer and within some approaches, even smaller than 1 nm [5, 18, 19].

3 Glassy Dynamics Under Nanoscale Confinement: Experimental Observations

3.1 Thermal Glass Transition

In the previous section, among the different aspects of glassy dynamics, those related to the presence of a typical length scale have been emphasized. As a consequence, if such a length scale really exists, it is possible to speculate *a-priori* that once the dimensions of the glass former approach those of such length scale, a modification of glassy dynamics must be expected.

In this context, pioneering studies of McKenna and co-workers [60] on several low molecular weight glass formers confined in nanopores with diameters as low as 4 nm showed clear depression of the calorimetric T_g. The first report of confinement effects in glass forming polymers was presented by Keddie et al. [62]. They employed ellipsometry to determine the temperature dependence of the thickness of thin polystyrene (PS) films supported on silicon wafers. They found significant T_g depression in films thinner than 50 nm. Since the work of Keddie et al. [62], a huge amount of work has been presented on the glass transition of thin polymer films in different configurations, included supported [4, 30, 35, 37, 38, 42, 44, 47, 51, 52, 107, 111, 117, 123, 124], capped [16, 46, 77, 98] and freestanding films [4, 10, 16, 41, 63, 64, 84, 102, 126]. Most of the results generally indicate that nanoscale confinement induces T_g depression. As in the work of Keddie et al. [62], effects on the T_g are visible at thicknesses smaller than 50 nm and the largest depression is of the order of 30 K for films thinner than 10 nm. Apart from this, it is important to remark that T_g determinations deliver considerable scattering of data. This implies that other factors, beyond the film thickness, are of importance in determining the magnitude of T_g depression in thin polymer films. This becomes immediately clear once T_g data of freestanding polymer films are considered [4, 10, 16, 41, 63, 64, 84, 102, 126]. In such a case, decreases of T_g in comparison to the corresponding bulk polymer as large as 70 K are observed and effects are visible already for thicknesses larger than 100 nm [10]. Furthermore, while for thin PS films the vast majority

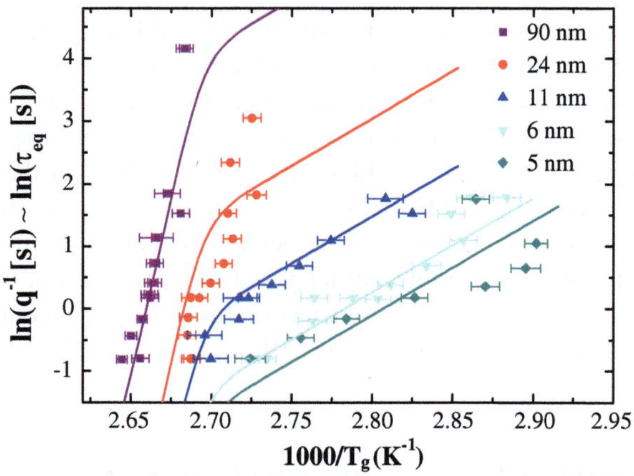

Fig. 1 Natural logarithm of the inverse cooling rate, q^{-1}, versus reciprocal glass transition temperature, $1000/T_g$, for supported thin PS films on platinum-coated silicon nitride with different films thicknesses. *Continuous lines* are the fits of the FVHD model to experimental data (Reprinted with permission from Ref. [12])

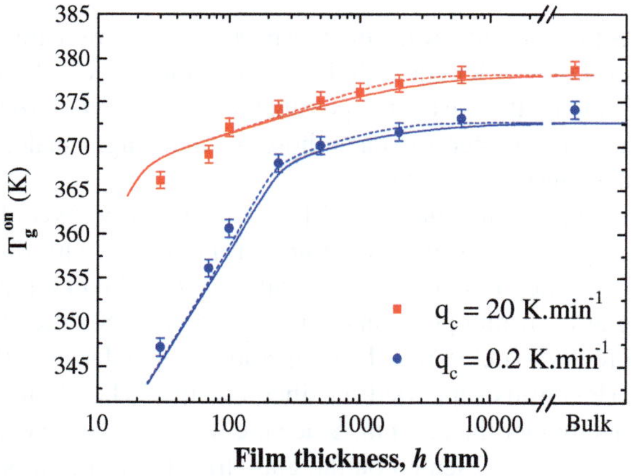

Fig. 2 T_g as a function of films thickness and applied cooling rate (q_c) for freestanding thin PS films. *Continuous* and *dashed lines* are the fits of the FVHD model to experimental data via two different approaches. For details regarding the difference between the two approaches see Ref. [10]. (Reprinted with permission from Ref. [10])

of studies generally provides evidence for T_g depression, there exists a significant number of studies on thin poly(methyl methacrylate) (PMMA) films supported on silicon-based substrates [52, 107], where an increase of the T_g in comparison to the bulk polymers is observed. These observations suggest that the nature of the interface is an important parameter affecting the magnitude of T_g depression.

An additional parameter that has been considered in the study of thin polymer films is the dependence of the magnitude of T_g depression on the applied cooling rate. This aspect has been considered after the introduction into the scientific debate of the study by fast calorimetry by Efremov et al. [35]. Studying thin PS films, they showed little effect, if any, of the film thickness down to several nanometers on the T_g measured at 9×10^4 K min^{-1} cooling rate. A systematic study on the cooling rate dependence of the T_g in thin PS films was later conducted by Fakhraai and Forrest [38] by ellipsometry, a technique delivering information of the temperature dependence of the film thickness. They showed pronounced thickness dependence of the T_g at low cooling rates, as in the vast majority of experiments. However, when the cooling rate is increased, the magnitude of the T_g depression reduces. This is shown in Fig. 1, where data of Fakhraai and Forrest [38] are summarized. Similar investigations have been performed by calorimetry at two cooling rates in freestanding thin PS films [10]. These results are presented in Fig. 2.

With regard to supported or capped thin films, very recently, the effect of polymer adsorption on the substrate has been investigated. In particular, it has been shown that extended annealing at temperatures substantially larger than the T_g of the bulk polymer (453 K for PS) induced a reduction of confinement effects on the T_g [87, 89, 106, 113, 129]. Furthermore, the annealing time dependence of such a reduction was shown to be significantly dependent on the polymer's molecular weight, as shown in Fig. 3. In particular, for films with molecular weight larger than 160 Kg mol^{-1} no shift in T_g were observed for annealing times as large as 10^5 s. Conversely, at similar annealing times lower molecular weights films could recover the bulk T_g. This indicates that the kinetic of adsorption is somehow related to chain dynamics. Further investigation on the impact of high temperature annealing on T_g's deviations revealed that, rather than the thickness of the adsorbed layer, the crucial parameter determining the magnitude of T_g depression is the amount of free interface [87].

Beside the huge scientific activity on thin polymer films, several studies on the effect on the thermal T_g in different kind of confinement have been carried out. These involve polymer nanocomposites and nanospheres, and glass formers confined in nanopores. With regard to polymer nanocomposites, those systems exhibiting weak interactions with the nanoparticles' surface generally exhibit T_g depression [7, 11, 15, 22, 105]. This result suggests a general analogy with thin polymer films [7]. Conversely, in those polymer nanocomposites where strong interactions, such as hydrogen bonding, are allowed, a T_g increase is observed [75, 100, 105].

Recent scientific activity has been devoted to the study of the thermal glass transition in polymer nanospheres. Here, results appears to be somewhat scattered even for the same type of interface. In particular, some calorimetric studies report no T_g dependence with the nanosphere diameter [48, 120]. Conversely, several works show that freestanding nanospheres with diameter of the order of several tens of nanometers generally exhibit increased T_g in comparison to the bulk [80, 83]. However, once the nanospheres diameter approaches 100 nm or larger a reduction in T_g is observed [31, 131, 132]. This nonmonotonic T_g dependence on the nanospheres diameter has been explained according to entropic arguments by Martinez-Tong et al. [80]. They pointed out that, when the nanospheres radius of curvature approaches the typical size of the macromolecules (e.g., the radius of gyration), a decrease of the number of configurational degrees of freedom occurs and, as a consequence,

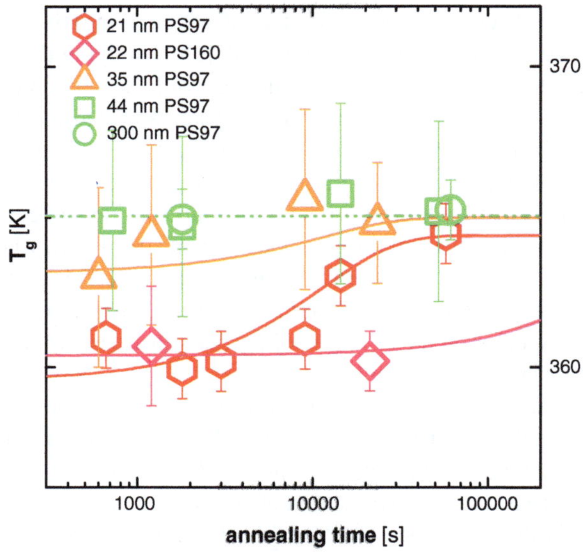

Fig. 3 T_g as a function of annealing time at 453 K for Al-capped thin PS films. PS97 and PS160 stand for polystyrenes with molecular weight 97 and 160 Kg mol^{-1}, respectively (Re-adapted from Ref. [89])

the molecular motion associated to the α relaxation is slowed down. As in the case of polymer thin films, an additional ingredient in determining the magnitude of T_g deviations from bulk behavior is the nature of the interface. Recent studies showed that PS nanospheres exhibit either decrease or no-change in T_g depending on the presence of surfactants and its nature at the interface [40] or the presence of silica capped on the surface [132].

Regarding the increase of T_g in polymer nanospheres with diameter in the sub-100 nm regime [80, 83], whatever the explanation, it is obvious that there exists a marked difference with thin polymer films with the same (equivalent) size. Conversely for nanospheres with diameter larger than 100 nm, analogy with thin polymer films can be put forward [131]. Hence, the difference between polymer nanospheres and thin films must originate from the curvature at the interface in the former geometry and its effect on the polymer conformation at such interface. This conclusion is corroborated by experiments in low molecular weight glass formers confined in spherical nanopores, which exhibit decrease [60] or no change in T_g.

3.2 *Out-of-Equilibrium Dynamics*

The glass transition marks the crossover from the equilibrium melt state to the out-of-equilibrium glass. Physical aging, that is the recovery of equilibrium of the glass,

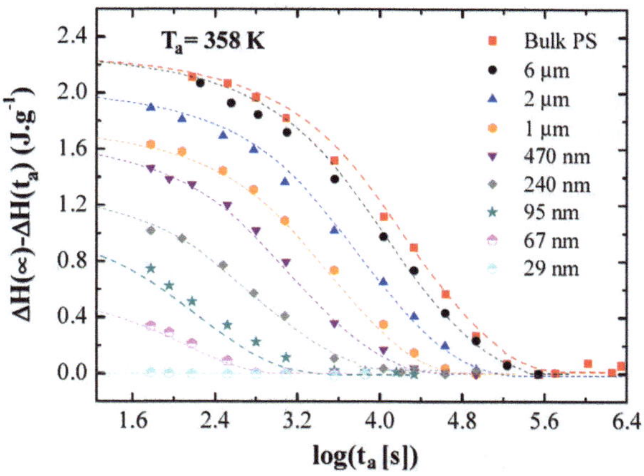

Fig. 4 Recovered enthalpy, expressed in terms of distance from equilibrium, for freestanding thin PS films and bulk PS at 358 K (Reprinted with permission from Ref. [10])

in nanoscale confinement has been deeply investigated in recent years [23, 96]. This phenomenon is intimately linked to the thermal glass transition. A T_g decrease indicates the ability of the glass to equilibrate more efficiently. Hence, if confinement-dependent T_g is observed, it is obvious to expect deep effects also in the physical aging behavior. This is actually found in several works in polymer thin films [10, 65], nanocomposites [3, 11, 14, 15, 22, 24] and nanospheres [53], where the entire enthalpy or volume recovery function is obtained in calorimetric and dilatometric experiments, respectively. All these studies show faster achievement of equilibrium in the nanostructured glass in comparison to the bulk counterpart. As an example, the enthalpy recovery during physical aging of freestanding thin PS films at 358 K, taken from Ref. [10], is presented in Fig. 4. Importantly, deviations from the bulk physical aging behavior can be clearly observed at thicknesses as large as several microns.

Other studies reported the aging time dependence of a given observable in a relatively limited time window. In such a case, an aging rate, defined as the slope of the decay in the observable as a function of the logarithm of the aging time [57], is determined. The studies delivering such information are based on the employment of ellipsometry [6, 43, 61, 101, 103, 104], fluorescent spectroscopy [97, 99, 100], dielectric methods [13, 26, 45, 100], permeability measurements [85, 108], dilatometry [24], and Positron annihilation lifetime spectroscopy (PALS) [109]. All of these studies point toward significant effects of confinement on the physical aging behavior. Interestingly, when the depth profile of the physical aging pattern is determined [97, 109], achievement of thermodynamic equilibrium appears to be faster the closer the free interface of the film is. Other factors, such as the presence of mechanical stress [103], the type of interface [104] and the chain architecture [43], have been shown to be of importance in determining the magnitude of the aging rate.

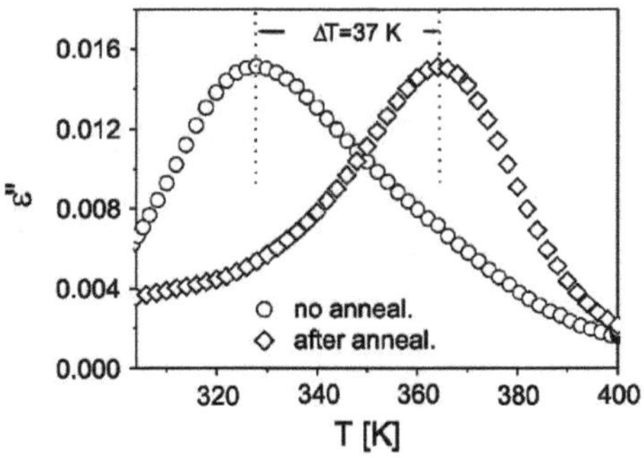

Fig. 5 Loss part of the dielectric permittivity as a function of temperature for PMMA at 12 kHz with thickness 57 nm before and after annealing for 12 h at 400 K in pure nitrogen (Reprinted with permission from Ref. [113])

3.3 Dynamic Glass Transition

Besides the investigation of the thermal glass transition, numerous studies have documented the effect of confinement on the rate of spontaneous fluctuations associated with the α process. With regard to polymer thin films, early experiments showed acceleration of the α relaxation with decreasing thickness. The first important contribution in the field was provided by Fukao and Miyamoto [46]. They showed that the molecular dynamics—probed by broadband dielectric spectroscopy (BDS)—of Al-capped PS films thinner than 20 nm was accelerated in comparison to the bulk. Similar results, also by BDS, were later reported by other authors [54, 116].

In subsequent studies, Kremer and co-workers emphasized that preparation and experimental conditions may have significant effect on BDS results [113]. In particular, they showed that annealing above T_g to remove the solvent of spin-coated thin polymer films and the environmental conditions of the experiments (nitrogen *verus* air) are key factors in determining the location of the most probable frequency of relaxation in BDS experiments. This is shown in Fig. 5, where the dielectric response as a function of the temperature at 12 kHz for a thin PMMA film with thickness 57 nm is presented. In this figure, it is shown how the typical temperature of the α relaxation shifts by more than 30 K once the films are annealed over extended time well above T_g. Beside these experiments, Perlich et al. [94] showed that solvent removal in supported thin PS films was considerably more difficult than in the bulk polymer. The important consequence of these observations was that, once measured in inert atmosphere (e.g., nitrogen) and prepared under drastic conditions for solvent removal, thin films exhibited identical molecular dynamics as the bulk counterpart. This result has been found measuring the α relaxation by different techniques, included BDS [16, 67, 79, 88, 89, 112, 121, 122, 128, 130], AC-calorimetry [58, 79, 121], shear modulation force microscopy (SMFM) [49], Near-edge X-ray absorption fine structure (NEXAFS) [72] and optical photobleaching [93]. In the latter case, pronounced bulk-like dynamics was observed for freestanding PS films as thin as 10 nm, as shown in Fig. 6.

Apart from thin polymer films, bulk-like dynamics has been observed in other type confinement. In particular, several polymer nanocomposites [11, 13, 15, 22]

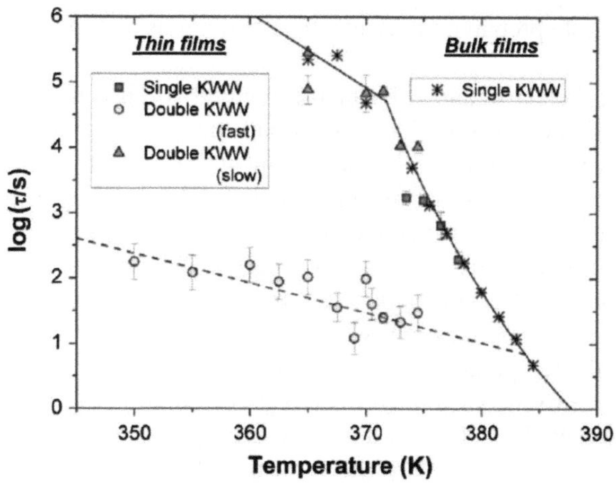

Fig. 6 Temperature dependence of the molecular relaxation time for 17 nm freestanding thin PS film (filled symbols) and bulk PS (stars) (Reprinted with permission from Ref. [93])

and nanospheres [131] exhibit filler size and diameter independent molecular dynamics, respectively. For polymer nanocomposites exhibiting strong interaction at the interface polymer/nanofillers, a slowing down of the segmental dynamics is rather observed [68, 100].

Despite the presence of bulk-like dynamics in different confining geometries, it is important to mention that several studies suggest that such systems display a rather complex relaxational behavior. This is generally found in confined systems with at least one free surface [39, 90, 92, 93, 131] and, therefore, is attributed to the existence of relatively fast dynamics in proximity of such surface. Within this explanation, the thickness of such layer is estimated to be of the order of nanometers [93].

4 Decoupling Between Equilibrium and Out-of-Equilibrium Dynamics

In the previous section of the chapter, the importance of sample preparation and the environmental conditions employed to measure glassy dynamics in thin polymer films have been discussed [113]. The important outcome within this context is that confinement effects on the α relaxation might be a consequence of the experimental conditions employed to determine such dynamics. These experimental facts generate an important question: is the observed thermal T_g depression (and beside it the acceleration of equilibrium recovery) a true confinement effect or does this originate from the employed experimental procedure? To answer this question it is vital to recall those studies where both the intrinsic molecular mobility, on the one hand, and the T_g and recovery of equilibrium in the physical aging regime, on the other, are probed in sample prepared under identical conditions and, possibly, in the same measurement.

In this sense, dielectric and calorimetric methods offer a unique possibility to probe the rate of spontaneous fluctuations and the out-of-equilibrium dynamics. The first study where simultaneous measurements of these two aspects of glassy dynamics were performed is that of Lupascu et al. [78]. In samples prepared under identical conditions, they measured the molecular dynamics in the linear regime and the thermal T_g simultaneously. The latter determination is based on the so-called capacitive dilatometry (CD) method. This consists in measuring the high frequency real part of the dielectric permittivity, where no relaxational contributions are present, as a function of the temperature. The value of such permittivity is connected to the density of the glass formers via the refractive index. In doing so, for Al-capped thin PS films, Lupascu et al. [78] found a weak speed-up of the molecular dynamics in comparison to the bulk. Conversely, the T_g of thin PS films exhibited slightly more pronounced thickness dependence.

A systematic investigation on the molecular dynamics and thermal T_g of thin PS films has been recently performed by Boucher et al. [16] employing BDS and calorimetry in samples subjected to the same preparation procedures and measured under identical environmental conditions. In doing so, they found a marked decoupling between the rate of spontaneous fluctuations and the thermal T_g. The former was found to be independent of the thickness and identical to that of bulk PS. Contrariwise, the thermal T_g exhibited clear depression, increasing with decreasing film thickness, and being more pronounced for freestanding films. In the case of BDS, exploiting the ability of this technique to achieve information on both aspects of glassy dynamics, this result was found in the same measurement. This result unequivocally indicates that T_g depression is a real feature of glassy dynamics in confinement and that this can be present in glass formers in confinement exhibiting bulk-like dynamics. The main outcome of this study is presented in Fig. 7, where the temperature dependence of the molecular relaxation time (τ) (upper panel) and the thermal T_g as a function of the thickness (lower panel) are reported. Apart from the study of Boucher et al. [16], other works report deviating results depending on the information delivered in thin polymer films [27, 59, 114, 115].

In ways analogous to thin polymer films, there exists a number of recent studies by BDS and calorimetry reporting apparently contrasting results regarding the intrinsic molecular dynamics and the thermal T_g in other type of confinement. In the Sects. 3.1 and 3.2, several examples of polymer nanocomposites exhibiting T_g depression have been reviewed [7, 11, 15, 22, 105]. In analogy to thin polymer films, in those nanocomposites where the molecular dynamics have been probed, identical rate of spontaneous fluctuations as those of the bulk polymer has been found [11, 15, 22]. This result also applies to PS nanospheres, where the T_g from calorimetry [132] and CD [131] have been found to be depressed with decreasing nanospheres diameters, whereas no shifts in the intrinsic molecular mobility were detected [131].

Regarding the simultaneous measurement, or at least in samples prepared under identical conditions, of the intrinsic molecular mobility and the rate of equilibrium recovery in the physical aging regime, this has been performed in several

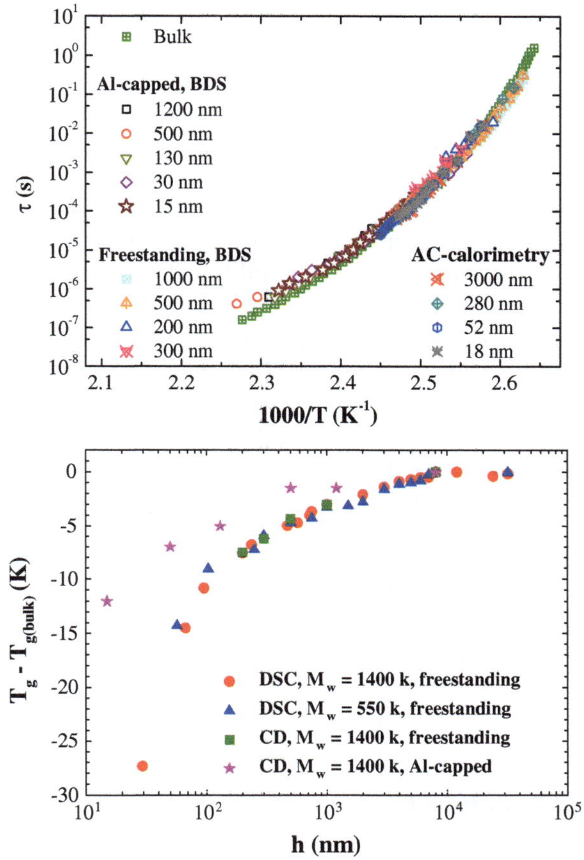

Fig. 7 *Upper panel* Temperature dependence of the molecular relaxation time of thin PS films in different confinement conditions; *Lower panel* Thickness dependence of T_g for the same systems of the *upper panel* (Reprinted with permission from Ref. [16])

confining geometries, including thin polymer films [10], polymer nanocomposites [11, 13, 15, 22] and nanospheres [53, 131]. In all cases, accelerated equilibrium recovery is found despite the unaltered molecular dynamics.

To understand these results, one way could be that the former is generally measured at temperatures somewhat larger than those relevant for the determination of the T_g and the recovery of equilibrium. Hence, one could hypothesize that, at lower temperatures, a drastic variation of the temperature dependence of the typical relaxation time occurs. However, there exists a number of studies which appear to contradict such scenario. First of all, the molecular dynamics of freestanding thin PS films have been shown to possess pronounced bulk-like dynamics even for temperatures of the order of the calorimetric T_g [93]. Furthermore, the molecular relaxation time monitored during the course of aging in poly(vinyl acetate) (PVAc)/silica nanocomposites has been shown to increase more rapidly in systems with larger nanofillers content. This implies that the instantaneous relaxation time, that is, the one at a given aging time, is larger in nanocomposites than in the pure polymer despite the faster evolution toward equilibrium [11]. Molecular dynamics simulations on a coarse-grained polymer nanocomposites show a significant acceleration of physical aging close to the polymer/filler interface accompanied by a slight speed-up of the molecular mobility seemingly insufficient to justify the increase of the aging rate [71]. Finally, some crucial experimental observations pointing toward the infeasibility of arguments exclusively based on the molecular mobility to describe the out-of-equilibrium dynamics are those reporting acceleration of physical aging in thin films [10, 81, 85, 95, 108] and polymer nanocomposites [9, 11, 13, 15, 22], with typical confinement length scale of the order of microns. Whatever the approach employed for the description of glassy dynamics, it is unphysical to attribute the observed acceleration to a modification of the rate of spontaneous fluctuation for systems exhibiting confinement length scale in the microns range.

According to the previous observations, it is possible to conclude that, differently from bulk glass formers [34, 55, 66, 125], the rate of spontaneous fluctuations and the out-of-equilibrium dynamics, that is that monitored after the application of a perturbation in the nonlinear regime are decoupled [16, 23]. In other words, the latter aspect of glassy dynamics in confinement is not exclusively determined by the molecular mobility and some confinement-specific features must be included to fully account for it.

5 Theoretical Frameworks

In the previous section of the chapter, the need to seek for an explanation to the peculiarities of glassy dynamics in confinement has been evidenced. In particular, in confinement the out-of-equilibrium dynamics, which could be expressed by an equilibration time τ_{eq}, must be connected to the molecular relaxation time τ plus an additional factor. This must depend exclusively on a confinement length. In particular, one can express the equilibration time as a function of the molecular relaxation time and a function only depending on such length: $\tau_{eq} = g(h)\tau$. Here h is the *confinement length scale*, which, as will be described, in the most trivial case of freestanding films is the film thickness. Given this very general approach to the problem, the main challenge is the search for the physics behind the function $g(h)$. In other words, whatever the approach employed to describe glassy dynamics in confinement, this *must* account for the experimental evidence that the rate of spontaneous fluctuations is not one-to-one related to the out-of-equilibrium dynamics.

Among the numerous theoretical approaches, most of them describe the T_g depression in terms of altered intrinsic molecular mobility [50, 70]. Hence, they do not account for the entire phenomenology of the glass transition in confinement. Others, such as those based on percolation arguments [69, 74], also rely on the effect of altered molecular mobility on the T_g. However, this kind of approach is based on the change of dimensionality of percolation under confinement, a purely geometric argument that could in principle be adapted to catch experimental observations.

The free volume hole diffusion (FVHD) model potentially represents a suitable candidate for the description of glassy dynamics in confinement. This model, rather than describing the different aspects of the relaxation in glass forming liquids, exclusively attempts to describe the way a glass recovers equilibrium in the physical aging regime (or try to maintain it on cooling). Alfrey et al. [2], who first tried to develop this idea, proposed that diffusion of free volume holes and their annihilation at the external surface of the sample could be responsible for the rapidity of achievement of equilibrium in the out-of-equilibrium glass. An important interference to the development of such idea is due to Braun and Kovacs [66]. They compared the volume recovery of milled PS, with typical size of several microns, with that of the bulk polymer and found no differences, at odds with the qualitative scenario expected from the FVHD model. To overcome the apparent size-independent evolution of out-of-equilibrium glasses, Curro et al. [28] assumed the presence of an internal length scale, where free volume holes can annihilate. In doing so, they could describe several aspects of PVAc volume recovery.

More recently, the model has been revitalized [21, 26, 82, 118, 119] after the finding of accelerated physical aging in numerous nanostructured glasses with typical confinement length scale of several microns or shorter [9–11, 13, 15, 22, 81, 85, 95, 108]. The basic equations to apply the model are: (i) the second equation of Fick:

$$\frac{\partial f_v(\mathbf{r},t)}{\partial t} = \nabla(D\nabla f_v(\mathbf{r},t)) \tag{1}$$

where f_v is the fractional free volume and D is the diffusion coefficient of free volume holes; and (ii) and the equation expressing the mean square displacement (MSD) ($\langle x^2\rangle$) as a function of time for one-dimensional linear diffusion[1]:

$$\langle x^2\rangle = 2Dt \tag{2}$$

The former equation can be applied to fit the evolution of magnitudes related to the free volume holes fraction during physical aging. The description of the T_g depression can be performed employing Eq. (2). This because, within the FVHD model, the glass transition occurs when $\langle x^2\rangle$ in the observation time $t \sim q^{-1}$, where q is the cooling rate of the experiment, is of the order of the square of half the film thickness: $(h/2)^2 = \langle x^2\rangle = 4Dq^{-1}$. A crucial point of the model is that, to describe accelerated physical aging and T_g depression, it does not requires any acceleration of the glass former molecular mobility, as suggested by several experiments.

The lines in Figs. 1 and 2 are the fits of the FVHD model to T_g data of thin PS films supported [12] and freestanding [10, 12], respectively. Interestingly, the FVHD model is able to account for both the thickness and cooling rate depen-

[1] The assumption of one-dimensional confinement is obviously true is thin films. For polymer nanocomposites and nanospheres it is approximately valid if the radius of curvature of nanoparticles and nanospheres, respectively, is considerably larger than the size of free volume holes.

dence of the T_g. In particular, the weak thickness dependence at high cooling rates [36, 38] is due to the VFT behavior of D in the temperature range relevant for such rates [10, 12, 15]. Conversely, at lower temperatures, in particular those relevant for determinations of the T_g at relatively low cooling rates, the diffusion coefficient exhibits weak Arrhenius temperature dependence [10, 12, 15]. This gives rise to large variations of the T_g with the film thickness. Successful fits of the model to the physical aging of freestanding thin PS films [10, 82, 118, 119] and several polymer nanocomposites [13, 15, 22] were also achieved, as shown by the lines in Fig. 2 for the former systems as a showcase.

With regard to supported or capped films, it is worth pointing out that only a portion of the polymer surface is available for elimination of free volume holes. This is due to irreversible chain adsorption [87, 89, 106, 129]. In this case, adsorbed chains constitute an infinitely high potential energy barrier to be overcome by free volume holes. Hence, these will be eliminated only at the free interface. A recent experimental study actually showed that the magnitude of T_g depression in capped thin PS films scales with the amount of free interface, in qualitative agreement with the FVHD model. Subsequently, such agreement was quantitatively tested and confirmed [86]. As such, the FVHD model, in combination with the experimental evidence reporting on chain adsorption at the interface, is able to account for the relatively limited T_g depression observed in supported and capped films in comparison to freestanding films. In relation to the presence of some energetic barrier at the interface, the FVHD model may provide an explanation to the contrasting results obtained for the deviation from bulk behavior of the T_g of polymer nanospheres exhibiting different kind of surfactants (if any) [40] or silica capped [132] at the polymer interface. Further investigation is required in this sense.

6 Conclusions and Perspectives

This chapter has emphasized the recent advancements in the understanding of the overall phenomenology of glassy dynamics in confinement. In view of recent results, it has been shown that, in the search for an explanation for the physics behind the observed T_g depression and accelerated recovery of equilibrium in confinement, arguments exclusively based on the alteration of the intrinsic molecular mobility are not sufficient. Therefore, any theoretical effort to describe glassy dynamics in confinement must account for this fact. In this context, the FVHD model potentially constitutes a suitable candidate to describe the physics of T_g depression and accelerated recovery of equilibrium, without need to invoke any change in the intrinsic molecular mobility. Obviously, such model needs to be tested for a variety of confinement configurations.

Apart from the search for a suitable framework to describe experimental results of glassy dynamics in confinement, the ability of nanostructured glasses to maintain equilibrium at temperatures lower than for bulk glass formers may open the perspective to so-far unexplored temperature ranges. This in view of the fact that in confinement the general features of the intrinsic molecular mobility, as discussed in this chapter, as well as the thermodynamics (at least for freestanding films thicker than 30 nm) [127] are essentially unaltered in comparison to the corresponding bulk glass formers. This implies that it is in principle possible to achieve information on the properties of equilibrium glasses down in the energy landscape. Within the context of the dynamics and thermodynamics of glass forming liquids, this implies that insight on the alleged divergence of the relaxation time and the vanishing of the configurational entropy at a finite temperature, that is, the Vogel and the Kauzmann temperatures respectively, can be obtained. Apart from this, a recent study on the enthalpy recovery of several polymer glasses in bulk over prolonged aging times (more than 1 year) revealed a complex scenario of both dynamics and thermodynamics [20]. In particular, it was found that recovery of equilibrium occurs in two stages, with partial and complete enthalpy recovery. One important implication of this result was that two equilibration mechanisms exist and, therefore, if a calorimetric experiment was performed at extremely low cooling rates, two jumps in the specific heat would be found. Alternatively, exploiting the more rapid equilibration of confined glass formers, such two jumps would be observed at cooling rates typical of experiments delivering a thermodynamic property as a function of temperature. Very recently, this has been actually found in freestanding thin PS films [102]. Such finding, in relation to the double equilibration mechanism found in the enthalpy recovery of bulk polymers, opens new perspective on the exploration of glasses low in the energy landscape exploiting the mentioned peculiarities of confined glasses.

Acknowledgments The author acknowledges the University of the Basque Country and Basque Country Government (Ref. No. IT-654-13 (GV)), Depto. Educación, Universidades e investigación; and Spanish Government (Grant No. MAT2012-31088) for their financial support.

References

1. Adam G, Gibbs JH (1965) On temperature dependence of cooperative relaxation properties in glass-forming liquids. J Chem Phys 43(1):139–146
2. Alfrey T, Goldfinger G, Mark H (1943) The apparent second-order transition point of polystyrene. J Appl Phys 14(12):700–705
3. Amanuel S, Gaudette AN, Sternstein SS (2008) Enthalpic relaxation of silicapolyvinyl acetate nanocomposites. J Polym Sci Pol Phys 46(24):2733–2740
4. Baeumchen O, McGraw JD, Forrest JA, Dalnoki-Veress K (2012) Reduced glass transition temperatures in thin polymer films: surface effect or artifact? Phys Rev Lett 109(5):055701
5. Bahar I, Erman B, Kremer F, Fischer E (1992) Segmental motions of cis-polyisoprene in the bulk state–interpretation of dielectric-relaxation data. Macromolecules 25(2):816–825
6. Baker EA, Rittigstein P, Torkelson JM, Roth CB (2009) Streamlined ellipsometry procedure for characterizing physical aging rates of thin polymer films. J Polym Sci Pt B-Polym Phys 47(24, SI):2509–2519
7. Bansal A, Yang H, Li C, Cho K, Benicewicz B, Kumar S, Schadler L (2005) Quantitative equivalence between polymer nanocomposites and thin polymer films. Nat Mater 4(9): 693–698

8. Berthier L, Biroli G, Bouchaud JP, Cipelletti L, Masri DE, L'Hôte D, Ladieu F, Pierno M (2005) Direct experimental evidence of a growing length scale accompanying the glass transition. Science 310(5755):1797–1800
9. Boucher VM, Cangialosi D, Alegría A, Colmenero J (2010) Enthalpy recovery of pmma/silica nanocomposites. Macromolecules 43(18):7594–7603
10. Boucher VM, Cangialosi D, Alegría A, Colmenero J (2012) Enthalpy recovery in nanometer to micrometer thick ps films. Macromolecules 45(12):5296–5206
11. Boucher VM, Cangialosi D, Alegría A, Colmenero J (2012) Time dependence of the segmental relaxation time of poly(vinyl acetate)-silica nanocomposites. Phys Rev E 86(4, Part 1): 041501
12. Boucher VM, Cangialosi D, Alegria A, Colmenero J (2014) Accounting for the thickness dependence of the Tg in supported PS films via the volume holes diffusion model. Thermochim Acta 575:233–237
13. Boucher VM, Cangialosi D, Alegría A, Colmenero J, Gonzalez-Irun J, Liz-Marzan LM (2010) Accelerated physical aging in pmma/silica nanocomposites. Soft Matter 6(14):3306–3317
14. Boucher VM, Cangialosi D, Alegría A, Colmenero J, Gonzalez-Irun J, Liz-Marzan LM (2011) Physical aging in pmma/silica nanocomposites: enthalpy and dielectric relaxation. J Non-Cryst Sol 357(2, SI): 605–609
15. Boucher VM, Cangialosi D, Alegría A, Colmenero J, Pastoriza-Santos I, Liz-Marzan LM (2011) Physical aging of polystyrene/gold nanocomposites and its relation to the calorimetric t(g) depression. Soft Matter 7(7):3607–3620
16. Boucher VM, Cangialosi D, Yin H, Schoenhals A, Alegría A, Colmenero J (2012) T-g depression and invariant segmental dynamics in polystyrene thin films. Soft Matter 8(19):5119–5122
17. Callen H, Greene R (1952) On a theorem of irreversible thermodynamics. Phys Rev 86(5): 702–710
18. Cangialosi D, Alegría A, Colmenero J (2007) Route to calculate the length scale for the glass transition in polymers. Phys Rev E 76(1, 1): 011514
19. Cangialosi D, Alegria A, Colmenero J (2007) "Self-concentration" effects on the dynamics of a polychlorinated biphenyl diluted in 1,4-polybutadiene. J Chem Phys 126(20)
20. Cangialosi D, Boucher V, Alegría A, Colmenero J (2013) Direct evidence of two equilibration mechanisms in glassy polymers. Phys Rev Lett 111:095701
21. Cangialosi D, Boucher VM, Alegría A, Colmenero J (2011) Free volume holes diffusion to describe physical aging in poly(mehtyl methacrylate)/silica nanocomposites. J Chem Phys 135(1):014901
22. Cangialosi D, Boucher VM, Alegría A, Colmenero J (2012) Enhanced physical aging of polymer nanocomposites: the key role of the area to volume ratio. Polymer 53(6):1362–1362
23. Cangialosi D, Boucher VM, Alegría A, Colmenero J (2013) Physical aging in polymers and polymer nanocomposites: recent results and open questions. Soft Matter 9(36):8619–8630
24. Cangialosi D, Boucher VM, Alegría A, Colmenero J (2013) Volume recovery of polystyrene/silica nanocomposites. J Polym Sci Part B: Polym Phys 51(10):847–853
25. Cangialosi D, Schwartz G, Alegria A, Colmenero J (2005) Combining configurational entropy and self-concentration to describe the component dynamics in miscible polymer blends. J Chem Phys 123(14)
26. Cangialosi D, Wübbenhorst M, Groenewold J, Mendes E, Schut H, van Veen A, Picken SJ (2004) Physical aging of polycarbonate far below the glass transition temperature: evidence for the diffusion mechanism. Phys Rev B 70:224213
27. Clough A, Peng D, Yang Z, Tsui OKC (2011) Glass transition temperature of polymer films that slip. Macromolecules 44(6):1649–1653
28. Curro JG, Lagasse RR, Simha R (1982) Diffusion model for volume recovery in glasses. Macromolecules 15(6):1621–1626
29. Debenedetti PG (1996) Metastable liquids: concepts and principles. Princeton University Press, Princeton
30. DeMaggio GB, Frieze WE, Gidley DW, Zhu M, Hristov HA, Yee AF (1997) Interface and surface effects on the glass transition in thin polystyrene films. Phys Rev Lett 78:1524–1527

31. Ding J, Xue G, Dai Q, Cheng R (1993) Glass-transition temperature of polystyrene microparticles. Polymer 34(15):3325–3327
32. Donati C, Douglas J, Kob W, Plimpton S, Poole P, Glotzer S (1998) Stringlike cooperative motion in a supercooled liquid. Phys Rev Lett 80(11):2338–2341
33. Donth E (1982) The size of cooperatively rearranging regions at the glass-transition. J Non Cryst Sol 53(3):325–330
34. Donth E, Korus J, Hempel E, Beiner M (1997) Comparison of dsc heating rate and hcs frequency at the glass transition. Thermochim Acta 305(6):239–239
35. Efremov M, Olson E, Zhang M, Zhang Z, Allen L (2003) Glass transition in ultrathin polymer films: calorimetric study. Phys Rev Lett 91(8):085703
36. Efremov MY, Olson EA, Zhang M, Zhang ZS, Allen LH (2004) Probing glass transition of ultrathin polymer films at a time scale of seconds using fast differential scanning calorimetry. Macromolecules 37(12):4607–4616
37. Ellison CJ, Torkelson JM (2003) The distribution of glass-transition temperatures in nanoscopically confined glass formers. Nat Mater 2(10):695–700
38. Fakhraai Z, Forrest JA (2005) Probing slow dynamics in supported thin polymer films. Phys Rev Lett 95(2):025701
39. Fakhraai Z, Forrest JA (2008) Measuring the surface dynamics of glassy polymers. Science 319(5863):600–604
40. Feng S, Li Z, Liu R, Mai B, Wu Q, Liang G, Gao H, Zhu F (2013) Glass transition of polystyrene nanospheres under different confined environments in aqueous dispersions. Soft Matter 9(18):4614–4620
41. Forrest J, Dalnoki-Veress K, Stevens J, Dutcher J (1996) Effect of free surfaces on the glass transition temperature of thin polymer films. Phys Rev Lett 77(10):2002–2005
42. Forrest J, DalnokiVeress K, Dutcher J (1997) Interface and chain confinement effects on the glass transition temperature of thin polymer films. Phys Rev E 56(5, B):5705–5716
43. Frieberg B, Glynos E, Sakellariou G, Green PF (2012) Physical aging of star-shaped macromolecules. ACS Macro Lett 1(5):636–640
44. Fryer D, Peters R, Kim E, Tomaszewski J, de Pablo J, Nealey P, White C, Wu W (2001) Dependence of the glass transition temperature of polymer films on interfacial energy and thickness. Macromolecules 34(16):5627–5634
45. Fukao K, Koizumi H (2008) Glassy dynamics in thin films of polystyrene. Phys Rev E 77(2, Part 1):021503
46. Fukao K, Miyamoto Y (2000) Glass transitions and dynamics in thin polymer films: dielectric relaxation of thin films of polystyrene. Phys Rev E 61(2):1743–1754
47. Gao S, Koh YP, Simon SL (2013) Calorimetric glass transition of single polystyrene ultrathin films. Macromolecules 46(2):562–570
48. Gaur U, Wunderlich B (1980) Study of microphase separation in block co-polymers of styrene and alpha-methylstyrene in the glass-transition region using quantitative thermal-analysis. Macromolecules 13(6):1618–1625
49. Ge S, Pu Y, Zhang W, Rafailovich M, Sokolov J, Buenviaje C, Buckmaster R, Overney R (2000) Shear modulation force microscopy study of near surface glass transition temperatures. Phys Rev Lett 85(11):2340–2343
50. de Gennes P (2000) Glass transitions in thin polymer films. Eur Phys J E 2(3):201–203
51. Glynos E, Frieberg B, Oh H, Liu M, Gidley DW, Green PF (2011) Role of molecular architecture on the vitrification of polymer thin films. Phys Rev Lett 106(12):128301
52. Grohens Y, Brogly M, Labbe C, David MO, Schultz J (1998) Glass transition of stereoregular poly(methyl methacrylate) at interfaces. Langmuir 14(11):2929–2932
53. Guo Y, Zhang C, Lai C, Priestley RD, D'Acunzi M, Fytas G (2011) Structural relaxation of polymer nanospheres under soft and hard confinement: isobaric versus isochoric conditions. ACS Nano 5(7):5365–5373
54. Hartmann L, Gorbatschow W, Hauwede J, Kremer F (2002) Molecular dynamics in thin films of isotactic poly(methyl methacrylate). Eur Phys J E 8(2):145–154

55. Hecksher T, Olsen NB, Niss K, Dyre JC (2010) Physical aging of molecular glasses studied by a device allowing for rapid thermal equilibration. J Chem Phys 133(17):174514
56. Hempel E, Hempel G, Hensel A, Schick C, Donth E (2000) Characteristic length of dynamic glass transition near t-g for a wide assortment of glass-forming substances. J Phys Chem B 104(11):2460–2466
57. Hutchinson JM (1995) Physical aging of polymers. Prog Pol Sci 20(4):703–760
58. Huth H, Minakov AA, Schick C (2006) Differential ac-chip calorimeter for glass transition measurements in ultrathin films. J Polym Sci Pt B-Polym Phys 44(20):2996–3005
59. Inoue R, Kanaya T, Nishida K, Tsukushi I, Telling MTF, Gabrys BJ, Tyagi M, Soles C, Wu WI (2009) Glass transition and molecular mobility in polymer thin films. Phys Rev E 80(3):031802
60. Jackson CL, McKenna GB (1991) The glass-transition of organic liquids confined to small pores. J Non-Cryst Sol 131(Part 1):221–224
61. Kawana S, Jones RAL (2003) Effect of physical ageing in thin glassy polymer films. Eur Phys J E 10(3):223–230
62. Keddie JL, Jones RAL, Cory RA (1994) Size-dependent depression of the glass-transition temperature in polymer-films. Europhys Lett 27(1):59–64
63. Kim JH, Jang J, Zin WC (2000) Estimation of the thickness dependence of the glass transition temperature in various thin polymer films. Langmuir 16(9):4064–4067
64. Koh YP, McKenna GB, Simon SL (2006) Calorimetric glass transition temperature and absolute heat capacity of polystyrene ultrathin films. J Polym Sci Pt B-Polym Phys 44(24):3518–3527
65. Koh YP, Simon SL (2008) Structural relaxation of stacked ultrathin polystyrene films. J Polym Sci Pt B-Polym Phys 46(24):2741–2753
66. Kovacs AJ (1963) Glass transition in amorphous polymers: a phenomenological study. Fortsch Hochpolym Fo 3(1/2):394–508
67. Labahn D, Mix R, Schoenhals A (2009) Dielectric relaxation of ultrathin films of supported polysulfone. Phys Rev E 79(1, Part 1):011801
68. Lee A, Lichtenhan JD (1998) Viscoelastic responses of polyhedral oligosilsesquioxane reinforced epoxy systems. Macromolecules 31(15):4970–4974
69. Lipson JEG, Milner ST (2009) Percolation model of interfacial effects in polymeric glasses. Eur Phys J B 72(1):133–137
70. Lipson JEG, Milner ST (2010) Local and average glass transitions in polymer thin films. Macromolecules 43
71. Liu AYH, Rottler J (2009) Physical aging and structural relaxation in polymer nanocomposites. J Polym Sci Pt B-Polym Phys 47(18):1789–1798
72. Liu Y, Russell T, Samant M, Stohr J, Brown H, Cossy-Favre A, Diaz J (1997) Surface relaxations in polymers. Macromolecules 30(25):7768–7771
73. Lodge T, McLeish T (2000) Self-concentrations and effective glass transition temperatures in polymer blends. Macromolecules 33(14):5278–5284
74. Long D, Lequeux F (2001) Heterogeneous dynamics at the glass transition in van der waals liquids, in the bulk and in thin films. Eur Phys J E 4(3):371–387
75. Lu H, Nutt S (2003) Restricted relaxation in polymer nanocomposites near the glass transition. Macromolecules 36(11):4010–4016
76. Lubchenko V, Wolynes PG (2007) Theory of structural glasses and supercooled liquids. Annu Rev Phys Chem 58:235–266
77. Lupascu V, Huth H, Schick C, Wubbenhorst M (2005) Specific heat and dielectric relaxations in ultra-thin polystyrene layers. Thermochim Acta 432(2):222–228
78. Lupascu V, Picken SJ, Wubbenhorst M (2006) Cooperative and non-cooperative dynamics in ultra-thin films of polystyrene studied by dielectric spectroscopy and capacitive dilatometry. J Non-Cryst Solids 352(52–54):5594–5600
79. Mapesa EU, Tress M, Schulz G, Huth H, Schick C, Reiche M, Kremer F (2013) Segmental and chain dynamics in nanometric layers of poly(cis-1,4-isoprene) as studied by broadband dielectric spectroscopy and temperature-modulated calorimetry. Soft Matter 9(44):10592–10598

80. Martinez-Tong DE, Soccio M, Sanz A, Garcia C, Ezquerra TA, Nogales A (2013) Chain arrangement and glass transition temperature variations in polymer nanoparticles under 3d-confinement. Macromolecules 46(11):4698–4705
81. McCaig MS, Paul DR (2000) Effect of film thickness on the changes in gas permeability of a glassy polyarylate due to physical aging part i. experimental observations. Polymer 41(2):629–637
82. McCaig MS, Paul DR, Barlow JW (2000) Effect of film thickness on the changes in gas permeability of a glassy polyarylate due to physical aging part i. experimental observations. Polymer 41(2):639–648
83. Ming W, Zhao J, Lu X, Wang C, Fu S (1996) Novel characteristics of polystyrene microspheres prepared by microemulsion polymerization. Macromolecules 29(24):7678–7682
84. Miyazaki T, Inoue R, Nishida K, Kanaya T (2007) X-ray reflectivity studies on glass transition of free standing polystyrene thin films. Eur Phys J Spec Top 141:203–206
85. Murphy TM, Langhe DS, Ponting M, Baer E, Freeman BD, Paul DR (2011) Physical aging of layered glassy polymer films via gas permeability tracking. Polymer 52(26):6117–6125
86. Napolitano S, Cangialosi D (2013) Interfacial free volume and vitrification: reduction in T_g in proximity of an adsorbing interface explained by the free volume holes diffusion model. Macromolecules 46(19):8051–8053
87. Napolitano S, Rotella C, Wübbenhorst M (2012) Can thickness and interfacial interactions univocally determine the behavior of polymers confined at the nanoscale? ACS Macro Lett 1(10):1189–1193
88. Napolitano S, Rotella C, Wuebbenhorst M (2011) Is the reduction in tracer diffusivity under nanoscopic confinement related to a frustrated segmental mobility? Macromol Rapid Commun 32(11):844–848
89. Napolitano S, Wübbenhorst M (2011) The lifetime of the deviations from bulk behaviour in polymers confined at the nanoscale. Nat Commun 2:260
90. Napolitano S, Wuebbenhorst M (2010) Structural relaxation and dynamic fragility of freely standing polymer films. Polymer 51(23):5309–5312
91. Nyquist H (1928) Thermal agitation of electric charge in conductors. Phys Rev 32(1):110–113
92. O'Connell P, McKenna G (2005) Rheological measurements of the thermoviscoelastic response of ultrathin polymer films. Science 307(5716):1760–1763
93. Paeng K, Swallen SF, Ediger MD (2011) Direct measurement of molecular motion in free-standing polystyrene thin films. J Am Chem Soc 133(22):8444–8447
94. Perlich J, Koerstgens V, Metwalli E, Schulz L, Georgii R, Mueller-Buschbaum P (2009) Solvent content in thin spin-coated polystyrene homopolymer films. Macromolecules 42(1): 337–344
95. Pfromm PH, Koros WJ (1995) Accelerated physical aging of thin glassy polymer-films–evidence from gas-transport measurements. Polymer 36(12):2379–2387
96. Priestley RD (2009) Physical aging of confined glasses. Soft Matter 5(5):919–926
97. Priestley RD, Broadbelt LJ, Torkelson JM (2005) Physical aging of ultrathin polymer films above and below the bulk glass transition temperature: effects of attractive vs neutral polymer-substrate interactions measured by fluorescence. Macromolecules 38(3):654–657
98. Priestley RD, Broadbelt LJ, Torkelson JM, Fukao K (2007) Glass transition and alpha-relaxation dynamics of thin films of labeled polystyrene. Phys Rev E 75(6, 1):061806
99. Priestley RD, Ellison CJ, Broadbelt LJ, Torkelson JM (2005) Structural relaxation of polymer glasses at surfaces, interfaces and in between. Science 309(5733):456–459
100. Priestley RD, Rittigstein P, Broadbelt LJ, Fukao K, Torkelson JM (2007) Evidence for the molecular-scale origin of the suppression of physical ageing in confined polymer: fluorescence and dielectric spectroscopy studies of polymer-silica nanocomposites. J Phys Condens Matt 19(20):2996–3005
101. Pye JE, Rohald KA, Baker EA, Roth CB (2010) Physical aging in ultrathin polystyrene films: evidence of a gradient in dynamics at the free surface and its connection to the glass transition temperature reductions. Macromolecules 43(19):8296–8303

102. Pye JE, Roth CB (2011) Two simultaneous mechanisms causing glass transition temperature reductions in high molecular weight freestanding polymer films as measured by transmission ellipsometry. Phys Rev Lett 107(23):235701
103. Pye JE, Roth CB (2013) Physical aging of polymer films quenched and measured free-standing via ellipsometry: controlling stress imparted by thermal expansion mismatch between film and support. Macromolecules 46(23):9455–9463
104. Rauscher PM, Pye JE, Baglay RR, Roth CB (2013) Effect of adjacent rubbery layers on the physical aging of glassy polymers. Macromolecules 46:9806–9817
105. Rittigstein P, Priestley RD, Broadbelt LJ, Torkelson JM (2007) Model polymer nanocomposites provide an understanding of confinement effects in real nanocomposites. Nat Mater 6(4):278–282
106. Rotella C, Wubbenhorst M, Napolitano S (2011) Probing interfacial mobility profiles via the impact of nanoscopic confinement on the strength of the dynamic glass transition. Soft Matter 7(11):5260–5266
107. Roth C, Dutcher J (2003) Glass transition temperature of freely-standing films of atactic poly(methyl methacrylate). Eur Phys J E 12(1):103–107
108. Rowe BW, Freeman BD, Paul DR (2009) Physical aging of ultrathin glassy polymer films tracked by gas permeability. Polymer 50(23):5565–5565
109. Rowe BW, Pas SJ, Hill AJ, Suzuki R, Freeman BD, Paul DR (2009) A variable energy positron annihilation lifetime spectroscopy study of physical aging in thin glassy polymer films. Polymer 50(25):6149–6156
110. Schmelzer JWP, Gutzow IS, Mazurin OV, Priven AI, Todorova SV, Petrov BP (2011) Glasses and the glass transition. Wiley-VCH Verlag GmbH & Co, KGaA, Weinheim
111. See Y, Cha J, Chang T, Ree M (2000) Glass transition temperature of poly(tert-butyl methacrylate) langmuir-blodgett film and spin-coated film by x-ray reflectivity and ellipsometry. Langmuir 16(5):2351–2355
112. Serghei A, Huth H, Schick C, Kremer F (2008) Glassy dynamics in thin polymer layers having a free upper interface. Macromolecules 41(10):3636–3639
113. Serghei A, Kremer F (2008) Metastable states of glassy dynamics, possibly mimicking confinement-effects in thin polymer films. Macromol Chem Phys 209(8):810–817
114. Sharp JS, Forrest JA (2003) Free surfaces cause reductions in the glass transition temperature of thin polystyrene films. Phys Rev Lett 91:235701
115. Soles C, Douglas J, Wu W, Peng H, Gidley D (2004) Comparative specular x-ray reflectivity, positron annihilation lifetime spectroscopy, and incoherent neutron scattering measurements of the dynamics in thin polycarbonate films. Macromolecules 37(8):2890–2800
116. Svanberg C (2007) Glass transition relaxations in thin suspended polymer films. Macromolecules 40(2):312–315
117. Tanaka Y, Yamamoto T (2012) Enthalpy relaxation of comb-like polymer analysed by combining activation energy spectrum and tnm models. J Non-Cryst Solids 358(14):1687–1698
118. Thornton AW, Hill AJ (2010) Vacancy diffusion with time-dependent length scale: an insightful new model for physical aging in polymers. Ind Eng Chem Res 49(23):12119–12124
119. Thornton AW, Nairn KM, Hill AJ, Hill JM, Huang Y (2009) New relation between diffusion and free volume: ii. predicting vacancy diffusion. J Membr Sci 338(1–2):38–42
120. Thurau CT, Ediger MD (2003) Change in the temperature dependence of segmental dynamics in deeply supercooled polycarbonate. J Chem Phys 118(4):1996–2004
121. Tress M, Erber M, Mapesa EU, Huth H, Mueller J, Serghei A, Schick C, Eichhorn KJ, Volt B, Kremer F (2010) Glassy dynamics and glass transition in nanometric thin layers of polystyrene. Macromolecules 43(23):9937–9944
122. Tress M, Mapesa EU, Kossack W, Kipnusu WK, Reiche M, Kremer F (2013) Glassy dynamics in condensed isolated polymer chains. Science 341(6152):1371–1374
123. Tsui O, Russell T, Hawker C (2001) Effect of interfacial interactions on the glass transition of polymer thin films. Macromolecules 34(16):5535–5539
124. Wallace W, Vanzanten J, Wu W (1995) Influence of an impenetrable interface on a polymer glass-transition temperature. Phys Rev E 52(4, A):R3329–R3332

125. Wang L, Velikov V, Angell C (2002) Direct determination of kinetic fragility indices of glass-forming liquids by differential scanning calorimetry: kinetic versus thermodynamic fragilities. J Chem Phys 117(22):10184–1019
126. Wang X, Zhou W (2002) Glass transition of microtome-sliced thin films. Macromolecules 35(18):6747–6750
127. White RP, Lipson JEG (2011) Thermodynamic treatment of polymer thin-film glasses. Phs Rev E 84(4, 1):041801
128. Yin H, Cangialosi D, Schoenhals A (2013) Glass transition and segmental dynamics in thin supported polystyrene films: the role of molecular weight and annealing. Thermochim Acta 566:186–192
129. Yin H, Napolitano S, Schoenhals A (2012) Molecular mobility and glass transition of thin films of poly(bisphenol a carbonate). Macromolecules 45(3):1652–1652
130. Yin H, Schoenhals A (2013) Calorimetric glass transition of ultrathin poly(vinyl methyl ether) films. Polymer 54(8, SI):2067–2070
131. Zhang C, Boucher VM, Cangialosi D, Priestley RD (2013) Mobility and glass transition temperature of polymer nanospheres. Polymer 54(1):230–235
132. Zhang C, Guo Y, Priestley RD (2011) Glass transition temperature of polymer nanoparticles under soft and hard confinement. Macromolecules 44(10):4001–4006

Index

F. Kremer (ed.), *Dynamics in Geometrical Confinement*, Advances in Dielectrics,
DOI: 10.1007/978-3-319-06100-9,

K

L

M

N

O

P

MIX
Papier aus verantwortungsvollen Quellen
Paper from responsible sources
FSC® C105338

If you have any concerns about our products,
you can contact us on
ProductSafety@springernature.com

In case Publisher is established outside the EU,
the EU authorized representative is:
Springer Nature Customer Service Center GmbH
Europaplatz 3, 69115 Heidelberg, Germany

Printed by Libri Plureos GmbH
in Hamburg, Germany